Oliver Goldsmith

A History of the Earth and Animated Nature

Oliver Goldsmith

A History of the Earth and Animated Nature

ISBN/EAN: 9783337025571

Printed in Europe, USA, Canada, Australia, Japan

Cover: Foto ©berggeist007 / pixelio.de

More available books at **www.hansebooks.com**

A HISTORY

OF

THE EARTH

AND

ANIMATED NATURE

BY

OLIVER GOLDSMITH, M.B.

A NEW EDITION,
WITH CORRECTIONS AND ALTERATIONS.

IN FOUR VOLUMES.

VOL. I.

NEW YORK:
R. WORTHINGTON, 770 BROADWAY.
1881

CONTENTS.

VOL. I.
PART I.

	Page
CHAP. I. A Sketch of the Universe	9
HAP. II. A short survey of the Globe, from the light of Astronomy, and Geography	11
CHAP. III. A view of the surface of the Earth	14
CHAP. IV. A review of the different Theories of the Earth	16
CHAP. V. Fossil-shells and other extraneous Fossils	23
CHAP. VI. The internal structure of the Earth	27
CHAP. VII. Caves and Subterraneous Passages that sink, but not perpendicularly, into the Earth	32
CHAP. VIII. Mines, Damps, and Mineral Vapours	36
CHAP. IX. Volcanoes and Earthquakes	41
CHAP. X. Earthquakes	47
CHAP. XI. The appearance of Islands and Tracts; and of the disappearing of others	55
CHAP XII. Mountains	59
CHAP XIII. Waters	69
CHAP. XIV. The origin of Rivers	81
CHAP. XV. The Ocean in general; and of its Saltness	94
CHAP. XVI. The Tides, Motion, and Currents of the Sea; with their effects	102
CHAP. XVII. The changes produced by the Sea upon the Earth	110
CHAP. XVIII. A summary account of the mechanical properties of the air	121
CHAP. XIX. An Essay towards a natural history of the Air	126
CHAP. XX. Winds regular and irregular	135
CHAP. XXI. Meteors, and such appearances as result from a combination of the Elements	147
CHAP. XXII. The conclusion	157

PART II.
ANIMALS.

CHAP. I. A comparison of Animals, with the inferior ranks of Creation	161
CHAP. II. The generation of Animals	166
CHAP. III. The infancy of Man	179
CHAP. IV. Puberty	186
CHAP. V. The Age of Manhood	189
CHAP. VI. Sleep and Hunger	206
CHAP. VII. Seeing	214
CHAP. VIII. Hearing	221
CHAP. IX. Smelling, Feeling, and Tasting	227
CHAP. X. Old age and Death	232
CHAP. XI. The varieties in the Human Race	239
CHAP. XII. Monsters	251
CHAP. XIII. Mummies, Wax-works, &c.	260
CHAP XIV. Animals	267
CHAP. XV. Quadrupeds in general, compared to Man	275

PREFACE.

Natural History, considered in its utmost extent, comprehends two objects. First, that of discovering, ascertaining, and naming all the various productions of Nature. Secondly, that of describing the properties, manners, and relations which they bear to us, and to each other. The first, which is the most difficult part of this science, is systematical, dry, mechanical, and incomplete. The second is more amusing, exhibits new pictures to the imagination, and improves our relish for existence, by widening the prospect of Nature around us.

Both, however, are necessary to those who would understand this pleasing science, in its utmost extent. The first care of every inquirer, no doubt, should be, to see, to visit, and examine every object, before he pretends to inspect its habitudes or its history. From seeing and observing the thing itself, he is most naturally led to speculate upon its uses, its delights, or its inconveniences.

Numberless obstructions, however, are found in this part of his pursuit, that frustrate his diligence, and retard his curiosity. The objects in Nature are so many, and even those of the same kind are exhibited in such a variety of forms, that the inquirer finds himself lost in the exuberance before him, and, like a man who attempts to count the stars, unassisted by Art, his powers are all distracted in the barren superfluity.

To remedy this embarrassment, artificial systems have been devised, which grouping into masses those parts of Nature more nearly resembling each other, refer the inquirer for the name of the single object he desires to know, to some one of those general distributions, where it is to be found by further examination.

If, for instance, a man should, in his walks, meet with an animal, the name, and consequently the history of which, he desires to know, he is taught by systematic writers of natural history, to examine its most obvious qualities, whether a quadruped, a bird, a fish, or an insect. Having determined it, for explanation sake, to be an insect, he examines whether it has wings; if he finds it possessed of these, he is taught to examine whether it has two or four; if possessed of four, he is taught to observe, whether the two upper wings are of a shelly hardness, and serve as cases to those under them; if he finds the wings composed in this manner, he is then taught to pronounce, that this insect is one of the beetle kind: of the beetle kind, there are three different classes, distinguished from each other by their feelers; he examines the insect before him, and finds that the feelers are clavated or knobbed at the ends; of beetles, with feelers thus formed, there are ten kinds; and, among those, he is taught to look for the precise name of that which is before him. If, for instance, the knob be divided at the ends, and the belly be streaked with white, it is no other than the Dor or the May-bug; an animal, the noxious qualities of which give it a very distinguished rank in the history of the insect creation. In this manner a system of natural history may, in some measure, be compared to a dictionary of words. Both are solely intended to explain the names of things;

PREFACE.

but with this difference, that in the dictionary of words we are led from the name of the thing to its definition; whereas in the system of natural history, we are led from the definition to find out the name.

Such are the efforts of writers, who have composed their works with great labour and ingenuity, to direct the learner in his progress through Nature, and to inform him of the name of every animal, plant, or fossil substance, that he happens to meet with; but it would be only deceiving the reader, to conceal the truth, which is, that books alone can never teach him this art in perfection; and the solitary student can never succeed. Without a master, and a previous knowledge of many of the objects of Nature, his book will only serve to confound and disgust him. Few of the individual plants or animals, that he may happen to meet with, are in that precise state of health, or that exact period of vegetation, from whence their descriptions were taken. Perhaps he meets the plant only with leaves, but the systematic writer has described it in a flower. Perhaps he meets the bird before it has moulted its first feathers, while the systematic description was made in its state of full perfection. He thus ranges without an instructor, confused, and with sickening curiosity from subject to subject, till at last he gives up the pursuit, in the multiplicity of his disappointments.

Some practice, therefore, much instruction, and diligent reading, are requisite to make a ready and expert naturalist, who shall be able, even by the help of a system, to find out the name of every object he meets with. But when this tedious, though requisite part of study is attained, nothing but delight and variety attend the rest of his journey. Wherever he travels, like a man in a country where he has many friends, he meets with nothing but acquaintances and allurements in all the stages of his way. The mere uninformed spectator passes on in gloomy solitude; but the naturalist, in every plant, in every insect, and every pebble, finds something to entertain his curiosity, and excite his speculation.

From hence it appears, that a system may be considered as a dictionary in the study of Nature. The ancients, however, who have all written most delightfully on this subject, seem entirely to have rejected those humble and mechanical helps to science. They contented themselves with seizing upon the great outlines of history, and passing over what was common, as not worth the detail, they only dwelt upon what was new, great and surprising, and sometimes even warmed the imagination at the expense of truth. Such of the moderns as revived this science in Europe, undertook the task more methodically, though not in a manner so pleasing. Aldrovandus, Gesner, and Johnson seemed desirous of uniting the entertaining and rich descriptions of the ancients with the dry and systematic arrangement, of which they were the first projectors. This attempt, however, was extremly imperfect, as the great variety of Nature was, as yet, but very inadequately known. Nevertheless, by attempting to carry on both objects at once, first of directing us to the name of the thing, and then giving the detail of its history, they drew out their works into a tedious and unreasonable length; and thus, mixing incompatible aims, they have left their labours rather to be occasionally consulted, than read with delight, by posterity.

The later moderns, with that good sense which they have carried into every other part of science, have taken a different method cultivating natural history. They have been content to give, not only the brevity, but also the dry and disgusting air of a dictionary to their systems. Ray, Klin, Brisson, and Linnæus, have had only one aim that of pointing out the object in Nature, of discovering its name, and where it was to be found in those authors that treated of it in a more prolix and satisfactory manner. Thus natural history, at present, is carried on in two distinct and separate channels, the one serving to lead us to the thing, the other conveying the history of the thing, as supposing it already known.

The following Natural History is written, with only such an attention to system as serves to remove the reader's embarrassments, and allure him to proceed. It can make no pretensions in directing him to the name of every object he meets with; that belongs to works of a very different kind, and written with very different aims. It will fully answer my design, if the reader, being already possessed of the name of any animal, shall find here a short, though satisfactory history of its habitudes, its subsistence, its manners, its friendships, and hostilities. My aim has been to carry on just as much method as was sufficient to shorten my description, by generalizing them, and never to follow order where the art of writing, which is but another name for good sense informed me that it would only contribute to the reader's embarrassment.

Still, however, the reader will perceive, that I have formed a kind of system in the history of every part of Animated Nature, directing myself by the great obvious distinctions that she herself seems to have made; which, though too few to point exactly to the name are yet sufficient to illuminate the subject, and remove the reader's perplexity. Mr. Buffon indeed, who has brought greater talents to this part of learning than any other man, has almost entirely rejected method in classing quadrupeds. This, with great deference to such a character, appears to me running into the opposite extreme; and, as some moderns have of late spent much time, great pains, and some learning, all to very little purpose, in systematic arrangement, he seems so much disgusted by their trifling, but ostentatious efforts, that he describes his animals almost in the order they happen to come before him. This want of method seems to be a fault; but he can lose little by a criticism which every dull man can make, or by an error in arrangement, from which the dullest are the most usually free.

In other respects, as far as this able philosopher has gone, I have taken him for my guide. The warmth of his style, and the brilliancy of his imagination, are inimitable. Leaving him, therefore, without a rival in these, and only availing myself of his information, I have been content to describe things in my own way; and though many of the materials are taken from him, yet I have added, retrenched, and altered, as I thought proper. It was my intention, at one time, whenever I differed from him, to have mentioned it at the bottom of the page; but this occurred so often, that I soon found it would look like envy, and might, perhaps, convict me of those very errors which I was wanting to lay upon him. I have, therefore, as being every way his debtor, concealed my dissent, where my opinion was different, but wherever I borrow from him, I take care at the bottom of the page to express my obligations. But though my obligations to this writer are many, they extend but to the smallest part of the work, as he has hitherto completed only the history of quadrupeds. I was therefore, left to my own reading alone, to make out the history of birds, fishes, and insects, of which the arrangement was so difficult, and the necessary information so widely diffused, and so obscurely related when found, that it proved by much the most laborious part of the undertaking. Thus having made use of Mr. Buffon's lights in the first part of the work, I may, with some share of confidence, recommend it to the public. But what shall I say to that part, where I have been entirely left without his assistance? As I would affect neither modesty nor confidence, it would be sufficient to say that my reading upon this part of the subject has been very extensive; and that I have taxed my scanty circumstances in procuring books, which are on this subject, of all others, the most expensive. In consequence of this industry, I here offer a work to the public, of a kind which has never been attempted in ours, or any other modern language that I know of. The ancients, indeed, and Pliny in particular, have anticipated me in the present manner of treating natural history. Like those historians who describe the events of a campaign, they have not condescended to give the private particulars of every individual that formed the army; they were content with characterizing the generals, and describing their operations, while they left it to meaner hands to carry the muster-roll. I have followed their manner, rejecting the numerous fables which they adopted, and adding the improvements of the moderns, which are so numerous, that they actually make up the bulk of natural history.

The delight which I found in reading Pliny, first inspired me with the idea of a work of this nature. Having a taste rather classical than scientific, and having but little employed myself in turning over the dry labours of modern system-makers, my earliest intention was to translate this agreeable writer, and by the help of a commentary to make my work as amusing as I could. Let us dignify natural history never so much with the grave appellation of *an useful science*, yet still we must confess that it is the occupation of the idle and the speculative, more than the busy, and the ambitious part of mankind. My intention, therefore, was to treat what I then conceived to be an idle subject, in an idle manner; and not to hedge round plain and simple narratives with hard words, accumulated distinctions, ostentatious learning, and disquisitions that produced no conviction. Upon the appearance, however, of Mr. Buffon's work, I dropped my former plan, and adopted the present, being convinced, by his manner, that the best imitation of the ancients was to write from our own feelings, and to imitate Nature.

It will be my chief pride, therefore, if this work may be found an innocent amusement for those who have nothing else to employ them or who require a relaxation from labour. Professed naturalists will, no doubt, find it superficial; and yet I should hope that even these will discover hints and remarks gleaned from various readings, not wholly trite or elementary. I would wish for their approbation. But my chief ambition is to drag up the obscure and gloomy learning of the cell to open inspection: to strip it from its garb of austerity and to show the beauties of that form, which only the industrious and the inquisitive have been hitherto permitted to approach.

A HISTORY OF THE EARTH.

CHAPTER I.

A SKETCH OF THE UNIVERSE.

The World may be considered as one vast mansion, where man has been admitted to enjoy, to admire, and to be grateful. The first desires of savage nature are merely to gratify the importunities of sensual appetite, and to neglect the contemplation of things, barely satisfied with their enjoyment: the beauties of nature, and all the wonders of creation, have but little charms for a being taken up in obviating the wants of the day, and anxious for precarious subsistence.

Philosophers, therefore, who have testified such surprise at the want of curiosity in the ignorant, seem not to consider that they are usually employed in making provisions of a more important nature; in providing rather for the necessities than the amusements of life. It is not till our more pressing wants are sufficiently supplied, that we can attend to the calls of curiosity; so that in every age scientific refinement has been the latest effort of human industry.

But human curiosity, though at first slowly excited, being at last possessed of leisure for indulging its propensity, becomes one of the greatest amusements of life, and gives higher satisfactions than what even the senses can afford. A man of this disposition turns all nature into a magnificent theatre, replete with objects of wonder and surprise, and fitted up chiefly for his happiness and entertainment: he industriously examines all things, from the minutest insect to the most finished animal; and, when his limited organs can no longer make the disquisition, he sends out his imagination upon new inquiries.

Nothing, therefore, can be more august and striking than the idea which his reason, aided by his imagination, furnishes of the universe around him. Astronomers tell us, that this earth which we inhabit forms but a very minute part in that great assemblage of bodies of which the world is composed. It is a million of times less than the sun, by which it is enlightened. The planets also, which, like it, are subordinate to the sun's influence, exceed the earth a thousand times in magnitude. These, which were at first supposed to wander in the heavens without any fixed path, and that took their name from their apparent deviations, have long been found to perform their circuits with great exactness and strict regularity. They have been discovered as forming, with our earth, a system of bodies circulating round the sun, all obedient to one law, and impelled by one common influence.

Modern philosophy has taught us to believe, that when the great Author of nature began the work of creation, he chose to operate by second causes; and that, suspending the constant exertion of his power, he endued matter with a quality, by which the universal economy of nature might be continued without his immediate assistance. This quality is called *attraction*; a sort of approximating influence, which all bodies, whether terrestrial or celestial, are found to possess; and which in all increases as the quantity of matter in each increases. The sun, by far the greatest body in our system, is, of consequence, possessed of much the greatest share of this attracting power; and all the planets, of which our earth is one, are, of course, entirely subject to its superior influence. Were this power, therefore, left uncontrolled by the other, the sun must quickly have attracted all the bodies of our celestial system to itself; but it is equally counteracted by another power of equal efficacy; namely, a progressive force, which each planet received when it was impelled forward by the divine Architect, upon its first formation. The heavenly bodies of our system being thus acted upon by two opposing powers; namely, by that of *attraction*, which draws them towards the sun; and that of *impulsion*, which drives them straight forward into the great void of space; they pursue a track between these contrary directions; and each, like a stone whirled about in a sling, obeying two opposite forces, circulates round its great centre of heat and motion.

In this manner, therefore, is the harmony of our planetary system preserved. The sun, in the midst, gives heat, and light, and circular motion to the planets which surround it: Mercury, Venus, the Earth, Mars, Jupiter, Saturn, and Herschel or the Georgium Sidus, perform their constant circuits at different distances, each taking up a time to complete its revolutions proportioned to the greatness of the circle which it is to describe. The lesser planets also, which are attendants upon some of the greater, are subject to the same laws; they circulate with the same exactness; and are, in the same manner, influenced by their respective centres of motion.

Besides those bodies which make a part of our peculiar system, and which may be said to reside within its great circumference, there are others that frequently come among us, from the most distant tracts of space, and that seem like dangerous intruders upon the beautiful simplicity of nature. These are comets, whose appearance was once so terrible to mankind; and the theory of which is so little understood at present: all we know is, that their number is much greater than that of the planets; and that, like these, they roll in orbits, in some measure obedient to solar influence. Astronomers have endeavoured to calculate the returning periods of many of them; but experience has not, as yet, confirmed the veracity of their investigations. Indeed, who can tell, when those wanderers have made their excursions into other worlds and distant systems, what obstacles may be found to oppose their progress, to accelerate their motions, or retard their return?

But what we have hitherto attempted to sketch, is but a small part of that great fabric in which the Deity has thought proper to manifest his wisdom and omnipotence. There are multitudes of other bodies, dispersed over the face of the heavens, that lie too remote for examination: these have no motion, such as the planets are found to possess, and are therefore called *fixed stars;* and from their extreme brilliancy, and their immense distance, philosophers have been induced to suppose them to be suns, resembling that which enlivens our system. As the imagination also, once excited, is seldom content to stop, it has furnished each with an attendant system of planets belonging to itself, and has even induced some to deplore the fate of those systems, whose imagined suns, which sometimes happens, have become no longer visible.

But conjectures of this kind, which no reasoning can ascertain, nor experiment reach, are rather amusing than useful. Though we see the greatness and wisdom of the Deity in all the seeming worlds that surround us, it is our chief concern to trace Him in that which we inhabit. The examination of the earth, the wonders of its contrivance, the history of its advantages, or of the seeming defects in its formation, are the proper business of the *natural historian.* A description of this *earth*, its *animals, vegetables,* and *minerals,* is the most delightful entertainment the mind can be furnished with, as it is the most interesting and useful. I would beg leave, therefore, to conclude these common-place speculations, with an observation, which, I hope, is not entirely so.

A use, hitherto not much insisted upon, that may result from the contemplation of celestial magnificence is, that it will teach us to make an allowance for the apparent irregularities we find below. Whenever we examine the works of the Deity at a proper point of distance, so as to take in the whole of his design, we see nothing but uniformity, beauty, and precision. The heavens present us with a plan, which, though inexpressibly magnificent, is yet regular beyond the power of invention. Whenever, therefore, we find any apparent defects in the Earth, which we are about to consider, instead of attempting to reason ourselves into an opinion that they are beautiful, it will be wiser to say, that we do not behold them at the proper point of distance, and that our eye is laid too close to the objects to take in the regularity of their connexion. In short, we may conclude, that God, who is regular in his GREAT productions, acts with equal uniformity in the LITTLE.

CHAPTER II.

A SHORT SURVEY OF THE GLOBE, FROM THE LIGHT OF ASTRONOMY AND GEOGRAPHY.

ALL the sciences are in some measure linked with each other, and before the one is ended, the other begins. In a natural history, therefore, of the earth, we must begin with a short account of its situation and form, as given by astronomers and geographers: it will be sufficient, however, upon this occasion, just to hint to the imagination, what they, by the most abstract reasonings, have forced upon the understanding. The earth which we inhabit is, as has been said before, one of those bodies which circulate in our solar system; it is placed at a happy middle distance from the centre; and even seems, in this respect, privileged beyond all other planets that depend upon our great luminary for their support. Less distant from the sun than Herschel or the Georgium Sidus, Saturn, Jupiter, and Mars, and yet less parched up than Venus and Mercury, that are situate too near the violence of its power, the Earth seems in a peculiar manner to share the bounty of the Creator: it is not, therefore, without reason, that mankind consider themselves as the peculiar objects of his providence and regard.

Besides that motion which the earth has round the sun, the circuit of which is performed in a year, it has another upon its own axle, which it performs in twenty-four hours. Thus, like a chariot-wheel, it has a compound motion; for while it goes forward on its journey, it is all the while turning round upon itself. From the first of these two arise the grateful vicissitude of the seasons; from the second, that of day and night.

It may be also readily conceived, that a body thus wheeling in circles will most probably be itself a sphere. The earth, beyond all possibility of doubt, is found to be so. Whenever its shadow happens to fall upon the moon, in an eclipse, it appears to be always circular, in whatever position it is projected: and it is easy to prove, that a body which in every position makes a circular shadow, must itself be round. The rotundity of the earth may be also proved from the meeting of two ships at sea: the top-masts of each are the first parts that are discovered by both, the under parts being hidden by the convexity of the globe which rises between them. The ships, in this instance, may be resembled to two men who approach each other on the opposite sides of a hill: their heads will first be seen, and gradually as they come nearer they will come entirely into view.

However, though the earth's figure is said to be spherical, we ought only to conceive it as being nearly so. It has been found in the last age to be rather flatted at both poles, so that its form is commonly resembled to that of a turnip. The cause of this swelling of the equator is ascribed to the greater rapidity of the motion with which the parts of the earth are there carried round; and which, consequently, endeavouring to fly off, act in opposition to central attraction. The twirling of a mop may serve as a homely illustration; which, as every one has seen, spreads and grows broader in the middle as it continues to be turned round.

As the earth receives light and motion from the sun, so it derives much of its warmth and power of vegetation from the same beneficent source. However, the different parts of the globe participate of these advantages in very different proportions, and accordingly put on very different appearances; a polar prospect, and a landscape at the equator, are as opposite in their appearances as in their situation.

The polar regions, that receive the solar beams in a very oblique direction, and that continue for one half of the year in night, receive but few of the genial comforts that other parts of the world enjoy. Nothing can be more mournful or hideous than the picture which

THE EARTH.

travellers present of those wretched regions. The ground,* which is rocky and barren, rears itself in every place in lofty mountains and inaccessible cliffs, and meets the mariner's eye at even forty leagues from shore. These precipices, frightful in themselves, receive an additional horror from being constantly covered with ice and snow, which daily seem to accumulate, and fill all the valleys with increasing desolation. The few rocks and cliffs, that are bare of snow, look at a distance of a dark brown colour, and quite naked. Upon a nearer approach, however, they are found replete with many different veins of coloured stone, here and there spread over with a little earth, and a scanty portion of grass and heath. The internal parts of the country are still more desolate and deterring. In wandering through these solitudes, some plains appear covered with ice, that, at first glance, seem to promise the traveller an easy journey.† But these are even more formidable and more unpassable than the mountains themselves, being cleft with dreadful chasms, and every where abounding with pits that threaten certain destruction. The seas that surround these inhospitable coasts, are still more astonishing, being covered with flakes of floating ice, that spread like extensive fields, or that rise out of the water like enormous mountains. These, which are composed of materials as clear and transparent as glass,‡ assume many strange and phantastic appearances. Some of them look like churches or castles, with pointed turrets; some like ships in full sail; and people have often given themselves the fruitless toil to attempt piloting the imaginary vessels into the harbour. There are still others that appear like large islands, with plains, valleys, and hills, which often rear their heads two hundred yards above the level of the sea, and although the height of these be amazing, yet their depth beneath is still more so; some of them being found to sink three hundred fathom under water.

The earth presents a very different appearance at the equator, where the sun-beams, darting directly downwards, burn up the lighter soils into extensive sandy deserts, or quicken all the moister tracts with incredible vegetation. In these regions, almost all the same inconveniences are felt from the proximity of the sun, that in the former were endured from its absence. The deserts are entirely barren, except where they are found to produce serpents, and that in such quantities, that some extensive plains seem almost entirely covered with them.§

It not unfrequently happens also, that this dry soil, which is so parched and comminuted by the force of the sun, rises with the smallest breeze of wind; and the sands being composed of parts, almost as small as those of water, they assume a similar appearance, rolling onward in waves like those of a troubled sea, and overwhelming all they meet with inevitable destruction. On the other hand, those tracts which are fertile teem with vegetation even to a noxious degree. The grass rises to such a height as often to require burning; the forests are impassable from underwoods, and so matted above, that even the sun, fierce as it is, can seldom penetrate.‖ These are so thick as

---

* Crantz's History of Greenland, p. 3. † Ibid, p. 22. ‡ Ibid, p. 27.
§ Adanson's Description of Senegal. ‖ Linnæi Amœnit. vol. vi. p. 67.

scarcely to be extirpated; for the tops being so bound together by the climbing plants that grow round them, though a hundred should be cut at the bottom, yet not one would fall, as they mutually support each other. In these dark and entangled forests, beasts of various kinds, insects in astonishing abundance, and serpents of surprising magnitude, find a quiet retreat from man, and are seldom disturbed except by each other.

In this manner the extremes of our globe seem equally unfitted for the comforts and conveniences of life: and although the imagination may find an awful pleasure in contemplating the frightful precipices of Greenland, or the luxurious verdure of Africa, yet true happiness can only be found in the more moderate climates, where the gifts of nature may be enjoyed, without incurring danger in obtaining them.

It is in the temperate zone, therefore, that all the arts of improving nature, and refining upon happiness, have been invented: and this part of the earth is, more properly speaking, the theatre of natural history. Although there be millions of animals and vegetables in the unexplored forests under the line, yet most of these may for ever continue unknown, as curiosity is there repressed by surrounding danger. But it is otherwise in these delightful regions which we inhabit, and where this art has had its beginning. Among us there is scarce a shrub, a flower, or an insect, without its particular history; scarce a plant that could be useful, which has not been propagated; nor a weed that could be noxious, which has not been pointed out.

CHAPTER III.

A VIEW OF THE SURFACE OF THE EARTH.

WHEN we take a slight survey of the surface of our globe, a thousand objects offer themselves, which, though long known, yet still demand our curiosity. The most obvious beauty that every where strikes the eye, is the verdant covering of the earth, which is formed by a happy mixture of herbs and trees of various magnitudes and uses. It has been often remarked, that no colour refreshes the sight so much as green: and it may be added, as a further proof of the assertion, that the inhabitants of those places where the fields are continually white with snow, generally become blind long before the usual course of nature.

This advantage, which arises from the verdure of the fields, is not a little improved by their agreeable inequalities. There are scarcely two natural landscapes that offer prospects entirely resembling each other; their risings and depressions, their hills and valleys, are never entirely the same, but always offer something new to entertain and refresh the imagination.

But to increase the beauties of the face of nature, the landscape is enlivened by springs and lakes, and intersected by rivulets. These lend a brightness to the prospect; give motion and coolness to the air; and, what is much more important, furnish health and subsistence to animated nature.

Such are the most obvious and tranquil objects that every where offer: but there are others of a more awful and magnificent kind; the *Mountain* rising above the clouds, and topped with snow; the *River* pouring down its sides, increasing as it runs, and losing itself, at last, in the ocean; the *Ocean* spreading its immense sheet of waters over one half of the globe, swelling and subsiding at well-known intervals, and forming a communication between the most distant parts of the earth.

If we leave those objects that seem to be natural to our earth, and keep the same constant tenor, we are presented with the great irregularities of nature. The burning mountain; the abrupt precipice; the unfathomable cavern; the headlong cataract; and the rapid whirlpool.

If we carry our curiosity a little further, and descend to the objects immediately below the surface of the globe, we shall there find wonders still as amazing. We first perceive the earth for the most part lying in regular beds or layers, every bed growing thicker in proportion as it lies deeper, and its contents more compact and heavy. We shall find, almost wherever we make our subterranean inquiry, an amazing number of shells that once belonged to aquatic animals. Here and there, at a distance from the sea, beds of oyster-shells, several yards thick, and many miles over; sometimes testaceous substances of various kinds on the tops of mountains, and often in the heart of the hardest marble. These, which are dug up by the peasants in every country, are regarded with little curiosity; for being so very common, they are considered as substances entirely terrene. But it is otherwise with the inquirer after nature, who finds them, not only in shape but in substance, every way resembling those that are found in the sea; and he, therefore, is at a loss to account for their removal.

Yet not one part of nature alone, but all her productions and varieties, become the object of the speculative man's inquiry: he takes different views of nature from the inattentive spectator; and scarce an appearance, how common soever, but affords matter for his contemplation: he inquires how and why the surface of the earth has those risings and depressions which most men call natural; he demands in what manner the mountains were formed, and in what consists their uses; he asks from whence springs arise, and how rivers flow round the convexity of the globe; he enters into an examination of the ebbings and flowings, and the other wonders of the deep; he acquaints himself with the irregularities of nature, and endeavours to investigate their causes; by which, at least, he will become better versed in their history. The internal structure of the globe becomes an object of his curiosity; and, although his inquiries can fathom but a very little way, yet, if possessed with a spirit of theory, his imagination will supply the rest. He will endeavour to account for the situation of the marine fossils that are found in the earth, and for the appearance of the different beds of which it is composed. These have been the inquiries that have splendidly employed many of the philosophers of the last and present age,* and, to a certain degree, they must be serviceable. But the worst of it is, that, as speculations amuse the

---

* Buffon, Woodward, Burnet, Whiston, Kircher, Bourguet, Leibnitz, Steno, Ray, &c.

writer more than facts, they may be often carried to an extravagant length; and that time may be spent in reasoning upon nature, which might be more usefully employed in writing her history.

Too much speculation in natural history is certainly wrong; but there is a defect of an opposite nature that does much more prejudice; namely, that of silencing all inquiry by alleging the benefits we receive from a thing, instead of investigating the cause of its production. If I inquire how a mountain came to be formed; such a reasoner, enumerating its benefits, answers, because God knew it would be useful. If I demand the cause of an earthquake, he finds some good produced by it, and alleges that as the cause of its explosion. Thus such an inquirer has constantly some ready reason for every appearance in nature, which serves to swell his periods, and give splendour to his declamation: every thing about him is, on some account or other, declared to be good; and he thinks it presumption to scrutinize into its defects, or to endeavour to imagine how it might be better. Such writers, and there are many such, add very little to the advancement of knowledge. It is finely remarked by Bacon, that the investigation of final causes* is a barren study; and, like a virgin dedicated to the Deity, brings forth nothing. In fact, those men who want to compel every appearance and every irregularity in nature into our service, and expatiate on their benefits, combat that very morality which they would seem to promote. God has permitted thousands of natural evils to exist in the world, because it is by their intervention that man is capable of moral evil; and he has permitted that we should be subject to moral evil, that we might do something to deserve eternal happiness, by showing that we had the rectitude to avoid it.

CHAPTER IV.

A REVIEW OF THE DIFFERENT THEORIES OF THE EARTH.

Human invention has been exercised for several ages to account for the various irregularities of the earth. While those philosophers, mentioned in the last chapter, see nothing but beauty, symmetry, and order; there are others, who look upon the gloomy side of nature, enlarge on its defects, and seem to consider the earth, on which they tread, as one scene of extensive desolation.† Beneath its surface they observe minerals and waters confusedly jumbled together; its different beds of earth irregularly lying upon each other; mountains rising from places that once were level;‡ and hills sinking into valleys; whole regions swallowed by the sea, and others again rising out of its bosom. All these they suppose to be but a few of the changes that have been wrought in our globe; and they send out the imagination to describe its primæval state of beauty.

Of those who have written theories describing the manner of the original formation of the earth, or accounting for its present appear

---

* Investigatio causarum finalium sterilis est, et veluti virgo Deo dicata nil parit.
† Buffon's second Discourse.   ‡ Senec. Quæst. lib. vi. cap. 21

ances, the most celebrated are Burnet, Whiston, Woodward, and Buffon. As speculation is endless, so it is not to be wondered that all these differ from each other, and give opposite accounts of the several changes, which they suppose our earth to have undergone. As the systems of each have had their admirers, it is, in some measure, incumbent upon the natural historian to be acquainted, at least, with their outlines; and, indeed, to know what others have even dreamed in matters of science, is very useful, as it may often prevent us from indulging similar delusions ourselves, which we should never have adopted, but because we take them to be wholly our own. However, as entering into a detail of these theories, is rather furnishing a history of opinions than things, I will endeavour to be as concise as I can.

The first who formed this amusement of earth-making into system, was the celebrated Thomas Burnet, a man of polite learning and rapid imagination. His *Sacred Theory*, as he calls it, describing the changes which the earth has undergone, or shall hereafter undergo, is well known for the warmth with which it is imagined, and the weakness with which it is reasoned; for the elegance of its style, and the meanness of its philosophy. " The earth," says he, " before the deluge, was very differently formed from what it is at present: it was at first a fluid mass; a chaos composed of various substances, differing both in density and figure: those which were most heavy sunk to the centre, and formed in the middle of our globe a hard solid body; those of a lighter nature remained next; and the waters, which were lighter still, swam upon its surface, and covered the earth on every side. The air, and all those fluids which were lighter than water, floated upon this also; and in the same manner encompassed the globe; so that between the surrounding body of waters, and the circumambient air, there was formed a coat of oil, and other unctuous substances, lighter than water. However, as the air was still extremely impure, and must have carried up with it many of those earthy particles with which it once was intimately blended, it soon began to defecate, and to depose these particles upon the oily surface already mentioned, which soon uniting, the earth and oil formed that crust, which soon became a habitable surface, giving life to vegetation, and dwelling to animals.

" This imaginary antideluvian abode was very different from what we see it at present. The earth was light and rich; and formed of a substance entirely adapted to the feeble state of incipient vegetation: it was a uniform plain, every where covered with verdure; without mountains, without seas, or the smallest inequalities. It had no difference of seasons, for its equator was in the plane of the ecliptic, or, in other words, it turned directly opposite to the sun, so that it enjoyed one perpetual and luxuriant spring. However, this delightful face of nature did not long continue in the same state; for, after a time, it began to crack and open in fissures; a circumstance which always succeeds when the sun exhales the moisture from rich or marshy situations. The crimes of mankind had been for some time preparing to draw down the wrath of Heaven; and they, at length, induced the Deity to defer repairing these breaches in nature. Thus the chasms of the earth every day became wider, and, at length, they penetrated to the great abyss of waters; and the whole earth, in a manner, fell in. Then ensued a total disorder in the uniform beauty of the first creation, the terrene surface of the globe being broken down; as it sunk the waters gushed out into its place; the deluge became universal; all mankind, except eight persons, were destroyed, and their posterity condemned to toil upon the ruins of desolated nature."

It only remains to mention the manner in which he relieves the earth from this universal wreck, which would seem to be as difficult as even its first formation: " These great masses of earth falling into the abyss, drew down with them vast quantities also of air; and, by dashing against each other, and breaking into small parts by the repeated violence of the shock, they, at length, left between them large cavities, filled with nothing but air. These cavities naturally offered a bed to receive the influent waters; and in proportion as they filled, the face of the earth became once more visible. The higher parts of its broken surface, now become the tops of the mountains, were the first that appeared; the plains soon after came forward, and, at length, the whole globe was delivered from the waters, except the places in the lowest situations; so that the ocean and the seas are still a part of the ancient abyss, that have not had a place to return. Islands and rocks are fragments of the earth's former crust; kingdoms and continents are larger masses of its broken substance; and all the inequalities that are to be found on the surface of the present earth, are owing to the accidental confusion into which both earth and waters were then thrown."

The next theorist was Woodward, who, in his Essay towards a Natural History of the Earth, which was only designed to precede a greater work, has endeavoured to give a more rational account of its appearance; and was, in fact, much better furnished for such an undertaking than any of his predecessors, being one of the most assiduous naturalists of his time. His little book, therefore, contains many important facts, relative to natural history, although his system may be weak and groundless.

He begins by asserting that all terrene substances are disposed in beds of various natures, lying horizontally one over the other, somewhat like the coats of an onion: that they are replete with shells, and other productions of the sea; these shells being found in the deepest cavities, and on the tops of the highest mountains. From these observations, which are warranted by experience, he proceeds to observe, that these shells and extraneous fossils are not productions of the earth, but are all actual remains of those animals which they are known to resemble; that all the beds of the earth lie under each other, in the order of their specific gravity; and that they are disposed as if they had been left there by subsiding waters. All these assertions he affirms with much earnestness, although daily experience contradicts him in some of them; particularly we find layers of stone often over the lightest soils, and the softest earth under the hardest bodies. However, having taken it for granted, that all the layers of the earth are found in the order of their specific gravity, the lightest at the top, and the heaviest next the centre, he consequently asserts, and it will not improbably follow, that all the substances of which the earth is composed were once in an actual state of dissolution. This universal dis solution he takes to have happened at the time of the flood. He supposes, that at that time a body of water, which was then in the centre of the earth, uniting with that which was found on the surface, so far separated the terrene parts as to mix all together in one fluid mass; the contents of which afterwards sinking according to their respective gravities, produced the present appearances of the earth. Being aware, however, of an objection, that fossil substances are not found dissolved, he exempts them from this universal dissolution, and, for that purpose, endeavours to show that the parts of animals have a stronger cohesion than those of minerals; and that, while even the hardest rocks may be dissolved, bones and shells may still continue entire.

So much for Woodward; but of all the systems which were published respecting the earth's formation, that of Whiston was most applauded, and most opposed. Nor need we wonder: for being supported with all the parade of deep calculation, it awed the ignorant, and produced the approbation of such as would be thought otherwise ; as it implied a knowledge of abstruse learning, to be even thought capable of comprehending what the writer aimed at. In fact, it is not easy to divest this theory of its mathematical garb; but those who have had leisure, have found the result of our philosopher's reasoning to be thus: He supposes the earth to have been originally a comet; and he considers the history of the creation, as given us in scripture, to have its commencement just when it was, by the hand of the Creator, to be more regularly placed as a planet in our solar system. Before that time he supposes it to have been a globe without beauty or proportion ; a world in disorder; subject to all the vicissitudes which comets endure ; some of which have been found, at different times, a thousand times hotter than melted iron; at others, a thousand times colder than ice. These alternations of heat and cold, continually melting and freezing the surface of the earth, he supposes to have produced, to a certain depth, a chaos entirely resembling that described by the poets, surrounding the solid contents of the earth, which still continued unchanged in the midst, making a great burning globe of more than two thousand leagues in diameter. This surrounding chaos, however, was far from being solid: he resembles it to a dense though fluid atmosphere, composed of substances mingled, agitated, and shocked against each other; and in this disorder he describes the earth to have been just at the eve of creation.

But upon its orbit being then changed, when it was more regularly wheeled round the sun, every thing took its proper place; every part of the surrounding fluid then fell into a situation, in proportion as it was light or heavy.

The middle, or central part, which always remained unchanged, still continued so, retaining a part of that heat which it received in its primeval approaches towards the sun: which heat, he calculates, may continue for about six thousand years. Next to this fell the heavier parts of the chaotic atmosphere, which serve to sustain the lighter: but as in descending they could not entirely be separated from many watery parts, with which they were intimately mixed, they drew down a part of these also with them;

and these could not mount again after the surface of the earth was consolidated: they, therefore, surrounded the heavy first descending parts, in the same manner as these surround the central globe. Thus the entire body of the earth is composed internally of a great burning globe; next which is placed a heavy terrene substance, that encompasses it; round which also is circumfused a body of water. Upon this body of water, the crust of earth, which we inhabit, is placed: so that, according to him, the globe is composed of a number of coats, or shells, one within the other, all of different densities. The body of the earth being thus formed, the air, which is the lightest substance of all, surrounded its surface; and the beams of the sun, darting through, produced that light which, we are told, first obeyed the Creator's command.

The whole economy of the creation being thus adjusted, it only remained to account for the risings and depressions on the surface of the earth, with the other seeming irregularities of its present appearance. The hills and valleys are considered by him as formed by their pressing upon the internal fluid, which sustains the outward shell of earth, with greater or less weight: those parts of the earth which are heaviest sink into the subjacent fluid more deeply, and become valleys: those that are lightest, rise higher upon the earth's surface, and are called mountains.

Such was the face of nature before the deluge; the earth was then more fertile and populous than it is at present; the life of man and animals was extended to ten times its present duration; and all these advantages arose from the superior heat of the central globe, which ever since has been cooling. As its heat was then in full power, the genial principle was also much greater than at present; vegetation and animal increase were carried on with more vigour; and all nature seemed teeming with the seeds of life. But these physical advantages were only productive of moral evil; the warmth which invigorated the body increased the passions and appetites of the mind; and, as man became more powerful he grew less innocent. It was found necessary to punish this depravity; and all living creatures were overwhelmed by the deluge in universal destruction.

This deluge, which simple believers are willing to ascribe to a miracle, philosophers have long been desirous to account for by natural causes: they have proved that the earth could never supply from any reservoir towards its centre, nor the atmosphere by any discharge from above, such a quantity of water as would cover the surface of the globe to a certain depth over the tops of our highest mountains. Where, therefore, was all this water to be found? Whiston has found enough, and more than a sufficiency, in the tail of a comet; for he seems to allot comets a very active part in the great operations of nature.

He calculates, with great seeming precision, the year, the month, and the day of the week, on which this comet (which has paid the earth some visits since, though at a kinder distance) involved our globe in its tail. The tail he supposed to be a vaporous fluid substance, exhaled from the body of the comet by the extreme heat of the sun, and increasing in proportion as it approached that great luminary. It was in this that our globe was involved at the time of the deluge; and, as the earth still acted by its natural attraction, it drew to itself all the watery vapours which were in the comet's tail; and the internal waters being also at the same time let loose, in a very short space the tops of the highest mountains were laid under the deep.

The punishment of the deluge being thus completed, and all the guilty destroyed, the earth, which had been broken by the eruption of the internal waters, was also enlarged by it; so that, upon the comet's recess, there was found room sufficient in the internal abyss for the recess of the superfluous waters; whither they all retired, and left the earth uncovered, but in some respects changed, particularly in its figure, which, from being round, was now become oblate. In this universal wreck of nature, Noah survived, by a variety of happy causes, to re-people the earth, and to give birth to a race of men slow in believing ill-imagined theories of the earth.

After so many theories of the earth, which had been published, applauded, answered, and forgotten, Mr. Buffon ventured to add one more to the number. This philosopher was, in every respect, better qualified than any of his predecessors for such an attempt, being furnished with more materials, having a brighter imagination to find new proofs, and a better style to clothe them in. However, if one so ill qualified as I am may judge, this seems the weakest part of his admirable work; and I could wish that he had been content with giving us facts instead of systems; that, instead of being a reasoner, he had contented himself with being merely a historian.

He begins his system by making a distinction between the first part of it and the last; the one being founded only on conjecture, the other depending entirely upon actual observation. The latter part of his theory may, therefore, be true, though the former should be found erroneous.

" The planets," says he, " and the earth among the number, might have been formerly (he only offers this as conjecture) a part of the body of the sun, and adherent to its substance. In this situation, a comet falling in upon that great body, might have given it such a shock, and so shaken its whole frame, that some of its particles might have been driven off like streaming sparkles from red-hot iron; and each of these streams of fire, small as they were in comparison of the sun, might have been large enough to have made an earth as great, nay, many times greater than ours. So that in this manner the planets, together with the globe which we inhabit, might have been driven off from the body of the sun by an impulsive force: in this manner also they would continue to recede from it for ever, were they not drawn back by its superior power of attraction; and thus, by the combination of the two motions, they are wheeled round in circles.

" Being in this manner detached at a distance from the body of the sun, the planets, from having been at first globes of liquid fire, gradually became cool. The earth also having been impelled obliquely forward, received a rotatory motion upon its axis at the very instant of its formation; and this motion being greatest at the equator, the parts there acting against the force of gravity, they must have swollen out, and given the earth an oblate or flatted figure.

THE HISTORY OF

"As to its internal substance, our globe, having once belonged to the sun, it continues to be a uniform mass of melted matter, very probably vitrified in its primeval fusion. But its surface is very differently composed. Having been in the beginning heated to a degree equal to, if not greater than what comets are found to sustain; like them it had an atmosphere of vapours floating round it, and which cooling by degrees, condensed and subsided upon its surface. These vapours formed, according to their different densities, the earth, the water, and the air; the heavier parts falling first, and the lighter remaining still suspended."

Thus far our philosopher is, at least, as much a system-maker as Whiston or Burnet; and, indeed, he fights his way with great perseverance and ingenuity, through a thousand objections that naturally arise. Having, at last, got upon the earth, he supposes himself on firmer ground, and goes forward with greater security. Turning his attention to the present appearance of things upon this globe, he pronounces from the view, that the whole earth was at first under water. This water he supposes to have been the lighter parts of its former evaporation, which, while the earthy particles sunk downwards by their natural gravity, floated on the surface, and covered it for a considerable space of time.

"The surface of the earth," says he,[*] "must have been in the beginning much less solid than it is at present; and, consequently, the same causes which at this day produce but very slight changes, must then, upon so complying a substance, have had very considerable effects. We have no reason to doubt but that it was then covered with the waters of the sea; and that those waters were above the tops of our highest mountains: since, even in such elevated situations, we find shells and other marine productions in very great abundance. It appears also that the sea continued for a considerable time upon the face of the earth: for as these layers of shells are found so very frequent at such great depths, and in such prodigious quantities, it seems impossible for such numbers to have been supported all alive at one time: so that they must have been brought there by successive depositions. These shells are also found in the bodies of the hardest rocks, where they could not have been deposited, all at once, at the time of the deluge, or at any such instant revolution; since that would be to suppose, that all the rocks in which they are found, were, at that instant, in a state of dissolution, which would be absurd to assert. The sea, therefore, deposited them wheresoever they are now to be found, and that by slow and successive degrees.

"It will appear also, that the sea covered the whole earth, from the appearance of its layers, which lying regularly one above the other, seem all to resemble the sediment formed at different times by the ocean. Hence, by the irregular force of its waves, and its currents driving the bottom into sand-banks, mountains must have been gradually formed within this universal covering of waters; and these successively raising their heads above its surface, must, in time, have formed the highest ridges of mountains upon land, together with con

[*] Theorie de la Terre, vol. i. p. 111.

tinents, islands, and low grounds, all in their turns. This opinion will receive additional weight by considering, that in those parts of the earth where the power of the ocean is greatest, the inequalities on the surface of the earth are highest. The ocean's power is greatest at the equator, where its winds and tides are most constant; and, in fact, the mountains at the equator are found to be higher than in any other part of the world. The sea, therefore, has produced the principal changes in our earth: rivers, volcanoes, earthquakes, storms, and rain, having made but slight alterations, and only such as have affected the globe to very inconsiderable depths."

This is but a very slight sketch of Mr. Buffon's Theory of the Earth; a theory which he has much more powerfully supported, than happily invented; and it would be needless to take up the reader's time from the pursuit of truth in the discussion of plausibilities. In fact, a thousand questions might be asked this most ingenious philosopher, which he would not find it easy to answer; but such is the lot of humanity, that a single Goth can in one day destroy the fabric which Cæsars were employed an age in erecting. We might ask, how mountains, which are composed of the most compact and ponderous substances, should be the first whose parts the sea began to remove? We might ask, how fossil-wood is found deeper even than shells? which argues, that trees grew upon the places he supposes once to have been covered with the ocean. But we hope this excellent man is better employed than to think of gratifying the petulance of incredulity, by answering endless objections.

CHAPTER V.

OF FOSSIL-SHELLS, AND OTHER EXTRANEOUS FOSSILS.

WE may affirm of Mr. Buffon, that which has been said of the chymists of old; though he may have failed in attaining his principal aim, of establishing a theory, yet he has brought together such a multitude of facts relative to the history of the earth, and the nature of its fossil productions, that curiosity finds ample compensation, even while it feels the want of conviction.

Before, therefore, I enter upon the description of those parts of the earth, which seem more naturally to fall within the subject, it will not be improper to give a short history of those animal productions that are found in such quantities, either upon its surface, or at different depths below it. They demand our curiosity; and, indeed, there is nothing in natural history that has afforded more scope for doubt, conjecture, and speculation. Whatever depths of the earth we examine, or at whatever distance within land we seek, we most commonly find a number of fossil-shells, which being compared with others from the sea, of known kinds, are found to be exactly of a similar shape and nature.* They are found at the very bottom of quarries and mines, in the retired and inmost parts of the most firm and solid rocks, upon

* Woodward's Essay towards a Natural History, p. 16.

the tops of even the highest hills and mountains, as well as in the valleys and plains: and this not in one country alone, but in all places where there is any digging for marble, chalk, or any other terrestrial matters, that are so compact as to fence off the external injuries of the air, and thus preserve these shells from decay.

These marine substances, so commonly diffused, and so generally to be met with, were for a long time considered by philosophers as productions, not of the sea, but of the earth. "As we find that spars," said they, "always shoot into peculiar shapes, so these seeming snail, cockle, and muscle-shells, are only sportive forms that nature assumes amongst others of its mineral varieties: they have the shape of fish, indeed, but they have always been terrestrial substances."*

With this plausible solution mankind were for a long time content; but, upon closer inquiry, they were obliged to alter their opinion. It was found that these shells had in every respect the properties of animal, and not of mineral nature. They were found exactly of the same weight with their fellow shells upon shore. They answered all the chymical trials in the same manner as sea-shells do. Their parts, when dissolved, had the same appearance to view, the same smell and taste. They had the same effects in medicine, when inwardly administered; and, in a word, were so exactly conformable to marine bodies, that they had all the accidental concretions growing to them, (such as pearls, corals, and smaller shells,) which are found in shells just gathered on the shore. They were, therefore, from these considerations, given back to the sea; but the wonder was, how to account for their coming so far from their own natural element upon land.†

As this naturally gave rise to many conjectures, it is not to be wondered that some among them have been very extraordinary. An Italian, quoted by Mr. Buffon, supposes them to have been deposited in the earth at the time of the crusades, by the pilgrims who returned from Jerusalem; who gathering them upon the sea-shore, in their return carried them to their different places of habitation. But this conjecturer seems to have but a very inadequate idea of their numbers. At Touraine, in France, more than a hundred miles from the sea, there is a plain of about nine leagues long, and as many broad, whence the peasants of the country supply themselves with marl for manuring their lands. They seldom dig deeper than twenty feet, and the whole plain is composed of the same materials, which are shells of various kinds, without the smallest portion of earth between them. Here then is a large space, in which are deposited millions of tons of shells, that pilgrims could not have collected, though their whole employment had been nothing else. England is furnished with its beds, which, though not quite so extensive, yet are equally wonderful. "Near Reading, in Berkshire, for many succeeding generations, a continued body of oyster-shells has been found through the whole circumference of five or six acres of ground. The foundation of these shells is a hard rocky chalk; and above this chalk, the oyster-shells lie in a bed of green sand, upon a level, as nigh as can possibly

* Lowth's Abridgment. Phil Trans. vol   p. 426   † Woodwd, p. 43

Persian Sheep, p. 44

Gnu Antelope, or Bubalus, p. 55

Nyl Ghau. p 11, vol. iii.

p 49. vol. ii.

be judged, and about two feet in thickness."* These shells are in their natural state, but they are found also petrified, and almost in equal abundance† in all the Alpine rocks, in the Pyrenees, on the hills of France, England, and Flanders. Even in all quarries from whence marble is dug, if the rocks be split perpendicularly downwards, petrified shells and other marine substances will be plainly discerned.

"About a quarter of a mile from the river Medway, in the county of Kent, after the taking off the coping of a piece of ground there, the workmen came to a blue marble, which continued for three feet and a half deep, or more, and then beneath appeared a hard floor, or pavement, composed of petrified shells crowded closely together. This layer was about an inch deep, and several yards over; and it could be walked upon as upon a beach. These stones, of which it was composed, (the describer supposes them to have always been stones,) were either wreathed as snails, or bivalvular like cockles. The wreathed kinds were about the size of a hazel-nut, and were filled with a stony substance of the colour of marl; and they themselves, also, till they were washed, were of the same colour: but when cleaned, they appeared of the colour of bezoar, and of the same polish. After boiling in water they became whitish, and left a chalkiness upon the fingers."‡

In several parts of Asia and Africa, travellers have observed these shells in great abundance. In the mountains of Castravan, which lie above the city Barut, they quarry out a white stone, every part of which contains petrified fishes in great numbers, and of surprising diversity. They also seem to continue in such preservation, that their fins, scales, and all the minutest distinctions of their make, can be perfectly discerned.§

From all these instances we may conclude, that fossils are very numerous; and, indeed, independent of their situation, they afford no small entertainment to observe them as preserved in the cabinets of the curious. The varieties of their kinds are astonishing. Most of the sea-shells which are known, and many others to which we are entirely strangers, are to be seen either in their natural state, or in various degrees of petrifaction.|| In the place of some we have mere spar, or stone, exactly expressing all the lineaments of animals, as having been wholly formed from them. For it has happened, that the shells dissolving by very slow degrees, and the matter having nicely and exactly filled all the cavities within, this matter, after the shells have perished, has preserved exactly and regularly the whole print of their internal surface. Of these there are various kinds found in our pits; many of them resembling those of our own shores; and many others that are only to be found on the coasts of other countries. There are some shells resembling those that are never stranded upon our coasts,¶ but always remain in the deep:** and many more there are which we can assimilate with no shells that are known amongst us. But we find not only shells in our pits, but also fishes and corals in great abundance; together with almost every sort of marine productions.

* Phil. Trans. vol. ii. p. 427. † Buffon, vol i. p. 407. ‡ Phil. Trans. p. 426
§ Buffon. vol. i. p. 408. || Hill, p. 646. ¶ Little ales. ** Pelagii.

It is extraordinary enough, however, that the common red coral, though so very frequent at sea, is scarcely seen in the fossil world, nor is there any account of its having ever been met with. But to compensate for this, there are all the kinds of the white coral now known, and many other kinds of that substance with which we are unacquainted. Of animals there are various parts: the vertebræ of whales, and the mouths of lesser fishes; these, with teeth also of various kinds, are found in the cabinets of the curious; where they receive long Greek names, which it is neither the intention nor the province of this work to enumerate. Indeed, few readers would think themselves much improved, should I proceed with enumerating the various classes of Conicthyodontes, Polyleptoginglimi, or the Orthoceratites. These names, which mean no great matter when they are explained, may serve to guide in furnishing a cabinet; but they are of very little service in furnishing the page of instructive history.

From all these instances we see in what abundance petrifactions are to be found; and, indeed, Mr. Buffon, to whose accounts we have added some, has not been sparing in the variety of his quotations, concerning the places where they are mostly to be found. However, I am surprised that he should have omitted the mention of one, which, in some measure, more than any of the rest, would have served to strengthen his theory. We are informed, by almost every traveller,* that has described the pyramids of Egypt, that one of them is entirely built of a kind of free-stone, in which these petrified shells are found in great abundance. This being the case, it may be conjectured, as we have accounts of these pyramids among the earliest records of mankind, and of their being built so long before the age of Herodotus, who lived but fifteen hundred years after the flood, that even the Egyptian priests could tell neither the time nor the cause of their erection; I say, it may be conjectured that they were erected but a short time after the flood. It is not very likely, therefore, that the marine substances found in one of them, had time to be formed into a part of the solid stone, either during the deluge, or immediately after it; and, consequently, their petrifaction must have been before that period. And this is the opinion Mr. Buffon has so strenuously endeavoured to maintain; having given specious reasons to prove, that such shells were laid in the beds where they are now found, not only before the deluge, but even antecedent to the formation of man, at the time when the whole earth, as he supposes, was buried beneath a covering of waters.

But while there are many reasons to persuade us that these extraneous fossils have been deposited by the sea, there is one fact that will abundantly serve to convince us, that the earth was habitable, if not inhabited, before these marine substances came to be thus deposited. For we find fossil-trees, which no doubt once grew upon the earth, as deep, and as much in the body of solid rocks, as these shells are found to be. Some of these fallen trees also have lain at least as long, if not longer, in the earth, than the shells, as they have been found sunk deep in a marly substance, composed of decayed shells and other marine productions. Mr. Buffon has proved, that

* Hasselquist, Sandys.

fossil-shells could not have been deposited in such quantities all at once by the flood; and I think, from the above instance, it is pretty plain, that, howsoever they were deposited, the earth was covered with trees before their deposition; and, consequently, that the sea could not have made a very permanent stay. How then shall we account for these extraordinary appearances in nature? A suspension of all assent is certainly the first, although the most mortifying conduct. For my own part, were I to offer a conjecture, and all that has been said upon this subject is but conjecture, instead of supposing them to be the remains of animals belonging to the sea, I would consider them rather as bred in the numerous fresh-water lakes, that, in primeval times, covered the face of uncultivated nature. Some of these shells we know to belong to fresh waters; some can be assimilated to none of the marine shells now known;* why, therefore, may we not as well ascribe the productions of all to fresh waters, where we do not find them, as we do that of the latter to the sea only, where we never find them? We know that lakes, and lands also, have produced animals that are now no longer existing; why, therefore, might not these fossil productions be among the number? I grant that this is making a very harsh supposition; but I cannot avoid thinking, that it is not attended with so many embarrassments as some of the former, and that it is much easier to believe that these shells were bred in fresh water, than that the sea had for a long time covered the tops of the highest mountains.

CHAPTER VI.

OF THE INTERNAL STRUCTURE OF THE EARTH.

HAVING, in some measure, got free from the regions of conjecture, let us now proceed to a description of the earth as we find it by examination, and observe its internal composition, as far as it has been the subject of experience, or exposed to human inquiry. These inquiries, indeed, have been carried but to a very little depth below its surface, and even in that disquisition men have been conducted more by motives of avarice than of curiosity. The deepest mine, which is that of Cotteberg in Hungary,† reaches not more than three thousand feet deep; but what proportion does that bear to the depth of the terrestrial globe, down to the centre, which is above four thousand miles? All, therefore, that has been said of the earth, to a deeper degree, is merely fabulous or conjectural; we may suppose with one, that it is a globe of glass;‡ with another, a sphere of heated iron;§ with a third, a great mass of waters;‖ and with a fourth, one dreadful volcano;¶ but let us at the same time show our consciousness, that all these are but suppositions.

Upon examining the earth, where it has been opened to any depth, the first thing that occurs, is the different layers or beds of which it is

* Hill's Fossils, p. 641.   † Boyle, vol. iii. p. 240.   ‡ Buffon   § Whiston.
‖ Burnet.   ¶ Kircher

composed; these all lying horizontally one over the other, like the leaves of a book, and each of them composed of materials that increase in weight in proportion as they lie deeper. This is, in general, the disposition of the different materials, where the earth seems to have remained unmolested; but this order is frequently inverted; and we cannot tell whether from its original formation, or from accidental causes. Of different substances, thus disposed, the far greatest part of our globe consists, from its surface downwards to the greatest depths we ever dig or mine [*]

The first layer that is most commonly found at the surface, is that light coat of blackish mould, which is called by some *garden earth*. With this the earth is every where invested, unless it be washed off by rains, or removed by some other external violence. This seems to have been formed from animal and vegetable bodies decaying, and thus turning into its substance. It also serves again as a storehouse, from whence animal and vegetable nature are renewed, and thus are all vital blessings continued with unceasing circulation. This earth, however, is not to be supposed entirely pure, but is mixed with much stony and gravelly matter from the layers lying immediately beneath it. It generally happens, that the soil is fertile in proportion to the quantity that this putrified mould bears to the gravelly mixture; and as the former predominates, so far is the vegetation upon it more luxuriant. It is this external covering that supplies man with all the true riches he enjoys. He may bring up gold and jewels from greater depths; but they are merely the toys of a capricious being, things upon which he has placed an imaginary value, and for which fools alone part with the more substantial blessings of life. "It is this earth," says Pliny,[†] "that, like a kind mother, receives us at our birth, and sustains us when born. It is this alone, of all the elements around us, that is never found an enemy to man. The body of waters deluge him with rains, oppress him with hail, and drown him with inundations. The air rushes in storms, prepares the tempest, or lights up the volcano; but the earth, gentle and indulgent, ever subservient to the wants of man, spreads his walks with flowers, and his table with plenty; returns with interest every good committed to her care; and though she produces the poison, she still supplies the antidote; though constantly teazed more to furnish the luxuries of man than his necessities, yet, even to the last, she continues her kind indulgence, and, when life is over, she piously covers his remains in her bosom."

This external and fruitful layer which covers the earth, is, as was said, in a state of continual change. Vegetables, which are naturally fixed and rooted to the same place, receive their adventitious nourishment from the surrounding earth and water; animals, which change from place to place, are supported by these, or by each other. Both, however, having for a time enjoyed a life adapted to their nature, give back to the earth those spoils, which they had borrowed for a very short space, yet still to be quickened again into fresh existence. But the deposits they make are of very dissimilar kinds, and the earth is very differently enriched by their continuance: those countries that

---

[*] Woodward, p. 9.   [†] Plinii Naturalis Historia. lib. ii. cap. 63.

have for a long time supported men and other animals, having been observed to become every day more barren; while, on the contrary, those desolate places, in which vegetables only are abundantly produced, are known to be possessed of amazing fertility. "In regions which are uninhabited,"* says Mr. Buffon, "where the forests are not cut down, and where animals do not feed upon the plants, the bed of vegetable earth is constantly increasing. In all woods, and even in those which are often cut, there is a layer of earth of six or eight inches thick, which has been formed by the leaves, branches, and bark, which fall and rot upon the ground. I have frequently observed on a Roman way, which crosses Burgundy for a long extent, that there is a bed of black earth, of more than a foot thick, gathered over the stony pavement, on which several trees, of a very considerable size, are supported. This I have found to be nothing else than an earth formed by decayed leaves and branches, which have been converted by time into a black soil. Now as vegetables draw much more of their nourishment from the air and water than they do from the earth, it must follow, that in rotting upon the ground, they must give more to the soil than they have taken from it. Hence, therefore, in woods kept a long time without cutting, the soil below increases to a considerable depth; and such we actually find the soil in those American wilds, where the forests have been undisturbed for ages. But it is otherwise where men and animals have long subsisted; for as they make a considerable consumption of wood and plants, both for firing and other uses, they take more from the earth than they return to it; it follows, therefore, that the bed of vegetable earth, in an inhabited country, must be always diminishing; and must at length resemble the soil of Arabia Petrea, and other provinces of the East, which having been long inhabited, are now become plains of salt and sand; the fixed salt always remaining while the other volatile parts have flown away."

If from this external surface we descend deeper, and view the earth cut perpendicularly downwards, either in the banks of great rivers, or steepy sea-shores; or, going still deeper, if we observe it in quarries or mines, we shall find its layers regularly disposed in their proper order. We must not expect, however, to find them of the same kind or thickness in every place, as they differ in different soils and situations. Sometimes marl is seen to be over sand, and sometimes under it. The most common disposition is, that under the first earth is found gravel or sand, then clay or marl, then chalk or coal, marbles, ores, sands, gravels, and thus an alternation of these substances, each growing more dense as it sinks deeper. The clay, for instance, found at the depth of a hundred feet, is usually more heavy than that found not far from the surface. In a well which was dug at Amsterdam, to the depth of two hundred and thirty feet, the following substances were found in succession:† seven feet of vegetable earth, nine of turf, nine of soft clay, eight of sand, four of earth, ten of clay, four of earth, ten of sand, two of clay, four of white sand, one of soft earth, fourteen of sand, eight of clay mixed with sand, four of sea-

---

* Buffon, vol. i p. 353.   † Varenius, as quoted by Mr. Buffon, p. 358.

sand mixed with shells, then a hundred and two feet of soft clay, and then thirty-one feet of sand.

In a well dug at Marly, to the depth of a hundred feet, Mr. Buffon gives us a still more exact enumeration of its layers of earth. "Thirteen of a reddish gravel, two of gravel mingled with a vitrifiable sand, three of mud or slime, two of marl, four of marly stone, five of marl in dust mixed with vitrifiable sand, six of very fine vitrifiable sand, three of earthy marl, three of hard marl, one of gravel, one of eglantine, a stone of the hardness and grain of marble, one of gravelly marl, one of stony marl, one of a coarser kind of stony marl, two of a coarser kind still, one of vitrifiable sand mixed with fossil-shells, two of fine gravel, three of stony marl, one of coarse powdered marl, one of stone, calcinable like marble, three of gray sand, two of white sand, one of red sand streaked with white, eight of gray sand with shells, three of very fine sand, three of a hard gray stone, four of red sand streaked with white, three of white sand, and fifteen of reddish vitrifiable sand."

In this manner the earth is every where found in beds over beds; and, what is still remarkable, each of them, as far as it extends, always maintains exactly the same thickness. It is found also, that as we proceed to considerable depths, every layer grows thicker. Thus in the adduced instances we might have observed, that the last layer was fifteen feet thick, while most of the others were not above eight, and this might have gone much deeper, for ought we can tell, as before they got through it the workmen ceased digging.

These layers are sometimes very extensive, and often are found to spread over a space of some leagues in circumference. But it must not be supposed that they are uniformly continued over the whole globe without any interruption: on the contrary, they are ever, at small intervals, cracked through as it were by perpendicular fissures; the earth resembling, in this respect, the muddy bottom of a pond, from whence the water has been dried off by the sun, and thus gaping in several chinks, which descend in a direction perpendicular to its surface. These fissures are many times found empty, but oftener closed up with adventitious substances, that the rain, or some other accidental causes, have conveyed to fill their cavities. Their openings are not less different than their contents, some being not above half an inch wide, some a foot, and some several hundred yards asunder. These last form those dreadful chasms that are to be found in the Alps, at the edge of which the traveller stands dreading to look down at the immeasurable gulf below. These amazing clefts are well known to such as have passed these mountains, where a chasm frequently presents itself several hundred feet deep, and as many over, at the edge of which the way lies. It often happens also, that the road leads along the bottom, and then the spectator observes on each side frightful precipices several hundred yards above him; the sides of which correspond so exactly with each other, that they evidently seem torn asunder.

But these chasms, to be found in the Alps, are nothing to what Ovalle tells us are to be seen in the Andes. These amazing mountains, in comparison of which the former are but little hills, have their fissures in proportion to their greatness. In some places they are a mile wide, and deep in proportion; and there are some others, that, running under ground, in extent resemble a province.

Of this kind also is that cavern called *Eldenhole*, in Derbyshire, which Dr. Plott tells us, was sounded by a line of eight and twenty hundred feet, without finding the bottom, or meeting with water: and yet the mouth at the top is not above forty yards over.* This immeasurable cavern runs perpendicularly downward; and the sides of it seem to tally so plainly as to show that they once were united. Those who come to visit the place, generally procure stones to be thrown into its mouth; and these are heard for several minutes, falling and striking against the sides of the cavern, producing a sound that resembles distant thunder, dying away as the stone goes deeper.

Of this kind also is that dreadful cavern described by Ælian; his account of which the reader may not have met with.† "In the country of the Arrian Indians, is to be seen an amazing chasm, which is called, *The Gulf of Pluto*. The depth and the recesses of this horrid place, are as extensive as they are unknown. Neither the natives, nor the curious who visit it, are able to tell how it was first made, or to what depths it descends. The Indians continually drive thither great multitudes of animals, more than three thousand at a time, of different kinds, sheep, horses, and goats; and, with an absurd superstition, force them into the cavity, from whence they never return. Their several sounds, however, are heard as they descend; the bleating of sheep, the lowing of oxen, and the neighing of horses, issuing up to the mouth of the cavern. Nor do these sounds cease, as the place is continually furnished with a fresh supply."

There are many more of these dreadful perpendicular fissures in different parts of the earth; with accounts of which Kircher, Gaffarellus, and others who have given histories of the wonders of the subterranean world, abundantly supply us. The generality of readers, however, will consider them with less astonishment, when they are informed of their being common all over the earth; that in every field, in every quarry, these perpendicular fissures are to be found, either still gaping, or filled with matter that has accidentally closed their interstices. The inattentive spectator neglects the inquiry, but their being common is partly the cause that excites the philosopher's attention to them; the irregularities of nature he is often content to let pass unexamined; but when a constant and common appearance presents itself, every return of the object is a fresh call to his curiosity; and the chink in the next quarry becomes as great a matter of wonder as the chasm in Eldenhole. Philosophers have long, therefore, endeavoured to find out the cause of these perpendicular fissures, which our own countrymen, Woodward and Ray, were the first that found to be so common and universal. Mr. Buffon supposes them to be cracks made by the sun, in drying up the earth immediately after its emersion from the deep. The heat of the sun is very probably a principal cause; but it is not right to ascribe to one only, what we find may be the result of many. Earthquakes, severe frosts, bursting waters, and storms, tearing up the roots of trees, have, in our own times, pro

---

* Phil. Trans. vol. ii. p. 370.   † Ælian, Var. Hist. lib. xvi. cap. 16.

duced them: and to this variety of causes, we must, at present, be content to assign those that have happened before we had opportunities for observation.

CHAPTER VII.

OF CAVES AND SUBTERRANEOUS PASSAGES, THAT SINK, BUT NOT PERPENDICULARLY, INTO THE EARTH.

In surveying the subterranean wonders of the globe, besides those fissures that descend perpendicularly, we frequently find others that descend but a litle way, and then spread themselves often to a great extent below the surface. Many of these caverns, it must be confessed, may be the production of art and human industry; retreats made to protect the oppressed, or shelter the spoiler. The famous labyrinth of Candia, for instance, is supposed to be entirely the work of art. Mr. Tournefort assures us, that it bears the impression of human industry, and that great pains have been bestowed upon its formation. The stone-quarry of Maestricht is evidently made by labour: carts enter at its mouth, and load within, then return and discharge their freight into boats that lie on the brink of the river Maese. This quarry is so large, that forty thousand people may take shelter in it: and it in general serves for this purpose, when armies march that way; becoming then an impregnable retreat to the people that live thereabout. Nothing can be more beautiful than this cavern, when lighted up with torches; for there are thousands of square pillars, in large level walks, about twenty feet high; and all wrought with much neatness and regularity. In this vast grotto there is very little rubbish; which shows both the goodness of the stone, and the carefulness of the workmen. To add to its beauty, there also are, in various parts of it, little pools of water, for the convenience of the men and cattle. It is remarkable also, that no droppings are seen to fall from the roof, nor are the walks any way wet under foot, except in cases of great rains, where the water gets in by the air-shafts. The salt-mines in Poland are still more spacious than these. Some of the catacombs, both in Egypt and Italy, are said to be very extensive. But no part of the world has a greater number of artificial caverns than Spain, which were made to serve as retreats to the Christians against the fury of the Moors, when the latter conquered that country. However, an account of the works of art does not properly belong to a natural history. It will be enough to observe, that though caverns be found in every country, far the greatest part of them have been fashioned by the hand of nature only. Their size is found beyond the power of man to have effected, and their forms but ill adapted to the conveniences of a human habitation. In some places, indeed, we find mankind still make use of them as houses; particularly in those countries where the climate is very severe;* but in general they are deserted by every race of meaner animals, except the bat; these noc-

---

* Phil. Trans. vol. ii. p 368.

turnal solitary creatures are usually the only inhabitants; and these only in such whose descent is sloping, or, at least, not directly perpendicular.

There is scarcely a country in the world without its natural caverns; and many new ones are discovered every day. Of those in England, Oakey-hole, the Devil's-hole, and Penpark-hole, have been often described. The former, which lies on the south side of Mendip-hills,* within a mile of the town of Wells, is much resorted to by travellers. To conceive a just idea of this, we must imagine a precipice of more than a hundred yards high, on the side of a mountain which shelves away a mile above it. In this is an opening not very large, into which you enter, going along upon a rocky uneven pavement, sometimes ascending, and sometimes descending. The roof of it, as you advance, grows higher: and, in some places, is fifty feet from the floor. In some places, however, it is so low that a man must stoop to pass. It extends itself, in length, about two hundred yards; and from every part of the roof, and the floor, there are formed sparry concretions of various figures, that by strong imaginations have been likened to men, lions, and organs. At the farthest part of this cavern rises a stream of water, well stored with fish, large enough to turn a mill, and which discharges itself near the entrance.

Penpark-hole, in Gloucestershire, is almost as remarkable as the former. Captain Sturmey descended into this by a rope, twenty-five fathoms perpendicular, and at the bottom found a very large vault in the shape of a horseshoe. The floors consisted of a kind of white stone enamelled with lead ore, and the pendent rocks were glazed with spar. Walking forward on this stony pavement, for some time, he came to a great river, twenty fathoms broad, and eight fathoms deep; and having been informed that it ebbed and flowed with the sea, he remained in this gloomy abode for five hours, to make an exact observation. He did not find, however, any alteration whatsoever in its appearance. But his curiosity was ill requited; for it cost this unfortunate gentleman his life: immediately after his return he was seized with an unusual and violent headach, which threw him into a fever, of which he died soon after.

But of all the subterranean caverns now known, the grotto of Antiparos is the most remarkable, as well for its extent, as for the beauty of its sparry incrustations. This celebrated cavern was first discovered by one Magni, an Italian traveller, about a hundred years ago, at Antiparos, an inconsiderable island of the Archipelago.† The account he gives of it is long and inflated, but upon the whole amusing. "Having been informed," says he, "by the natives of Paros, that in the little island of Antiparos, which lies about two miles from the former, of a gigantic statue that was to be seen at the mouth of a cavern in that place, it was resolved that we (the French consul and himself) should pay it a visit. In pursuance of this resolution, after we had landed on the island, and walked about four miles through the midst

---

* Phil. Trans. vol. ii. p. 368.
† Kircher Mund. sub. 112. I have translated a part of Kircher's description, rather than Tournefort's, as the latter was written to support a hypothesis.

of beautiful plains, and sloping woodlands, we at length came to a little hill, on the side of which yawned a most horrid cavern, that with its gloom at first struck us with terror, and almost repressed curiosity. Recovering the first surprise, however, we entered boldly; and had not proceeded above twenty paces, when the supposed statue of the giant presented itself to our view. We quickly perceived, that what the ignorant natives had been terrified at as a giant, was nothing more than a sparry concretion, formed by the water dropping from the roof of the cave, and by degrees hardening into a figure that their fears had formed into a monster. Incited by this extraordinary appearance, we were induced to proceed still farther, in quest of new adventures in this subterranean abode. As we proceeded, new wonders offered themselves; the spars, formed into trees and shrubs, presented a kind of petrified grove; some white, some green; and all receding in due perspective. They struck us with the more amazement, as we knew them to be mere productions of nature, who, hitherto in solitude, had, in her playful moments, dressed the scene as if for her own amusement.

" But we had as yet seen but a few of the wonders of the place; and were introduced only into the portico of this amazing temple. In one corner of this half-illuminated recess, there appeared an opening of about three feet wide, which seemed to lead to a place totally dark, and that, one of the natives assured us, contained nothing more than a reservoir of water. Upon this we tried, by throwing down some stones, which rumbled along the sides of the descent for some time, the sound seemed at last quashed in a bed of water. In order, however, to be more certain, we sent in a Levantine mariner, who, by the promise of a good reward, with a flambeau in his hand, ventured into this narrow aperture. After continuing within it for about a quarter of an hour, he returned, carrying some beautiful pieces of white spar in his hand, which art could neither imitate nor equal. Upon being informed by him that the place was full of these beautiful incrustations, I ventured in once more with him for about fifty paces, anxiously and cautiously descending by a steep and dangerous way. Finding, however, that we came to a precipice which led into a spacious amphitheatre, if I may so call it, still deeper than any other part, we returned, and being provided with a ladder, flambeaux, and other things to expedite our descent, our whole company, man by man, ventured into the same opening, and descending one after another, we at last saw ourselves all together in the most magnificent part of the cavern.

" Our candles being now all lighted up, and the whole place completely illuminated, never could the eye be presented with a more glittering, or a more magnificent scene. The roof all hung with solid icicles, transparent as glass, yet solid as marble. The eye could scarcely reach the lofty and noble ceiling: the sides were regularly formed with spars; and the whole presented the idea of a magnificent theatre, illuminated with an immense profusion of lights. The floor consisted of solid marble; and in several places magnificent columns, thrones, altars, and other objects, appeared, as if nature had designed to mock the curiosities of art. Our voices, upon speaking or singing, were redoubled to an astonishing loudness; and upon the firing of a gun, the noise and reverberations were almost deafening. In the midst of this grand amphitheatre rose a concretion of about fifteen feet high, that, in some measure, resembled an altar; from which taking the hint, we caused mass to be celebrated there. The beautiful columns that shot up round the altar, appeared like candlesticks; and many other natural objects represented the customary ornaments of this sacrament.

"Below even this spacious grotto there seemed another cavern; down which I ventured with my former mariner, and descended about fifty paces by means of a rope. I at last arrived at a small spot of level ground, where the bottom appeared different from that of the amphitheatre, being composed of soft clay, yielding to the pressure, and in which I thrust a stick to about six feet deep. In this however, as above, numbers of the most beautiful crystals were formed; one of which particularly resembled a table. Upon our egress from this amazing cavern, we perceived a Greek inscription upon a rock at the mouth, but so obliterated by time, that we could not read it. It seemed to import that one Antipater, in the time of Alexander, had come thither; but whether he had penetrated into the depths of the cavern, he does not think fit to inform us."

Such is the account of this beautiful scene, as communicated in a letter to Kircher. We have another, and a more copious description of it, by Tournefort, which is in every body's hands; but I have given the above, both because it was communicated by the first discoverer, and because it is a simple narrative of facts, without any reasoning upon them. According to Tournefort's account, indeed, we might conclude, from the rapid growth of the spars in this grotto, that it must every year be growing narrower, and that it must in time be choked up with them entirely; but no such thing has happened hitherto, and the grotto at this day continues as spacious as we ever knew it.

This is not a place for an inquiry into the seeming vegetation of those stony substances, with which this and almost every cavern are incrusted; it is enough to observe in general, that they are formed by an accumulation of that little gritty matter, which is carried thither by the waters, and which in time acquires the hardness of marble. What in this place more imports us to know is, how these amazing hollows in the earth came to be formed. And I think, in the three instances above mentioned, it is pretty evident, that their excavation has been owing to water. These, finding subterraneous passages under the earth, and by long degrees hollowing the beds in which they flowed, the ground above them has slipped down closer to their surface, leaving the upper layers of the earth or stone still suspended: the ground that sinks upon the face of the waters forming the floor of the cavern; the ground, or rock, that keeps suspended, forming the roof: and indeed there are but few of these caverns found without water, either within them, or near enough to point out their formation.

CHAPTER VIII.

OF MINES, DAMPS, AND MINERAL VAPOURS.

The caverns, which we have been describing, generally carry us but a very little way below the surface of the earth. Two hundred feet, at the utmost, is as much as the lowest of them is found to sink. The perpendicular fissures run much deeper; but few persons have been bold enough to venture down to their deepest recesses; and some few who have tried, have been able to bring back no tidings of the place, for unfortunately they left their lives below. The excavations of art have conducted us much further into the bowels of the globe. Some mines in Hungary are known to be a thousand yards perpendicularly downwards; and I have been informed, by good authority, of a coal mine in the north of England, a hundred yards deeper still.

It is beside our present purpose to inquire into the peculiar construction and contrivance of these, which more properly belongs to the history of fossils. It will be sufficient to observe in this place, that as we descend into the mines, the various layers of earth are seen as we have already described them; and in some of these are always found the metals or minerals for which the mine has been dug. Thus frequently gold is found dispersed and mixed with clay and gravel;[*] sometimes it is mixed with other metallic bodies, stones, or bitumens;[†] and sometimes united with that most obstinate of all substances, platina, from which scarce any art can separate it. Silver is sometimes found quite pure,[‡] sometimes mixed with other substances and minerals. Copper is found in beds mixed with various substances, marbles, sulphurs, and pyrites. Tin, the ore of which is heavier than that of any other metal, is generally found mixed with every kind of matter:[§] lead is also equally common; and iron we well know can be extracted from all the substances upon earth.

The variety of substances which are thus found in the bowels of the earth, in their native state, have a very different appearance from what they are afterwards taught to assume by human industry. The richest metals are very often less glittering and splendid than the most useless marcasites; and the basest ores are in general the most beautiful to the eye.

This variety of substances, which compose the internal parts of our globe, is productive of equal varieties, both above and below its surface. The combination of the different minerals with each other, the heats which arise from their mixture, the vapours they diffuse, the fires which they generate, or the colds which they sometimes produce, are all either noxious or salutary to man; so that in this great elaboratory of nature, a thousand benefits and calamities are forging, of which we are wholly unconscious; and it is happy for us that we are so.

Upon our descent into mines of considerable depth, the cold seems to increase from the mouth as we descend;[‖] but after passing very low down, we begin by degrees to come into a warmer air, which sensibly

---

[*] Ulloa, vol. ii. p. 470. [†] Ulloa, ibid. [‡] Macquer's Chymistry, vol. i. p. 316.
[§] Hill's Fossils, p. 628. [‖] Boyle, vol. iii. p. 232.

grows hotter as we go deeper, till, at last, the labourers can scarce bear any covering as they continue working.

This difference in the air was supposed by Boyle to proceed from magazines of fire that lay nearer the centre, and that diffused their heat to the adjacent regions. But we now know that it may be ascribed to more obvious causes. In some mines, the composition of the earth all around is of such a nature, that upon the admission of water or air, it frequently becomes hot, and often bursts out into eruptions. Besides this, as the external air cannot readily reach the bottom, or be renewed there, an observable heat is perceived below, without the necessity of recurring to the central heat for an explanation.

Hence, therefore, there are two principal causes of the warmth at the bottom of mines: the heat of the substances of which the sides are composed; and the want of renovation in the air below. Any sulphureous substance, mixed with iron, produces a very great heat, by the admission of water. If, for instance, a quantity of sulphur be mixed with a proportionable share of iron filings, and both kneaded together into a soft paste, with water, they will soon grow hot, and at last produce a flame. This experiment, produced by art, is very commonly effected within the bowels of the earth by nature. Sulphurs and irons are intimately blended together, and want only the mixture of water or air to excite their heat; and this, when once raised, is communicated to all bodies that lie within the sphere of their operation. Those beautiful minerals called *marcasites* and *pyrites*, are often of this composition; and wherever they are found, either by imbibing the moisture of the air, or having been by any means combined with water, they render the mine considerably hot.*

The want of fresh air also, at these depths, is, as we have said, another reason for their being found much hotter. Indeed, without the assistance of art, the bottom of most mines would, from this cause, be insupportable. To remedy this inconvenience, the miners are often obliged to sink, at some convenient distance from the mouth of the pit, where they are at work, another pit, which joins the former below, and which, in Derbyshire, is called an *air-shaft*. Through this the air circulates; and thus the workmen are enabled to breathe freely at the bottom of the place; which becomes, as Mr. Boyle affirms, very commodious for respiration, and also very temperate as to heat and cold.† Mr. Locke, however, who has left us an account of the Mendip mines, seems to present a different picture. "The descent into these is exceedingly difficult and dangerous; for they are not sunk like wells, perpendicularly, but as the crannies of the rocks happen to run. The constant method is to swing down by a rope placed under the arms, and clamber along by applying both feet and hands to the sides of the narrow passage. The air is conveyed into them through a little passage that runs along the sides from the top, where they set up some turfs, on the lea-side of the hole, to catch and force it down. These turfs being removed to the windy side, or laid over the mouth of the hole, the miners below presently want breath, and faint; and if sweet-smelling flowers chance to be

* Kircher Mund. Subt. vol. ii. p. 216. † Boyle, vol. iii. p. 238.

placed there, they immediately lose their fragrancy, and stink like carrion." An air so putrifying can never be very commodious for respiration.

Indeed, if we examine the complexion of most miners, we shall be very well able to form a judgment of the unwholesomeness of the place where they are confined. Their pale and sallow looks show how much the air is damaged by passing through those deep and winding ways, that are rendered humid by damps, or warmed with noxious exhalations. But although every mine is unwholesome, all are not equally so. Coal-mines are generally less noxious than those of tin; tin than those of copper; but of all, none are so dreadfully destructive as those of quicksilver. At the mines near the village of Idra, nothing can adequately describe the deplorable infirmities of such as fill the hospital there; emaciated and crippled, every limb contracted or convulsed, and some in a manner transpiring quicksilver at every pore. There was one man, says Dr. Pope,[*] who was not in the mines above half a year, and yet whose body was so impregnated with this mineral, that putting a piece of brass money in his mouth, or rubbing it between his fingers, it immediately became as white as if it had been washed over with quicksilver. In this manner all the workmen are killed sooner or later; first becoming paralytic, and then dying consumptive: and all this they sustain for the trifling reward of sevenpence a day.

But these metallic mines are not so noxious from their own vapours, as from those of the substances with which the ores are usually united, such as arsenic, cinnabar, bitumen, or vitriol. From the fumes of these, variously combined, and kept inclosed, are produced those various damps, that put on so many dreadful forms, and are usually so fatal. Sometimes those noxious vapours are perceived by the delightful fragrance of their smell,[†] somewhat resembling the pea-blossom in bloom, from whence one kind of damp has its name. The miners are not deceived, however, by its flattering appearances, but as they thus have timely notice of its coming, they avoid it while it continues, which is generally during the whole summer season. Another shows its approach by the burning of the candles, which seem to collect their flame into a globe of light, and thus gradually lessen, till they are quite extinguished. From this also, the miners frequently escape; however, such as have the misfortune to be caught in it, either swoon away, and are suffocated, or slowly recover in excessive agonies. Here also is a third, called *the fulminating damp*, much more dangerous than either of the former, as it strikes down all before it like a flash of gunpowder, without giving any warning of its approach. But there is still another, more deadly than all the rest, which is found in those places where the vapour has been long confined, and has been, by some accident, set free. The air rushing out from thence, always goes upon deadly errands; and scarce any escape to describe the symptoms of its operations.

Some colliers in Scotland, working near an old mine that had been long closed up, happened, inadvertently, to open a hole into it, from

---

[*] Phil. Trans. vol. ii. p. 578.   [†] Phil. Trans. vol. ii. p. 375.

the pit where they were then employed. By great good fortune, they at that time perceived their error, and instantly fled for their lives The next day, however, they were resolved to renew their work in the same pit, and eight of them ventured down, without any great apprehensions; but they had scarce got to the bottom of the stairs that led to the pit, when, coming within the vapour, they all instantly dropped down dead, as if they had been shot. Amongst these unfortunate poor men, there was one whose wife was informed he was stifled in the mine; and, as he happened to be next the entrance, she so far ventured down as to see where he lay. As she approached the place, the sight of her husband inspired her with a desire to rescue him, if possible, from that dreadful situation; though a little reflection might have shown her it was then too late. But nothing could deter her; she ventured forward, and had scarce touched him with her hand, when the damp prevailed, and the misguided, but faithful creature, fell dead by his side.

Thus the vapours found beneath the surface of the earth, are very various in their effects upon the constitution: and they are not less in their appearances. There are many kinds that seemingly are no way prejudicial to health, but in which the workmen breathe freely; and yet in these, if a lighted candle be introduced, they immediately take fire, and the whole cavern at once becomes one furnace of flame. In mines, therefore, subject to damps of this kind, they are obliged to have recourse to a very peculiar contrivance to supply sufficient light for their operations. This is by a great wheel; the circumference of which is beset with flints, which, striking against steels placed for that purpose at the extremity, a stream of fire is produced, which affords light enough, and yet which does not set fire to the mineral vapour.

Of this kind are the vapours of the mines about Bristol: on the contrary, in other mines, a single spark struck out from the collision of flint and steel, would set the whole shaft in a flame. In such, therefore, every precaution is used to avoid collision; the workmen making use only of wooden instruments in digging; and being cautious, before they enter the mine, to take out even the nails from their shoes. Whence this strange difference should arise, that the vapours of some mines catch fire with a spark, and others only with a flame, is a question that we must be content to leave in obscurity, till we know more of the nature both of mineral vapour and of fire. This only we may observe, that gunpowder will readily fire with a spark, but not with the flame of a candle: on the other hand, spirits of wine will flame with a candle, but not with a spark; but even here the cause of this difference as yet remains a secret.

As from this account of mines, it appears that the internal parts of the globe are filled with vapours of various kinds, it is not surprising that they should, at different times, reach the surface, and there put on various appearances. In fact, much of the salubrity, and much of the unwholesomeness of climates and soils, is to be ascribed to these vapours, which make their way from the bowels of the earth upwards, and refresh or taint the air with their exhalations. San mines, being naturally cold,* send forth a degree of coldness to the external air, to comfort and refresh it: on the contrary, metallic mines are known, not only to warm it with their exhalations, but often to destroy all kinds of vegetation by their volatile, corrosive fumes. In some mines dense vapours are plainly perceived issuing from their mouths, and sensibly warm to the touch. In some places, neither snow nor ice will continue on the ground that covers a mine; and over others the fields are found destitute of verdure.† The inhabitants, also, are rendered dreadfully sensible of these subterraneous exhalations, being affected with such a variety of evils proceeding entirely from this cause, that books have been professedly written upon this class of disorders.

Nor are these vapours, which thus escape to the surface of the earth, entirely unconfined; for they are frequently, in a manner, circumscribed to a spot. The grotto Del Cane, near Naples, is an instance of this; the noxious effects of which have made that cavern so very famous. This grotto, which has so much employed the attention of travellers, lies within four miles of Naples, and is situated near a large lake of clear and wholesome water.‡ Nothing can exceed the beauty of the landscape which this lake affords; being surrounded with hills covered with forests of the most beautiful verdure, and the whole bearing a kind of amphitheatrical appearance. However, this region, beautiful as it appears, is almost entirely uninhabited; the few peasants that necessity compels to reside there, looking quite consumptive and ghastly, from the poisonous exhalations that rise from the earth. The famous grotto lies on the side of a hill, near which place a peasant resides, who keeps a number of dogs for the purpose of showing the experiment to the curious. These poor animals always seem perfectly sensible of the approach of a stranger, and endeavour to get out of the way. However, their attempts being perceived, they are taken and brought to the grotto; the noxious effects of which they have so frequently experienced. Upon entering this place, which is a little cave, or hole rather, dug into the hill, about eight feet high, and twelve feet long, the observer can see no visible mark of its pestilential vapour; only to about a foot from the bottom, the wall seems to be tinged with a colour resembling that which is given by stagnant waters. When the dog, this poor philosophical martyr, as some have called him, is held above this mark, he does not seem to feel the smallest inconvenience; but when his head is thrust down lower, he struggles to get free for a little; but in the space of four or five minutes he seems to lose all sensation, and is taken out seemingly without life. Being plunged in the neighbouring lake, he quickly recovers, and is permitted to run home, seemingly without the smallest injury

This vapour, which thus for a time suffocates, is of the humid kind, as it extinguishes a torch, and sullies a looking-glass; but there are other vapours perfectly inflammable, and that only require the approach of a candle to set them blazing. Of this kind was the burning well at Brosely, which is now stopped up; the vapour of which,

---

* Phil. Trans. vol. ii. p. 523.   † Boyle, vol. iii. p. 228
‡ Kircher, Mund. Subt. vol. i. p. 191.

when a candle was brought within about a foot of the surface of the water, caught flame like spirit of wine, and continued blazing for several hours after. Of this kind, also, are the perpetual fires in the kingdom of Persia. In that province, where the worshippers of fire hold their chief mysteries, the whole surface of the earth, for some extent, seems impregnated with inflammable vapours. A reed stuck into the ground continues to burn like a flambeau; a hole made beneath the surface of the earth, instantly becomes a furnace, answering all the purposes of a culinary fire. There they make lime by merely burying the stones in the earth, and watch with veneration the appearances of a flame that has not been extinguished for times immemorial. How different are men in various climates! This deluded people worship the vapours as a deity, which, in other parts of the world are considered as one of the greatest evils.

CHAPTER IX.

OF VOLCANOES AND EARTHQUAKES.

MINES and caverns, as we have said, reach but a very little way under the surface of the earth, and we have hitherto had no opportunities of exploring farther. Without all doubt, the wonders that are still unknown surpass those that have been represented, as there are depths of thousands of miles which are hidden from our inquiry. The only tidings we have from these unfathomable regions are by means of volcanoes, those burning mountains that seem to discharge their materials from the lowest abysses of the earth.* A volcano may be considered as a cannon of immense size, the mouth of which is often near two miles in circumference. From this dreadful aperture are discharged torrents of flame and sulphur, and rivers of melted metal. Whole clouds of smoke and ashes, with rocks of enormous size, are discharged to many miles distance; so that the force of the most powerful artillery, is but as a breeze agitating a feather in comparison. In the deluge of fire and melted matter which runs down the sides of the mountain, whole cities are sometimes swallowed up and consumed. Those rivers of liquid fire are sometimes two hundred feet deep; and when they harden, frequently form considerable hills. Nor is the danger of these confined to the eruption only; but the force of the internal fire struggling for vent, frequently produces earthquakes through the whole region where the volcano is situated. So dreadful have been these appearances, that men's terrors have added new horrors to the scene, and they have regarded as prodigies, what we know to be the result of natural causes. Some philosophers have considered them as vents communicating with the fires of the centre; and the ignorant, as the mouths of hell itself. Astonishment produces fear, and fear superstition: the inhabitants of Iceland believe the bellowings of Hecla are nothing else but the cries of the damned, and that its eruptions are contrived to increase their tortures.

* Buffon, vol i. p. 291.

But if we regard this astonishing scene of terror with a more tranquil and inquisitive eye, we shall find that these conflagrations are produced by very obvious and natural causes. We have already been apprised of the various mineral substances in the bosom of the earth, and their aptness to burst out into flames. Marcasites and pyrites, in particular, by being humified with water or air, contract this heat, and often endeavour to expand with irresistible explosion. These, therefore, being lodged in the depths of the earth, or in the bosom of mountains, and being either washed by the accidental influx of waters below, or fanned by air, insinuating itself through perpendicular fissures from above, take fire at first by only heaving in earthquakes, but at length by bursting through every obstacle, and making their dreadful discharge in a volcano.

These volcanoes are found in all parts of the earth: in Europe there are three that are very remarkable: Ætna in Sicily, Vesuvius in Italy, and Hecla in Iceland. Ætna has been a volcano for ages immemorial. Its eruptions are very violent, and its discharge has been known to cover the earth sixty-eight feet deep. In the year 1537, an eruption of this mountain produced an earthquake through the whole island for twelve days, overturned many houses, and at last formed a new aperture, which overwhelmed all within five leagues round. The cinders thrown up were driven even into Italy, and its burnings were seen at Malta at the distance of sixty leagues. "There is nothing more awful," says Kircher, "than the eruptions of this mountain, nor nothing more dangerous than attempting to examine its appearances, even long after the eruption has ceased. As we attempt to clamber up its steepy sides, every step we take upward, the feet sink back half way. Upon arriving near the summit, ashes and snow, with an ill-assorted conjunction, present nothing but objects of desolation. Nor is this the worst, for, as all places are covered over, many caverns are entirely hidden from the sight, into which if the inquirer happens to fall, he sinks to the bottom, and meets inevitable destruction. Upon coming to the edge of the great crater, nothing can sufficiently represent the tremendous magnificence of the scene. A gulf two miles over, and so deep that no bottom can be seen; on the sides pyramidical rocks starting out between apertures that emit smoke and flame; all this accompanied with a sound that never ceases, louder than thunder, strikes the bold with horror, and the religious with veneration for HIM that has power to control its burnings."

In the descriptions of Vesuvius, or Hecla, we shall find scarcely any thing but a repetition of the same terrible objects, but rather lessened, as these mountains are not so large as the former. The crater of Vesuvius is but a mile across, according to the same author; whereas that of Ætna is two. On this particular, however, we must place no dependence, as these caverns every day alter; being lessened by the mountain's sinking at one eruption, and enlarged by the fury of another. It is not one of the least remarkable particulars respecting Vesuvius, that Pliny the naturalist was suffocated in one of its eruptions; for his curiosity impelling him too near, he found himself involved in smoke and cinders when it was too late to retire; and his companions hardly escaped to give an account of the misfortune. It was in that dread ful eruption that the city of Herculaneum was overwhelmed; the ruins of which have lately been discovered at sixty feet distance below the surface, and, what is still more remarkable, forty feet below the bed of the sea. One of the most remarkable eruptions of this mountain was in the year 1707, which is finely described by Valetta; a part of whose description I shall beg leave to translate.

"Towards the latter end of summer, in the year 1707, the Mount Vesuvius, that had for a long time been silent, now began to give some signs of commotion. Little more than internal murmurs at first were heard, that seemed to contend within the lowest depths of the mountain; no flame, nor even any smoke, was as yet seen. Soon after some smoke appeared by day, and a flame by night, which seemed to brighten all the campania. At intervals, also, it shot off substances with a sound very like that of artillery, but which, even at so great a distance as we were at, infinitely exceeded them in greatness. Soon after it began to throw up ashes, which becoming the sport of the winds, fell at great distances, and some many miles. To this succeeded showers of stones, which killed many of the inhabitants of the valley, but made a dreadful ravage among the cattle. Soon after a torrent of burning matter began to roll down the sides of the mountain, at first with a slow and gentle motion, but soon with increased celerity. The matter thus poured out, when cold, seemed upon inspection to be of vitrified earth, the whole united into a mass of more than stony hardness. But what was particularly observable was, that upon the whole surface of these melted materials, a light spongy stone seemed to float, while the lower body was of the hardest substance of which our roads are usually made. Hitherto there were no appearances but what had been often remarked before; but on the third or fourth day, seeming flashes of lightning were shot forth from the mouth of the mountain, with a noise far exceeding the loudest thunder. These flashes, in colour and brightness, resembled what we usually see in tempests, but they assumed a more twisted and serpentine form. After this followed such clouds of smoke and ashes, that the whole city of Naples, in the midst of the day, was involved in nocturnal darkness, and the nearest friends were unable to distinguish each other in this frightful gloom. If any person attempted to stir out without torch-light, he was obliged to return, and every part of the city was filled with supplications and terror. At length, after a continuance of some hours, about one o'clock at midnight, the wind blowing from the north, the stars began to be seen; the heavens, though it was night, began to grow brighter; and the eruptions, after a continuance of fifteen days, to lessen. The torrent of melted matter was seen to extend from the mountain down to the shore; the people began to return to their former dwellings, and the whole face of nature to resume its former appearance."

The famous Bishop Berkeley gives an account of one of these eruptions in a manner something different from the former.* "In the year 1717, and the middle of April, with much difficulty I reached the top of Mount Vesuvius, in which I saw a vast aperture full of smoke

* Phil. Trans. vol. II. p. 209.

which hindered me from seeing its depth and figure. I heard within that horrid gulf certain extraordinary sounds, which seemed to proceed from the bowels of the mountain; a sort of murmuring, sighing, dashing sound; and, between whiles, a noise like that of thunder or cannon, with a clattering like that of tiles falling from the tops of houses into the streets. Sometimes, as the wind changed, the smoke grew thinner, discovering a very ruddy flame, and the circumference of the crater streaked with red and several shades of yellow. After an hour's stay, the smoke, being moved by the wind, gave us short and partial prospects of the great hollow; in the flat bottom of which I could discern two furnaces almost contiguous; that on the left seeming about three yards over, glowing with ruddy flame, and throwing up red-hot stones with a hideous noise, which, as they fell back, caused the clattering already taken notice of.—May 8, in the morning, I ascended the top of Vesuvius a second time, and found a different face of things. The smoke ascending upright, gave a full prospect of the crater, which, as I could judge, was about a mile in circumference, and a hundred yards deep. A conical mount had been formed, since my last visit, in the middle of the bottom, which I could see was made by the stones, thrown up and fallen back again into the crater. In this new hill remained the two furnaces already mentioned. The one was seen to throw up every three or four minutes, with a dreadful sound, a vast number of red-hot stones, at least three hundred feet higher than my head, as I stood upon the brink; but as there was no wind, they fell perpendicularly back from whence they had been discharged. The other was filled with red-hot liquid matter, like that in the furnace of a glass-house, raging and working like the waves of the sea, with a short abrupt noise. This matter would sometimes boil over, and run down the side of the conical hill, appearing at first red-hot, but changing colour as it hardened and cooled. Had the wind driven in our faces, we had been in no small danger of stifling by the sulphureous smoke, or being killed by the masses of melted minerals that were shot from the bottom. But as the wind was favourable, I had an opportunity of surveying this amazing scene for above an hour and a half together. On the fifth of June, after a horrid noise, the mountain was seen at Naples to work over; and about three days after, its thunders were renewed so, that not only the windows in the city, but all the houses, shook. From that time it continued to overflow, and sometimes at night were seen columns of fire shooting upward from its summit. On the tenth, when all was thought to be over, the mountain again renewed its terrors, roaring and raging most violently. One cannot form a juster idea of the noise, in the most violent fits of it, than by imagining a mixed sound made up of the raging of a tempest, the murmur of a troubled sea, and the roaring of thunder and artillery confused all together. Though we heard this at the distance of twelve miles, yet it was very terrible. I therefore resolved to approach nearer to the mountain; and, accordingly, three or four of us got into a boat, and were set ashore at a little town situated at the foot of the mountain. From thence we rode about four or five miles, before we came to the torrent of fire that was descending from the side of the volcano; and here the roaring grew exceedingly loud and

THE EARTH.

terrible as we approached. I observed a mixture of colours in the cloud, above the crater, green, yellow, red, and blue. There was likewise a ruddy dismal light in the air, over that tract where the burning river flowed. These circumstances, set off and augmented by the horror of the night, made a scene the most uncommon and astonishing I ever saw; which still increased as we approached the burning river. Imagine a vast torrent of liquid fire, rolling from the top, down the side of the mountain, and with irresistible fury bearing down and consuming vines, olives, and houses; and divided into different channels, according to the inequalities of the mountain. The largest stream seemed half a mile broad at least, and five miles long. I walked so far before my companions up the mountain, along the side of the river of fire, that I was obliged to retire in great haste, the sulphureous steam having surprised me, and almost taken away my breath. During our return, which was about three o'clock in the morning, the roaring of the mountain was heard all the way, while we observed it throwing up huge spouts of fire and burning stones, which falling, resembled the stars in a rocket. Sometimes I observed two or three distinct columns of flame, and sometimes one only that was large enough to fill the whole crater. These burning columns, and fiery stones, seemed to be shot a thousand feet perpendicular above the summit of the volcano; and in this manner the mountain continued raging for six or eight days after. On the 18th of the same month, the whole appearance ended, and the mountain remained perfectly quiet, without any visible smoke or flame."

The matter which is found to roll down from the mouth of all volcanoes in general, resembles the dross that is thrown from a smith's forge. But it is different, perhaps, in various parts of the globe; for, as we have already said, there is not a quarter of the world that has not its volcanoes. In Asia, particularly in the islands of the Indian Ocean, there are many. One of the most famous is that of Albouras, near Mount Taurus, the summit of which is continually on fire, and covers the whole adjacent country with ashes. In the island of Ternate there is a volcano, which, some travellers assert, burns most furiously in the times of the equinoxes, because of the winds which then contribute to increase the flames. In the Molucca islands, there are many burning mountains; they are also seen in Japan, and the islands adjacent; and in Java and Sumatra, as well as in other of the Philippine islands. In Africa there is a cavern, near Fez, which continually sends forth either smoke or flames. In the Cape de Verde islands, one of them, called *the Island del Fuego*, continually burns; and the Portuguese, who frequently attempted a settlement there, have as often been obliged to desist. The Peak of Teneriffe is, as every body knows, a volcano, that seldom desists from eruptions. But of all parts of the earth, America is the place where those dreadful irregularities of nature are the most conspicuous. Vesuvius, and Ætna itself, are but mere fire-works in comparison to the burning mountains of the Andes; which, as they are the highest mountains of the world, so also are they the most formidable for their eruptions. The mountain of Arequipa in Peru, is one of the most celebrated; Tarassa, and Malahallo, are very considerable; but that of Cotopaxi, in the province of Quito, exceeds any thing we have hitherto read or heard of The mountain of Cotopaxi, as described by Ulloa,* is more than three miles perpendicular from the sea; and it became a volcano at the time of the Spaniards' first arrival in that country. A new eruption of it happened in the year 1743, having been some days preceded by a continual roaring in its bowels. The sound of one of these mountains is not, like that of the volcanoes in Europe, confined to a province, but is heard at a hundred and fifty miles distance.† "An aperture was made in the summit of this immense mountain; and three more about equal heights, near the middle of its declivity, which was at that time buried under prodigious masses of snow. The ignited substances ejected on that occasion, mixed with a prodigious quantity of ice and snow, melting amidst the flames, were carried down with such astonishing rapidity, that in an instant the valley from Callo to Latacunga was overflowed; and besides its ravages in bearing down the houses of the Indians, and other poor inhabitants, great numbers of people lost their lives. The river of Latacunga was the channel of this terrible flood; till being too small for receiving such a prodigious current, it overflowed the adjacent country, like a vast lake near the town, and carried away all the buildings within its reach. The inhabitants retired into a spot of higher ground behind the town, of which those parts which stood within the limits of the current were totally destroyed. The dread of still greater devastations did not subside for three days; during which the volcano ejected cinders, while torrents of melted ice and snow poured down its sides. The eruption lasted several days, and was accompanied with terrible roarings of the wind, rushing through the volcano, still louder than the former rumblings in its bowels. At last all was quiet, neither fire nor smoke to be seen, nor noise to be heard; till, in the ensuing year, the flames again appeared with recruited violence, forcing their passage through several other parts of the mountain, so that in clear nights the flames being reflected by the transparent ice, formed an awfully magnificent illumination."

Such is the appearance and the effect of those fires which proceed from the more inward recesses of the earth: for that they generally come from deeper regions than man has hitherto explored, I cannot avoid thinking, contrary to the opinion of Mr. Buffon, who supposes them rooted but a very little way below the bed of the mountain. "We can never suppose," says this great naturalist, "that these substances are ejected from any great distance below, if we only consider the great force already required to fling them up to such vast heights above the mouth of the mountain; if we consider the substances thrown up, which we shall find upon inspection to be the same with those of the mountain below; if we take into our consideration, that air is always necessary to keep up the flame; but, most of all, if we attend to one circumstance, which is, that if these substances were exploded from a vast depth below, the same force required to shoot them up so high, would act against the sides of the volcano, and tear the whole mountain in pieces." To all this specious reasoning, par-

---

* Ulloa, vol. i. p. 442.   † Ulloa, vol. i. p. 442

ticular answers might be easily given; as, that the length of the funnel increases the force of the explosion; that the sides of the funnel are actually often burst with the great violence of the flame; that air may be supposed at depths at least as far as the perpendicular fissures descend. But the best answer is a well-known fact; namely, that the quantity of matter discharged from Ætna alone, is supposed, upon a moderate computation, to exceed twenty times the original bulk of the mountain.* The greatest part of Sicily seems covered with its eruptions. The inhabitants of Catanea have found, at the distance of several miles, streets and houses sixty feet deep, overwhelmed by the lava or matter it has discharged. But what is still more remarkable, the walls of these very houses have been built of materials evidently thrown up by the mountain. The inference from all this is very obvious; that the matter thus exploded cannot belong to the mountain itself, otherwise it would have been quickly consumed; it cannot be derived from moderate depths, since its amazing quantity evinces, that all the places near the bottom must have long since been exhausted; nor can it have an extensive, and, if I may so call it, a superficial spread, for then the country round would be quickly undermined; it must, therefore, be supplied from the deeper regions of the earth; those undiscovered tracts where the Deity performs his wonders in solitude, satisfied with self-approbation!

CHAPTER X.

OF EARTHQUAKES.

HAVING given the theory of volcanoes, we have, in some measure, given also that of earthquakes. They both seem to proceed from the same cause, only with this difference, that the fury of the volcano is spent in the eruption; that of an earthquake spreads wider, and acts more fatally by being confined. The volcano only affrights a province, earthquakes have laid whole kingdoms in ruin.

Philosophers† have taken some pains to distinguish between the various kinds of earthquakes, such as the tremulous, the pulsative, the perpendicular, and the inclined; but these are rather the distinctions of art than of nature, mere accidental differences arising from the situation of the country or of the cause. If, for instance, the confined fire acts directly under a province or a town, it will heave the earth perpendicularly upward, and produce a *perpendicular* earthquake. If it acts at a distance, it will raise that tract obliquely, and thus the inhabitants will perceive an *inclined* one.

Nor does it seem to me that there is much greater reason for Mr. Buffon's distinction of earthquakes. One kind of which he supposes‡ to be produced by fire in the manner of volcanoes, and confined to but a very narrow circumference. The other kind he ascribes to the struggles of confined air, expanded by heat in the bowels of the earth, and endeavouring to get free. For how do these two causes differ?

* Kircher. Mund. Subt. vol. i. p. 202. †Aristotle, Agricola, Buffon. ‡ Buffon, vol. ii. p. 328

Fire is an agent of no power whatsoever without air. It is the air, which being at first compressed, and then dilated in a cannon, that drives the ball with such force. It is the air struggling for vent in a volcano, that throws up its contents to such vast heights. In short, it is the air confined in the bowels of the earth, and acquiring elasticity by heat, that produces all those appearances which are generally ascribed to the operation of fire. When, therefore, we are told that there are two causes of earthquakes, we only learn that a greater or smaller quantity of heat produces those terrible effects; for air is the only active operator in either.

Some philosophers, however, have been willing to give the air as great a share in producing these terrible effects as they could; and, magnifying its powers, have called in but a very moderate degree of heat to put it in action. Although experience tells us that the earth is full of inflammable materials, and that fires are produced wherever we descend; although it tells us that those countries where there are volcanoes are most subject to earthquakes; yet they step out of their way, and so find a new solution. These only allow but just heat enough to produce the most dreadful phenomena, and backing their assertions with long calculations, give theory an air of demonstration Mr. Amontons* has been particularly sparing of the internal heat in this respect; and has shown, perhaps accurately enough, that a very moderate degree of heat may suffice to give the air amazing powers of expansion.

It is amusing enough, however, to trace the progress of a philosophical fancy let loose in imaginary speculations. They run thus: "A very moderate degree of heat may bring the air into a condition capable of producing earthquakes; for the air, at the depth of forty-three thousand five hundred and twenty-eight fathom below the surface of the earth, becomes almost as heavy as quicksilver. This, however, is but a very slight depth in comparison of the distance to the centre, and is scarce a seventieth part of the way. The air, therefore, at the centre, must be infinitely heavier than mercury, or any body that we know of. This granted, we shall take something more, and say, that it is very probable there is nothing but air at the centre. Now let us suppose this air heated, by some means, even to the degree of boiling water, as we have proved that the density of the air is here very great, its elasticity must be in proportion; a heat, therefore, which at the surface of the earth would have produced but a slight expansive force, must, at the centre, produce one very extraordinary, and, in short, be perfectly irresistible. Hence this force may, with great ease, produce earthquakes; and, if increased, it may convulse the globe; it may (by only adding figures enough to the calculation) destroy the solar system, and even the fixed stars themselves." These reveries generally produce nothing; for, as I have ever observed, increased calculations, while they seem to tire the memory, give the reasoning faculty perfect repose.

However, as earthquakes are the most formidable ministers of nature, it is not to be wondered that a multitude of writers have been

---

* Memoires de l'Academie des Sciences. An. 1703.

curiously employed in their consideration. Woodward has ascribed the cause to a stoppage of the waters below the earth's surface by some accident. These being thus accumulated, and yet acted upon by fires, which he supposes still deeper, both contribute to heave up the earth upon their bosom. This, he thinks, accounts for the lakes of water produced in an earthquake, as well as for the fires that sometimes burst from the earth's surface upon those dreadful occasions. There are others still who have supposed that the earth may be itself the cause of its own convulsions. "When," say they, "the roots or basis of some large tract is worn away by a fluid underneath, the earth sinking therein, its weight occasions a tremor of the adjacent parts, sometimes producing a noise, and sometimes an inundation of water." Not to tire the reader with a history of opinions instead of facts, some have ascribed them to electricity, and some to the same causes that produce thunder.

It would be tedious, therefore, to give all the various opinions that have employed the speculative upon this subject. The activity of the internal heat seems alone sufficient to account for every appearance that attends these tremendous irregularities of nature. To conceive this distinctly, let us suppose, at some vast distance under the earth, large quantities of inflammable matter, pyrites, bitumens, and marcasites, disposed, and only waiting for the aspersion of water, or the humidity of the air, to put their fires in motion: at last, this dreadful mixture arrives; waters find their way into those depths, through the perpendicular fissures; or air insinuates itself through the same minute apertures: instantly new appearances ensue; those substances, which for ages before lay dormant, now conceive new apparent qualities; they grow hot, produce new air, and only want room for expansion. However, the narrow apertures by which the air or water had at first admission, are now closed up; yet as new air is continually generated, and as the heat every moment gives this air new elasticity, it at length bursts, and dilates all round; and, in its struggles to get free, throws all above it into similar convulsions. Thus an earthquake is produced, more or less extensive, according to the depth or the greatness of the cause.

But before we proceed with the causes, let us take a short view of the appearances which have attended the most remarkable earthquakes. By these we shall see how far the theorist corresponds with the historian. The greatest we find in antiquity is that mentioned by Pliny,* in which twelve cities in Asia Minor were swallowed up in one night: he tells us also of another, near the lake Thrasymene, which was not perceived by the armies of the Carthaginians and Romans, that were then engaged near that lake, although it shook the greatest part of Italy. In another place† he gives the following account of an earthquake of an extraordinary kind. "When Lucius Marcus and Sextus Julius were consuls, there appeared a very strange prodigy of the earth, (as I have read in the books of Ætruscan discipline) which happened in the province of Mutina. Two mountains shocked against each other, approaching and retiring with the most

* Pl'n lib. ii. cap. 86. † Ibid. lib. iii. cap. 85.

dreadful noise. They, at the same time, and in the midst of the day appeared to cast forth fire and smoke, while a vast number of Roman knights and travellers from the Æmilian Way, stood and continued amazed spectators. Several towns were destroyed by this shock; and all the animals that were near them were killed." In the time of Trajan, the city of Antioch, and a great part of the adjacent country, was buried by an earthquake. About three hundred years after, in the times of Justinian, it was once more destroyed, together with forty thousand inhabitants; and, after an interval of sixty years, the same ill-fated city was a third time overturned, with the loss of not less than sixty thousand souls. In the year 1182, most of the cities of Syria, and the kingdom of Jerusalem, were destroyed by the same accident. In the year 1594, the Italian historians describe an earthquake at Puteoli, which caused the sea to retire two hundred yards from its former bed.

But one of those most particularly described in history, is that of the year 1693; the damages of which were chiefly felt in Sicily, but its motion perceived in Germany, France, and England. It extended to a circumference of two thousand six hundred leagues; chiefly affecting the sea-coasts, and great rivers; more perceivable also upon the mountains than in the valleys. Its motions were so rapid, that those who lay at their length were tossed from side to side, as upon a rolling billow.* The walls were dashed from their foundations; and no less than fifty-four cities, with an incredible number of villages, were either destroyed or greatly damaged. The city of Catanea, in particular, was utterly overthrown. A traveller, who was on his way thither, at the distance of some miles, perceived a black cloud, like night, hanging over the place. The sea, all of a sudden, began to roar; Mount Ætna to send forth great spires of flame; and soon after a shock ensued, with a noise as if all the artillery in the world had been at once discharged. Our traveller, being obliged to alight, instantly felt himself raised a foot from the ground; and turning his eyes to the city, he, with amazement, saw nothing but a thick cloud of dust in the air. The birds flew about astonished; the sun was darkened; the beasts ran howling from the hills; and, although the shock did not continue above three minutes, yet near nineteen thousand of the inhabitants of Sicily perished in the ruins.—Catanea, to which city the describer was travelling, seemed the principal scene of ruin; its place only was to be found; and not a footstep of its former magnificence was to be seen remaining.

The earthquake which happened in Jamaica, in 1692, was very terrible, and its description sufficiently minute. "In two minutes' time it destroyed the town of Port-Royal, and sunk the houses in a gulf forty fathoms deep. It was attended with a hollow rumbling noise, like that of thunder; and, in less than a minute, three parts of the houses, and their inhabitants, were all sunk quite under water. While they were thus swallowed up on one side of the street, on the other the houses were thrown into heaps; the sand of the street rising like the waves of the sea, lifting up those that stood upon it, and immedi-

* Phil. Trans.

ately overwhelming them in pits. All the wells discharged their waters with the most vehement agitation. The sea felt an equal share of turbulence, and, bursting over its mounds, deluged all that came in its way. The fissures of the earth were, in some places, so great, that one of the streets appeared twice as broad as formerly. In many places, however, it opened and closed again, and continued this agitation for some time. Of these openings, two or three hundred might be seen at a time; in some whereof the people were swallowed up; in others, the earth closing, caught them by the middle, and thus crushed them instantly to death. Other openings, still more dreadful than the rest, swallowed up whole streets; and others, more formidable still, spouted up whole cataracts of water, drowning such as the earthquake had spared. The whole was attended with the most noisome stench; while the thundering of the distant falling mountains, the whole sky overcast with a dusky gloom, and the crash of falling habitations, gave unspeakable horror to the scene. After this dreadful calamity was over, the whole island seemed converted into a scene of desolation; scarce a planter's house was left standing; almost all were swallowed up; houses, people, trees, shared one universal ruin; and in their places appeared great pools of water, which, when dried up by the sun, left only a plain of barren sand, without any vestige of former inhabitants. Most of the rivers, during the earthquake, were stopt up by the falling in of the mountains; and it was not till after some time that they made themselves new channels. The mountains seemed particularly attacked by the force of the shock; and it was supposed that the principal seat of the concussion was among them. Those who were saved got on board ships in the harbour, where many remained above two months; the shocks continuing, during that interval, with more or less violence every day."

As this description seems to exhibit all the appearances that usually make up the catalogue of terrors belonging to an earthquake, I will suppress the detail of that which happened at Lisbon, in our own times, and which is too recent to require a description. In fact, there are few particulars, in the accounts of those who were present at that scene of desolation, that we have not more minutely and accurately transmitted to us by former writers, whose narratives I have for that reason preferred. I will therefore close this description of human calamities, with the account of the dreadful earthquake at Calabria, in 1638. It is related by the celebrated Father Kircher, as it happened while he was on his journey to visit Mount Ætna, and the rest of the wonders that lie towards the south of Italy. I need scarce inform the reader, that Kircher is considered, by scholars, as one of the greatest prodigies of learning.

"Having hired a boat, in company with four more, two friars of the order of St. Francis, and two seculars, we launched, on the twenty-fourth of March, from the harbour of Messina, in Sicily, and arrived the same day at the promontory of Pelorus. Our destination was for the city of Euphæmia, in Calabria, where we had some business to transact, and where we designed to tarry for some time. However, Providence seemed willing to cross our design; for we were obliged to continue for three days at Pelorus, upon account of the weather;

and though we often put out to sea, yet we were as often driven back. At length, however, wearied with the delay, we resolved to prosecute our voyage; and, although the sea seemed more than usually agitated, yet we ventured forward. The gulf of Charybdis, which we approached, seemed whirled round in such a manner, as to form a vast hollow, verging to a point in the centre. Proceeding onward, and turning my eyes to Ætna, I saw it cast forth large volumes of smoke, of mountainous sizes, which entirely covered the whole island, and blotted out the very shores from my view. This, together with the dreadful noise, and the sulphureous stench, which was strongly perceived, filled me with apprehensions that some more dreadful calamity was impending. The sea itself seemed to wear a very unusual appearance; those who have seen a lake in a violent shower of rain covered all over with bubbles, will conceive some idea of its agitations. My surprise was still increased by the calmness and serenity of the weather; not a breeze, not a cloud, which might be supposed to put all nature thus into motion. I therefore warned my companions that an earthquake was approaching; and, after some time, making for the shore with all possible diligence, we landed at Tropæa, happy and thankful for having escaped the threatening dangers of the sea.

"But our triumphs at land were of short duration; for we had scarce arrived at the Jesuits' College in that city, when our ears were stunned with a horrid sound, resembling that of an infinite number of chariots driven fiercely forward, the wheels rattling, and the thongs cracking. Soon after this, a most dreadful earthquake ensued, so that the whole tract upon which we stood seemed to vibrate, as if we were in the scale of a balance that continued wavering. This motion, however, soon grew more violent; and being no longer able to keep my legs, I was thrown prostrate upon the ground. In the mean time, the universal ruin round me redoubled my amazement. The crash of falling houses, the tottering of towers, and the groans of the dying, all contributed to raise my terror and despair. On every side of me I saw nothing but a scene of ruin, and danger threatening wherever I should fly. I commended myself to God, as my last great refuge. At that hour, O how vain was every sublunary happiness! wealth, honour, empire, wisdom, all mere useless sounds, and as empty as the bubbles in the deep. Just standing on the threshold of eternity, nothing but God was my pleasure; and the nearer I approached, I only loved him the more.—After some time, however, finding that I remained unhurt amidst the general concussion, I resolved to venture for safety, and running as fast as I could, reached the shore, but almost terrified out of my reason. I did not search long here till I found the boat in which I had landed, and my companions also, whose terrors were even greater than mine. Our meeting was not of that kind where every one is desirous of telling his own happy escape; it was all silence, and a gloomy dread of impending terrors.

"Leaving this seat of desolation, we prosecuted our voyage along the coast; and the next day came to Rochetta, where we landed, although the earth still continued in violent agitations. But we were scarce arrived at our inn, when we were once more obliged to return to the boat, and in about half an hour we saw the greatest part of the town, and the inn at which we had set up, dashed to the ground, and burying all its inhabitants beneath its ruins.

"In this manner, proceeding onward in our little vessel, finding no safety at land, and yet, from the smallness of our boat, having but a very dangerous continuance at sea, we at length landed at Lopizium, a castle midway between Tropæa and Euphæmia, the city to which, as I said before, we were bound. Here, wherever I turned my eyes, nothing but scenes of ruin and horror appeared; towns and castles levelled to the ground; Strombalo, though at sixty miles distance, belching forth flames in an unusual manner, and with a noise which I could distinctly hear. But my attention was quickly turned from more remote to contiguous danger. The rumbling sound of an approaching earthquake, which we by this time were grown acquainted with, alarmed us for the consequences; it every moment seemed to grow louder, and to approach more near. The place on which we stood now began to shake most dreadfully, so that being unable to stand, my companions and I caught hold of whatever shrub grew next us, and supported ourselves in that manner.

"After some time, this violent paroxysm ceasing, we again stood up, in order to prosecute our voyage to Euphæmia, that lay within sight. In the mean time, while we were preparing for this purpose, I turned my eyes towards the city, but could see only a frightful dark cloud that seemed to rest upon the place. This the more surprised us, as the weather was so very serene. We waited, therefore, till the cloud was past away; then turning to look for the city, it was totally sunk. Wonderful to tell! nothing but a dismal and putrid lake was seen where it stood. We looked about to find some one that could tell us of its sad catastrophe, but could see none! All was become a melancholy solitude! a scene of hideous desolation! Thus proceeding pensively along, in quest of some human being that could give us some little information, we at length saw a boy sitting by the shore, and appearing stupified with terror. Of him, therefore, we inquired concerning the fate of the city, but he could not be induced to give us an answer. We intreated him with every expression of tenderness and pity to tell us: but his senses were quite wrapt up in contemplation of the danger he had escaped. We offered him some victuals, but he seemed to loathe the sight. We still persisted in our offices of kindness; but he only pointed to the place of the city, like one out of his senses; and then running up into the woods, was never heard of after. Such was the fate of the city of Euphæmia! and as we continued our melancholy course along the shore, the whole coast, for the space of two hundred miles, presented nothing but the remains of cities, and men scattered, without a habitation, over the fields. Proceeding thus along, we at length ended our distressful voyage by arriving at Naples, after having escaped a thousand dangers both at sea and land."

The reader, I hope, will excuse me for this long translation from a favourite writer, and that the sooner, as it contains some particulars relative to earthquakes not to be found elsewhere. From the whole of these accounts we may gather, that the most concomitant circumstances are these:

A rumbling sound before the earthquake. This proceeds from the air or fire, or both, forcing their way through the chasms of the earth, and endeavouring to get free, which is also heard in volcanoes.

A violent agitation or heaving of the sea, sometimes before and sometimes after that at land. This agitation is only a similar effect produced on the waters with that at land, and may be called, for the sake of perspicuity, a *seaquake;* and this also is produced by volcanoes.

A spouting up of waters to great heights. It is not easy to describe the manner in which this is performed; but volcanoes also perform the same: Vesuvius being known frequently to eject a vast body of water.

A rocking of the earth to and fro, and sometimes a perpendicular bouncing, if it may be so called, of the same. This difference chiefly arises from the situation of the place with respect to the subterranean fire. Directly under, it lifts; at a farther distance, it rocks.

Some earthquakes seem to travel onward, and are felt in different countries at different hours the same day. This arises from the great shock being given to the earth at one place, and that, being communicated onward by an undulatory motion, successively affects different regions in its progress. As the blow given by a stone falling in a lake, is not perceived at the shores till some time after the first concussion.

The shock is sometimes instantaneous, like the explosion of gunpowder; and sometimes tremulous, and continuing for several minutes. The nearer the place where the shock is first given, the more instantaneous and simple it appears. At a greater distance, the earth redoubles the first blow with a sort of vibratory continuation.

As waters have generally so great a share in producing earthquakes, it is not to be wondered that they should generally follow those breaches made by the force of fire, and appear in the great chasms which the earthquake has opened.

These are some of the most remarkable phenomena of earthquakes, presenting a frightful assemblage of the most terrible effects of air, earth, fire, and water.

The valley of Solfatara, near Naples, seems to exhibit, in a minuter degree, whatever is seen of this horrible kind on the great theatre of nature. This plain, which is about twelve hundred feet long, and a thousand broad, is embosomed in mountains, and has in the middle of it a lake of noisome blackish water, covered with a bitumen, that floats upon its surface. In every part of this plain, caverns appear smoking with sulphur, and often emitting flames. The earth, wherever we walk over it, trembles beneath the feet. Noises of flames, and the hissing of waters, are heard at the bottom. The water sometimes spouts up eight or ten feet high. The most noisome fumes, fœtid water, and sulphureous vapours, offend the smell. A stone thrown into any of the caverns, is ejected again with considerable violence. These appearances generally prevail when the sea is any way disturbed; and the whole seems to exhibit the appearance of an earthquake in miniature. However, in this smaller scene of wonders, as well as in the greater, there are many appearances for which, perhaps we shall never account; and many questions may be asked, which no conjectures can thoroughly resolve. It was the fault of the philosophers of the last age, to be more inquisitive after the causes of things than after the things themselves. They seemed to think that a confession of ignorance cancelled their claims to wisdom; they, therefore, had a solution for every demand. But the present age has grown, if not more inquisitive, at least more modest; and none are now ashamed of that ignorance, which labour can neither remedy nor remove.

CHAPTER XI.

OF THE APPEARANCE OF NEW ISLANDS AND TRACTS; AND OF THE DISAPPEARANCE OF OTHERS.

Hitherto we have taken a survey only of the evils which are produced by subterranean fires, but we have mentioned nothing of the benefits they may possibly produce. They may be of use in warming and cherishing the ground, in promoting vegetation, and giving a more exquisite flavour to the productions of the earth. The imagination of a person who has never been out of our own mild region, can scarcely reach to that luxuriant beauty with which all nature appears clothed in those very countries that we have just now described as desolated by earthquakes, and undermined by subterranean fires. It must be granted, therefore, that though in those regions they have a greater share in the dangers, they have also a larger proportion in the benefits of nature.

But there is another advantage arising from subterranean fires, which, though hitherto disregarded by man, yet may one day become serviceable to him; I mean, that while they are found to swallow up cities and plains in one place, they are also known to produce promontories and islands in another. We have many instances of islands being thus formed in the midst of the sea, which though for a long time barren, have afterwards become fruitful seats of happiness and industry.

New islands are formed in two ways; either suddenly, by the action of subterranean fires; or more slowly, by the deposition of mud, carried down by rivers, and stopped by some accident.* With respect particularly to the first, ancient historians, and modern travellers, give us such accounts as we can have no room to doubt of. Seneca assures us, that in his time the island of Therasia appeared unexpectedly to some mariners, as they were employed in another pursuit. Pliny assures us, that thirteen islands in the Mediterranean appeared at once emerging from the water; the cause of which he ascribes rather to the retiring of the sea in those parts, than to any subterraneous elevation. However, he mentions the island of Hiera, near that of Therasia, as formed by subterraneous explosions; and adds to his list several others formed in the same manner. In one of which he relates that

* Buffon. vol. ii. p. 347.

fish in great abundance were found, and that all those who eat of them died shortly after.

"On the twenty-fourth of May,* in the year 1707, a slight earthquake was perceived at Santorin; and the day following, at sun-rising, an object was seen by the inhabitants of that island, at two or three miles distance at sea, which appeared like a floating rock. Some persons, desirous either of gain, or incited by curiosity, went there, and found, even while they stood upon this rock, that it seemed to rise beneath their feet. They perceived also, that its surface was covered with pumice-stones and oysters, which it had raised from the bottom. Every day after, until the fourteenth of June, this rock seemed considerably to increase; and then was found to be half a mile round, and about thirty feet above the sea. The earth of which it was composed seemed whitish, with a small proportion of clay. Soon after this the sea again appeared troubled, and steams arose which were very offensive to the inhabitants of Santorin. But on the sixteenth of the succeeding month, seventeen or eighteen rocks more were seen to rise out of the sea, and at length to join together. All this was accompanied with the most terrible noise, and fires that proceeded from the island that was newly formed. The whole mass, however, of all this new formed earth, uniting, increased every day, both in height and breadth, and, by the force of its explosions, cast forth rocks to seven miles distance. This continued to bear the same dreadful appearances till the month of November in the same year; and it is at present a valcano which sometimes renews its explosions. It is about three miles in circumference; and more than from thirty-five to forty feet high."

It seems extraordinary, that about this place in particular, islands have appeared at different times, particularly that of Hiera, mentioned above, which has received considerable additions in succeeding ages. Justin tells us,† that at the time the Macedonians were at war with the Romans, a new island appeared between those of Theramenes and Therasia, by means of an earthquake. We are told that this became half as large again about a thousand years after, another island rising up by its side, and joining to it, so as scarce at present to be distinguished from the former.

A new island was formed, in the year 1720, near that of Tercera, near the continent of Africa, by the same causes. In the beginning of December, at night, there was a terrible earthquake at that place, and the top of a new island appeared, which cast forth smoke in vast quantities. The pilot of a ship, who approached it, sounded on one side of this island, and could not find ground at sixty fathom: at the other side the sea was totally tinged of a different colour, exhibiting a mixture of white, blue, and green; and was very shallow. This island, on its first appearance, was larger than it is at present; for it has since that time sunk in such a manner, as to be scarcely above water.

A traveller, whom these appearances could not avoid affecting, speaks of them in this manner:‡ "What can be more surprising than

* Hist. del Acad an. 1708, p. 23. † Justin, lib. xxx. cap. 4 ‡ Phil. Tran. vol. ? p. 197.

to see fire not only break out of the bowels of the earth, but also make itself a passage through the waters of the sea! What can be more extraordinary, or foreign to our common notions of things, than to see the bottom of the sea rise up into a mountain above the water, and become so firm an island as to be able to resist the violence of the greatest storms! I know that subterraneous fires, when pent in a narrow passage, are able to raise up a mass of earth as large as an island: but that this should be done in so regular and exact a manner that the water of the sea should not be able to penetrate and extinguish those fires; that after having made so many passages, they should retain force enough to raise the earth; and, in fine, after having been extinguished, that the mass of earth should not fall down, or sink again with its own weight, but still remain in a manner suspended over the great arch below! This is what to me seems more surprising than any thing that has been related of Mount Ætna, Vesuvius, or any other volcano."

Such are his sentiments; however, there are few of these appearances any way more extraordinary than those attending volcanoes and earthquakes in general. We are not more to be surprised that inflammable substances should be found beneath the bottom of the sea, than at similar depths at land. These have all the force of fire giving expansion to air, and tending to raise the earth at the bottom of the sea, till it at length heaves above water. These marine volcanoes are not so frequent; for, if we may judge of the usual procedure of nature- it must very often happen, that, before the bottom of the sea is elevated above the surface, a chasm is opened in it, and then the water pressing in, extinguishes the volcano before it has time to produce its effects. This extinction, however, is not effected without very great resistance from the fire beneath. The water, upon dashing into the cavern, is very probably at first ejected back with great violence; and thus some of those amazing water-spouts are seen, which have so often astonished the mariner, and excited curiosity. But of these in their place.

Besides the production of those islands by the action of fire, there are others, as was said, produced by rivers or seas carrying mud, earth, and such like substances, along with their currents; and at last depositing them in some particular place. At the mouths of most great rivers, there are to be seen banks, thus formed by the sand and mud carried down with the stream, which have rested at that place, where the force of the current is diminished by its junction with the sea. These banks, by slow degrees, increase at the bottom of the deep: the water in those places, is at first found by mariners to grow more shallow; the bank soon heaves up above the surface; it is considered, for a while, as a tract of useless and barren sand: but the seeds of some of the more hardy vegetables are driven thither by the wind, take root, and thus binding the sandy surface, the whole spot is clothed in time with a beautiful verdure. In this manner there are delightful and inhabited islands at the mouths of many rivers, particularly the Nile, the Po, the Mississippi, the Ganges, and the Senegal. There has been, in the memory of man, a beautiful and large island formed in this manner, at the mouth of the river Nanquin, in China made from depositions of mud at its opening: it is not less than sixty miles long, and about twenty broad. La Loubere informs us,* in his voyage to Siam, that these sand-banks increase every day, at the mouths of all the great rivers in Asia: and hence, he asserts, that the navigation up these rivers becomes every day more difficult, and will, at one time or other, be totally obstructed. The same may be remarked with regard to the Wolga, which has at present seventy openings into the Caspian sea; and of the Danube, which has seven into the Euxine. We have had an instance of the formation of a new island not very long since at the mouth of the Humber, in England. "It is yet within the memory of man," says the relator,† "since it began to raise its head above the ocean. It began its appearance at low water, for the space of a few hours, and was buried again till the next tide's retreat. Thus successively it lived and died, until the year 1666, when it began to maintain its ground against the insult of the waves, and first invited the aid of human industry. A bank was thrown about its rising grounds, and being thus defended from the incursions of the sea, it became firm and solid, and, in a short time, afforded good pasturage for cattle. It is about nine miles in circumference, and is worth to the proprietor about eight hundred pounds a year." It would be endless to mention all the islands that have been thus formed, and the advantages that have been derived from them. However, it is frequently found, that new islands may be often considered as only turning the rivers from their former beds; so that in proportion as land is gained at one part, it is lost by the overflowing of some other.

Little, therefore, is gained by such accessions; nor is there much more by the new islands which are sometimes formed from the spoils of the continent. Mariners assure us, that there are sometimes whole plains unrooted from the main lands, by floods and tempests. These being carried out to sea, with all the trees and animals upon them, are frequently seen floating in the ocean, and exhibiting a surprising appearance of rural tranquillity in the midst of danger. The greatest part, however, having the earth at their roots at length washed away, are dispersed, and their animals drowned; but now and then some are found to brave the fury of the ocean, till being struck either among rocks or sands, they again take firm footing, and become permanent islands.

As different causes have thus concurred to produce new islands, so we have accounts of others that the same causes have contributed to destroy. We have already seen the power of earthquakes exerted in sinking whole cities, and leaving lakes in their room. There have been islands, and regions also, that have shared the same fate; and have sunk with their inhabitants never more to be heard of. Thus Pausanias‡ tells us of an island, called Chryses, that was sunk near Lemnos. Pliny mentions several; among others, the island Cea, for thirty miles, having been washed away, with several thousands of its inhabitants. But of all the noted devastations of this kind, the total

---

* Lettres Curieuses et Edifiantes, sec. xi. p. 234. † Phil. Trans. vol. iv. p. 251.
‡ Pausanias, l. 8. in Arcad. p. 503.

submersion of the island of Atalantis, as mentioned by Plato, has been most the subject of speculation. Mankind, in general, now consider the whole of his description as an ingenious fable; but when fables are grown famous by time and authority, they become an agreeable, if not a necessary part of literary information.

"About nine thousand years are passed," says Plato,* " since the island of Atalantis was in being. The priests of Egypt were well acquainted with it: and the first heroes of Athens gained much glory in their wars with the inhabitants. This island was as large as Asia Minor and Syria united; and was situated beyond the Pillars of Hercules, in the Atlantic ocean. The beauty of the buildings, and the fertility of the soil, were far beyond any thing a modern imagination can conceive; gold and ivory were every where common; and the fruits of the earth offered themselves without cultivation. The arts and courage of the inhabitants were not inferior to the happiness of their situation; and they were frequently known to make conquests, and overrun the continents of Europe and Asia." The imagination of the poetical philosopher riots in the description of the natural and acquired advantages, which they long enjoyed in this charming region. "If," says he, " we compare that country to our own, ours will appear a mere wasted skeleton, when opposed to it. The mountains to the very tops were clothed with fertility, and poured down rivers to enrich the plains below."

However, all these beauties and benefits were destroyed in one day by an earthquake sinking the earth, and the sea overwhelming it. At present not the smallest vestiges of such an island are to be found; Plato remains as the only authority for its existence: and philosophers dispute about its situation. It is not for me to enter into the controversy, when there appears but little probability to support the fact; and, indeed, it would be useless to run back nine thousand years in search of difficulties, as we are surrounded with objects that more closely affect us, and that demand admiration at our very doors. When I consider, as Lactantius suggests, the various vicissitudes of nature; lands swallowed by yawning earthquakes, or overwhelmed in the deep; rivers and lakes disappearing, or dried away; mountains levelled into plains; and plains swelling up into mountains; I cannot help regarding this earth as a place of very little stability; as a transient abode of still more transitory beings.

CHAPTER XII.

OF MOUNTAINS.

HAVING at last, in some measure, emerged from the deeps of the earth, we come to a scene of greater splendour; the contemplation of its external appearance. In this survey, its mountains are the first objects that strike the imagination, and excite our curiosity. There is not, perhaps, any thing in all nature that impresses an unaccustomed

* Plato in Critia.

spectator with such ideas of awful solemnity, as these immense piles of Nature's erecting, that seem to mock the minuteness of human magnificence.

In countries where there are nothing but plains, the smallest elevations are apt to excite wonder. In Holland, which is all a flat, they show a little ridge of hills, near the sea-side, which Boerhaave generally marked out to his pupils, as being mountains of no small consideration. What would be the sensations of such an auditory, could they at once be presented with a view of the heights and precipices of the Alps or the Andes! Even among us in England, we have no adequate ideas of a mountain-prospect; our hills are generally sloping from the plain, and clothed to the very top with verdure; we can scarce, therefore, lift our imaginations to those immense piles, whose tops peep up behind intervening clouds, sharp and precipitate, and reach to heights that human avarice or curiosity have never been able to ascend.

We, in this part of the world, are not, for that reason, so immediately interested in the question which has so long been agitated among philosophers, concerning what gave rise to these inequalities on the surface of the globe. In our own happy region, we generally see no inequalities but such as contribute to use and beauty; and we, therefore, are amazed at a question, inquiring how such necessary inequalities came to be formed, and seeming to express a wonder how the globe comes to be so beautiful as we find it. But though with us there may be no great cause for such a demand, yet in those places where mountains deform the face of nature, where they pour down cataracts, or give fury to tempests, there seems to be good reason for inquiry either into their causes or their uses. It has been, therefore, asked by many, in what manner mountains have come to be formed; or for what uses they are designed?

To satisfy curiosity in these respects, much reasoning has been employed, and very little knowledge propagated. With regard to the first part of the demand, the manner in which mountains were formed, we have already seen the conjectures of different philosophers on that head. One supposing that they were formed from the earth's broken shell at the time of the deluge; another, that they existed from the creation, and only acquired their deformities in process of time; a third, that they owed their original to earthquakes; and still a fourth, with much more plausibility than the rest, ascribing them entirely to the fluctuations of the deep, which he supposes in the beginning to have covered the whole earth. Such as are pleased with disquisitions of this kind, may consult Burnett, Whiston, Woodward, or Buffon. Nor would I be thought to decry any mental amusements, that at worst keep us innocently employed; but for my own part, I cannot help wondering how the opposite demand has never come to be made; and why philosophers have never asked how we come to have plains? Plains are sometimes more prejudicial to man than mountains. Upon plains, an inundation has greater power; the beams of the sun are often collected there with suffocating fierceness; they are sometimes found desert for several hundred miles together, as in the country east of the Caspian sea, although otherwise fruitful, merely because there are no risings nor depressions to form reservoirs, or collect the smallest rivulet of water. The most rational answer, therefore, why either mountains or plains were formed, seems to be that they were thus fashioned by the hand of Wisdom, in order that pain and pleasure should be so contiguous, as that mortality might be exercised either in bearing the one, or communicating the other.

Indeed, the more I consider this dispute respecting the formation of mountains, the more I am struck with the futility of the question. There is neither a straight line nor an exact superficies, in all nature. If we consider a circle, even with mathematical precision, we shall find it formed of a number of small right lines, joining at angles together. These angles, therefore, may be considered in a circle as mountains are upon our globe; and to demand the reason for the one being mountainous, or the other angular, is only to ask, why a circle is a circle, or a globe is a globe. In short, if there be no surface without inequality in nature, why should we be surprised that the earth has such? It has often been said, that the inequalities of its surface are scarce distinguishable, if compared to its magnitude; and I think we have every reason to be content with the answer.

Some, however, have avoided the difficulty by urging the final cause. They allege that mountains have been formed merely because they are useful to man. This carries the inquirer but a part of the way; for no one can affirm, that in all places they are useful. The contrary is known, by horrid experience, in those valleys that are subject to their influence. However, as the utility of any part of our earthly habitation is a very pleasing and flattering speculation to every philosopher, it is not to be wondered that much has been said to prove the usefulness of these. For this purpose, many conjectures have been made, that have received a degree of assent even beyond their evidence; for men were unwilling to become more miserably wise.

It has been alleged, as one principal advantage that we derive from them, that they serve like hoops or ribs, to strengthen our earth, and to bind it together. In consequence of this theory, Kircher has given us a map of the earth, in this manner hooped with its mountains; which might have a much more solid foundation, did it entirely correspond with truth.

Others have found a different use for them, especially when they run surrounding our globe; which is, that they stop the vapours which are continually travelling from the equator to the poles; for these being urged by the heat of the sun, from the warm regions of the line, must all be accumulated at the poles, if they were not stopped in their way by those high ridges of mountains which cross their direction. But an answer to this may be, that all the great mountains in America lie lengthwise, and therefore do not cross their direction.

But to leave these remote advantages, others assert, that not only the animal but vegetable part of the creation would perish for want of convenient humidity, were it not for their friendly assistance. Their summits are, by these, supposed to arrest, as it were, the vapours which float in the regions of the air. Their large inflections and channels are considered as so many basins prepared for the reception of those thick vapours, and impetuous rains, which descend into them. The huge caverns beneath are so many magazines or conservatories of water for the peculiar service of man; and those orifices by which the water is discharged upon the plain, are so situated as to enrich and render them fruitful, instead of returning through subterraneous channels to the sea, after the performance of a tedious and fruitless circulation.*

However this be, certain it is, that almost all our great rivers find their source among mountains; and, in general, the more extensive the mountain, the greater the river: thus the river Amazons, the greatest in the world, has its source among the Andes, which are the highest mountains on the globe; the river Niger travels a long course of several hundred miles from the Mountains of the Moon, the highest in all Africa; and the Danube and the Rhine proceed from the Alps, which are probably the highest mountains of Europe.

It need scarce be said, that, with respect to height, there are many sizes of mountains, from the gentle rising upland, to the tall craggy precipice. The appearance is in general different in those of different magnitudes. The first are clothed with verdure to the very tops, and only seem to ascend to improve our prospects, or supply us with a purer air: but the lofty mountains of the other class have a very different aspect. At a distance their tops are seen, in wavy ridges, of the very colour of the clouds, and only to be distinguished from them by their figure; which, as I have said, resemble the billows of the sea.† As we approach, the mountain assumes a deeper colour: it gathers upon the sky, and seems to hide half the horizon behind. Its summits also are become more distinct, and appear with a broken and perpendicular line. What at first seemed a single hill, is now found to be a chain of continued mountains, whose tops running along in ridges, are embosomed in each other; so that the curvatures of one are fitted to the prominences of the opposite side, and form a winding valley between, often of several miles in extent; and all the way continuing nearly of the same breadth.

Nothing can be finer, or more exact, than Mr. Pope's description of a traveller straining up the Alps. Every mountain he comes to he thinks will be the last; he finds, however, an unexpected hill rise before him; and that being scaled, he finds the highest summit almost at as great a distance as before. Upon quitting the plain, he might have left a green and fertile soil, and a climate warm and pleasing. As he ascends, the ground assumes a more russet colour; the grass becomes more mossy, and the weather more moderate. Still as he ascends, the weather becomes more cold, and the earth more barren. In this dreary passage he is often entertained with a little valley of surprising verdure, caused by the reflected heat of the sun collected into a narrow spot on the surrounding heights. But it much more frequently happens that he sees only frightful precipices beneath, and lakes of amazing depths, from whence rivers are formed, and fountains derive their original. On those places next the highest summits, vegetation is scarcely carried on; here and there a few plants of the most hardy kind appear. The air is intolerably cold; either continually refrige-

* Nature Displayed, vol. iii. p. 82.   † Lettres Philosophiques sur la Formation, &c. p. 190

rated with frosts, or disturbed with tempests. All the ground here wears an eternal covering of ice, and snows that seem constantly accumulating. Upon emerging from this war of the elements, he ascends into a purer and serener region, where vegetation is entirely ceased; where the precipices, composed entirely of rocks, rise perpendicularly above him; while he views beneath him all the combat of the elements; clouds at his feet, and thunders darting upwards from their bosoms below.* A thousand meteors, which are never seen on the plain, present themselves. Circular rainbows;† mock suns; the shadow of the mountain projected upon the body of the air;‡ and the traveller's own image, reflected as in a looking-glass, upon the opposite cloud.§

Such are, in general, the wonders that present themselves to a traveller in his journey either over the Alps or the Andes. But we must not suppose that this picture exhibits either a constant or an invariable likeness of those stupendous heights. Indeed, nothing can be more capricious or irregular than the forms of many of them. The tops of some run in ridges for a considerable length, without interruption; in others, the line seems indented by great valleys to an amazing depth. Sometimes a solitary and a single mountain rises from the bosom of the plain; and sometimes extensive plains, and even provinces, as those of Savoy and Quito, are found embosomed near the tops of mountains. In general, however, those countries that are most mountainous, are the most barren and uninhabitable.

If we compare the heights of mountains with each other, we shall find that the greatest and highest are found under the line.‖ It is thought by some, that the rapidity of the earth's motion in these parts, together with the greatness of the tides there, may have thrown up those stupendous masses of earth. But, be the cause as it may, it is a remarkable fact, that the inequalities of the earth's surface are greatest there. Near the poles, the earth, indeed, is craggy and uneven enough; but the heights of the mountains there, are very inconsiderable. On the contrary, at the equator, where nature seems to sport in the amazing size of all her productions, the plains are extensive, and the mountains remarkably lofty. Some of them are known to rise three miles perpendicular above the bed of the ocean.

To enumerate the most remarkable of these, according to their size, we shall begin with the Andes, of which we have an excellent description by Ulloa, who went thither by command of the king of Spain, in company with the French Academicians, to measure a degree of the meridian. His journey up these mountains is too curious not to give an extract from it.

After many incommodious days, sailing up the river Guayquil, he arrived at Caracol, a town situated at the foot of the Andes. Nothing could exceed the inconveniences which he experienced in his voyage, from the flies and moschitoes (an animal resembling our gnat.) "We were the whole day," says he, "in continual motion to keep them off; but at night our torments were excessive. Our gloves, indeed, were some defence to our hands; but our faces were entirely exposed; nor

---

* Ulloa, vol. i. † Ibid. ‡ Phil. Trans. vol. v. p. 152. § Ulloa, vol. i. ‖ Buffon, passim

were our clothes a sufficient defence for the rest of our bodies; for their stings penetrating through the cloth, caused a very painful and fiery itching. One night, in coming to an anchor near a large and handsome house that was uninhabited, we had no sooner seated ourselves in it, than we were attacked on all sides by swarms of moschitoes, so that it was impossible to have one moment's quiet. Those who had covered themselves with clothes made for this purpose, found not the smallest defence; wherefore, hoping to find some relief in the open fields, they ventured out, though in danger of suffering in a more terrible manner from the serpents. But both places were equally obnoxious. On quitting this inhospitable retreat, we the next night took up our quarters in a house that was inhabited; the host of which being informed of the terrible manner we had past the night before, he gravely told us, that the house we so greatly complained of, had been forsaken on account of its being the purgatory of a soul. But we had more reason to believe that it was quitted on account of its being the purgatory of the body. After having journeyed for upwards of three days, through boggy roads, in which the mules at every step sunk up to their bellies, we began at length to perceive an alteration in the climate; and having been long accustomed to heat, we now began to feel it grow sensibly colder.

"It is remarkable, that at Tariguagua we often see instances of the effects of two opposite temperatures, in two persons happening to meet; one of them leaving the plains below, and the other descending from the mountain. The former thinks the cold so severe, that he wraps himself up in all the garments he can procure; while the latter finds the heat so great, that he is scarce able to bear any clothes whatsoever. The one thinks the water so cold, that he avoids being sprinkled by it; the other is so delighted with its warmth, that he uses it as a bath. Nor is the case very different in the same person, who experiences the same diversity of sensation upon his journey up, and upon his return. This difference only proceeds from the change naturally felt at leaving a climate to which one has been accustomed, and coming into another of an opposite temperature.

"The ruggedness of the road from Tariguagua, leading up the mountain, is not easily described. In some parts, the declivity is so great, that the mules can scarce keep their footing; and in others, the acclivity is equally difficult. The trouble of having people going before to mend the road, the pains arising from the many falls and bruises, and the being constantly wet to the skin, might be supported, were not these inconveniences augmented by the sight of such frightful precipices, and deep abysses, as must fill the mind with ceaseless terror. There are some places where the road is so steep, and yet so narrow, that the mules are obliged to slide down, without making any use of their feet whatsoever. On one side of the rider, in this situation, rises an eminence of several hundred yards; and on the other, an abyss of equal depth; so that if he in the least checks his mule, so as to destroy the equilibrium, they both must unavoidably perish.

"After having travelled about nine days in this manner, slowly winding along the side of the mountain, we began to find the whole country covered with a hoar frost: and a hut, in which we lay, had ice on it. Having escaped many perils, we at length, after a journey of fifteen days, arrived upon the plain, on the extremity of which stands the city of Quito, the capital of one of the most charming regions upon earth. Here, in the centre of the torrid zone, the heat is not only very tolerable, but in some places the cold also is painful. Here they enjoy all the temperature and advantages of perpetual spring; their fields being always covered with verdure, and enamelled with flowers of the most lively colours. However, although this beautiful region be higher than any other country in the world, and although it took up so many days of painful journey in the ascent, it is still overlooked by tremendous mountains; their sides covered with snow, and yet flaming with volcanoes at the top. These seemed piled one upon the other, and rise to a most astonishing height, with great coldness. However, at a determined point above the surface of the sea, the congelation is found at the same height in all the mountains. Those parts which are not subject to a continual frost, have here and there growing upon them a rush, resembling the genista, but much more soft and flexible. Towards the extremity of the part where the rush grows, and the cold begins to increase, is found a vegetable, with a round bulbous head, which, when dried, becomes of amazing elasticity. Higher up, the earth is entirely bare of vegetation, and seems covered with eternal snow. The most remarkable mountains are, that of Cotopaxi (already described as a volcano,) Chimborazo, and Pichincha. Cotopaxi is more than three geographical miles above the surface of the sea: the rest are not much inferior. On the top of the latter was my station for measuring a degree of the meridian; where I suffered particular hardships, from the intenseness of the cold, and the violence of the storms. The sky around was, in general, involved in thick fogs, which, when they cleared away, and the clouds, by their gravity, moved nearer to the surface of the earth, they appeared surrounding the foot of the mountain, at a vast distance below, like a sea, encompassing an island in the midst of it. When this happened, the horrid noises of tempests were heard from beneath, then discharging themselves on Quito, and the neighbouring country. I saw the lightnings issue from the clouds, and heard the thunders roll far beneath me. All this time, while the tempest was raging below, the mountain top, where I was placed, enjoyed a delightful serenity; the wind was abated; the sky clear; and the enlivening rays of the sun moderated the severity of the cold. However, this was of no very long duration, for the wind returned with all its violence, and with such velocity as to dazzle the sight; whilst my fears were increased by the dreadful concussions of the precipice, and the fall of enormous rocks; the only sounds that were heard in this frightful situation."

Such is the animated picture of these mountains, as given us by this ingenious Spaniard: and I believe the reader will wish that I had made the quotation still longer. A passage over the Alps, or a journey across the Pyrennees, appear petty trips or excursions in the comparison; and yet these are the most lofty mountains we know of in Europe.

If we compare the Alps with the mountains already described, we shall find them but little more than one half of the height of the former. The Andes, upon being measured by the barometer, are found above three thousand one hundred and thirty-six toises or fathoms above the surface of the sea.* Whereas the highest points of the Alps is not above sixteen hundred. The one, in other words, is above three miles high; the other about a mile and a half. The highest mountains in Asia are, Mount Taurus, Mount Immaus, Mount Caucasus, and the mountains of Japan. Of these, none equals the Andes in height: although Mount Caucasus, which is the highest of them, makes very near approaches. Father Verbiest tells of a mountain in China, which he measured, and found a mile and a half high.† In Africa, the Mountains of the Moon, famous for giving source to the Niger and the Nile, are rather more noted than known. Of the Peak of Teneriffe, one of the Canary Islands that lie off this coast, we have more certain information. In the year 1727, it was visited by a company of English merchants, who travelled up to the top, where they observed its height, and the volcano on its very summit.‡ They found it a heap of mountains, the highest of which rises over the rest like a sugar-loaf, and gives a name to the whole mass. It is computed to be a mile and a half perpendicular from the surface of the sea. Kircher gives us an estimate of the heights of most of the other great mountains in the world; but as he has taken his calculations in general from the ancients, or from modern travellers, who had not the art of measuring them, they are quite incredible. The art of taking the heights of places by the barometer, is a new and an ingenious invention. As the air grows lighter as we ascend, the fluid in the tube rises in due proportion: thus the instrument being properly marked, gives the height with a tolerable degree of exactness; at least enough to satisfy curiosity.

Few of our great mountains have been estimated in this manner; travellers having, perhaps, been deterred, by a supposed impossibility of breathing at the top. However, it has been invariably found, that the air in the highest that our modern travellers have ascended, is not at all too fine for respiration. At the top of the Peak of Teneriffe, there was found no other inconvenience from the air, except its coldness; at the top of the Andes, there was no difficulty of breathing perceived. The accounts, therefore, of those who have asserted that they were unable to breathe, although at much less heights, are greatly to be suspected. In fact, it is very natural for mankind to paint those obstacles as insurmountable, which they themselves have not had the fortitude or perseverance to surmount.

The difficulty and danger of ascending to the tops of mountains, proceeds from other causes, not the thinness of the air. For instance, some of the summits of the Alps have never yet been visited by man. But the reason is, that they rise with such a rugged and precipitate ascent, that they are utterly inaccessible. In some places they appear like a great wall of six or seven hundred feet high; in others, there stick out enormous rocks, that hang upon the brow of the steep, and every moment threaten destruction to the traveller below.

* Ulloa, vol. . p. 442   † Verbiest, alla Chine.   ‡ Phil. Trans. vol. v.

In this manner almost all the tops of the highest mountains are bare and pointed. And this naturally proceeds from their being so continually assaulted by thunder and tempests. All the earthy substances with which they might have been once covered, have for ages been washed away from their summits; and nothing is left remaining but immense rocks, which no tempest has hitherto been able to destroy.

Nevertheless, time is every day, and every hour, making depredations; and huge fragments are seen tumbling down the precipice, either loosened from the summit by frost or rains, or struck down by lightning. Nothing can exhibit a more terrible picture than one of these enormous rocks, commonly larger than a house, falling from its height, with a noise louder than thunder, and rolling down the side of the mountain. Doctor Plot tells us of one in particular, which being loosened from its bed, tumbled down the precipice, and was partly shattered into a thousand pieces. Notwithstanding, one of the largest fragments of the same, still preserving its motion, travelled over the plain below, crossed a rivulet in the midst, and at last stopped on the other side of the bank! These fragments, as was said, are often struck off by lightning, and sometimes undermined by rains; but the most usual manner in which they are disunited from the mountain, is by frost: the rains insinuating between the interstices of the mountain, continue there until there comes a frost, and then, when converted into ice, the water swells with an irresistible force, and produces the same effect as gunpowder, splitting the most solid rocks, and thus shattering the summits of the mountain.

But not rocks alone, but whole mountains are, by various causes, disunited from each other. We see in many parts of the Alps, amazing clefts, the sides of which so exactly correspond with the opposite, that no doubt can be made of their having been once joined together. At Cajeta,* in Italy, a mountain was split in this manner by an earthquake; and there is a passage opened through it, that appears as if elaborately done by the industry of man. In the Andes these breaches are frequently seen. That at Thermopylæ, in Greece, has been long famous. The mountain of the Troglodytes, in Arabia, has thus a passage through it: and that in Savoy, which Nature began, and which Victor Amadeus completed, is an instance of the same kind.

We have accounts of some of these disruptions, immediately after their happening. "In the month of June,† in the year 1714, a part of the mountain of Diableret, in the district of Valais, in France, suddenly fell down, between two and three o'clock in the afternoon, the weather being very calm and serene. It was of a conical figure, and destroyed fifty-five cottages in the fall. Fifteen persons, together with about a hundred beasts, were also crushed beneath its ruins, which covered an extent of a good league square. The dust it occasioned instantly covered all the neighbourhood in darkness. The heaps of rubbish were more than three hundred feet high. They stopped the current of a river that ran along the plain, which now is formed into several new and deep lakes. There appeared through the whole of this rubbish, none of those substances that seemed to indicate that

* Buffon. vol. ii. p 364.   † Hist. de l'Academie des Sciences, p. 1 an. 1715

this disruption had been made by means of subterraneous fires. Most probably, the base of this rocky mountain was rotted and decayed; and thus fell, without any extraneous violence." In the same manner, in the year 1618, the town of Pleurs, in France, was buried beneath a rocky mountain, at the foot of which it was situated.

These accidents, and many more that might be enumerated of the same kind, have been produced by various causes: by earthquakes, as in the mountain at Cajeta; or by being decayed at the bottom, as at Diableret. But the most general way is, by the foundation of one part of the mountain being hollowed by waters, and thus wanting a support, breaking from the other. Thus it generally has been found in the great chasms in the Alps; and thus it almost always is known in those disruptions of hills, which are known by the name of landslips. These are nothing more than the slidings down of a higher piece of ground, disrooted from its situation by subterraneous inundations, and settling itself upon the plain below.

There is not an appearance in all nature that so much astonished our ancestors, as these land-slips. In fact, to behold a large upland, with its houses, its corn, and cattle, at once loosened from its place, and floating, as it were, upon the subjacent water; to behold it quitting its ancient situation, and travelling forward like a ship in quest of new adventures; this is certainly one of the most extraordinary appearances that can be imagined; and to a people, ignorant of the powers of nature, might well be considered as a prodigy. Accordingly, we find all our old historians mentioning it as an omen of approaching calamities. In this more enlightened age, however, its cause is very well known; and, instead of exciting ominous apprehensions in the populace, it only gives rise to some very ridiculous law-suits among them, about whose the property shall be; whether the land which has thus slipt, shall belong to the original possessor, or to him upon whose grounds it has encroached and settled. What has been the determination of the judges, is not so well known, but the circumstances of the slips have been minutely and exactly described.

In the lands of Slatberg,* in the kingdom of Iceland, there stood a declivity, gradually ascending for near half a mile. In the year 1713, and on the 10th of March, the inhabitants perceived a crack on its side, somewhat like a furrow made with a plough, which they imputed to the effects of lightning, as there had been thunder the night before. However, on the evening of the same day, they were surprised to hear a hideous confused noise issuing all round from the side of the hill; and their curiosity being raised, they resorted to the place. There, to their amazement, they found the earth, for near five acres, all in gentle motion, and sliding down the hill upon the subjacent plain. This motion continued the remaining part of the day, and the whole night; nor did the noise cease during the whole time; proceeding, probably, from the attrition of the ground beneath. The day following, however, this strange journey down the hill ceased entirely; and above an acre of the meadow below was found covered with what before composed a part of the declivity.

* Phil. Trans. vol. iv. p. 250.

However, these slips, when a whole mountain's side seems to descend, happen but very rarely. There are some of another kind, however, much more common; and, as they are always sudden, much more dangerous. These are snow-slips, well known, and greatly dreaded by travellers. It often happens, that when snow has long been accumulated on the tops and on the sides of mountains, it is borne down the precipice, either by means of tempests, or its own melting. At first, when loosened, the volume in motion is but small; but gathers as it continues to roll; and, by the time it has reached the habitable parts of the mountain, is generally grown of enormous bulk. Wherever it rolls, it levels all things in its way; or buries them in unavoidable destruction. Instead of rolling, it sometimes is found to slide along from the top; yet even thus it is generally as fatal as before. Nevertheless, we have had an instance, a few years ago, of a small family in Germany, that lived for above a fortnight beneath one of these snow-slips. Although they were buried during that whole time, in utter darkness, and under a bed of some hundred feet deep, yet they were luckily taken out alive; the weight of the snow being supported by a beam that kept up the roof; and nourishment being supplied them by the milk of an ass, if I remember right, that was buried under the same ruin.

But it is not the parts alone that are thus found to subside, whole mountains have been known totally to disappear. Pliny tells us,* that in his own time, the lofty mountain of Cybotus, together with the city of Eurites, were swallowed by an earthquake. The same fate, he says, attended Phlegium, one of the highest mountains in Ethiopia; which, after one night's concussion, was never seen more. In more modern times, a very noted mountain in the Molucca islands, known by the name of the *Peak*, and remarkable for being seen at a very great distance from sea, was swallowed by an earthquake; and nothing but a lake was left in the place where it stood. Thus, while storms and tempests are levelled against mountains above, earthquakes and waters are undermining them below. All our histories talk of their destruction; and a very few new ones (if we except Mount Cenere, and one or two such heaps of cinders) are produced. If mountains, therefore, were of such great utility as some philosophers make them to mankind, it would be a very melancholy consideration that such benefits were diminishing every day. But the truth is, the valleys are fertilized by that earth which is washed from their sides; and the plains become richer, in proportion as the mountains decay.

CHAPTER XIII.

OF WATER.

In contemplating nature, we shall often find the same substances possessed of contrary qualities, and producing opposite effects. Air, which liquifies one substance, dries up another. That fire which is

* Plin. l. ii. cap. 93.

seen to burn up the desert, is often found, in other places, to assist the luxuriance of vegetation; and water, which, next to fire, is the most fluid substance upon earth, nevertheless, gives all other bodies their firmness and durability; so that every element seems to be a powerful servant, capable either of good or ill, and only awaiting external direction, to become the friend or the enemy of mankind. These opposite qualities, in this substance in particular, have not failed to excite the admiration and inquiry of the curious.

That water is the most fluid penetrating body, next to fire, and the most difficult to confine, is incontestibly proved by a variety of experiments. A vessel through which water cannot pass, may be said to retain any thing. It may be objected indeed, that syrups, oils, and honey, leak through some vessels that water cannot pass through; but this is far from being the result of the greater tenuity and fineness of their parts; it is owing to the rosin wherewith the wood of such vessels abounds, which oils and syrups have a power of dissolving; so that these fluids, instead of finding their way, may more properly be said to eat their way through the vessels that contain them. However, water will at last find its way even through these; for it is known to escape through vessels of every substance, glass only excepted. Other bodies may be found to make their way out more readily indeed; as air, when it finds a vent, will escape at once; and quicksilver, because of its weight, quickly penetrates through whatever chinky vessel confines it: but water, though it operates more slowly, yet always finds a more certain issue. As, for instance, it is well known that air will not pass through leather; which water will very readily penetrate. Air also may be retained in a bladder; but water will quickly ooze through. And those who drive this to the greatest degree of precision, pretend to say, that it will pass through pores ten times smaller than air can do. Be this as it may, we are very certain that its parts are so small that they have been actually driven through the pores of gold. This has been proved by the famous Florentine experiment, in which a quantity of water was shut up in a hollow ball of gold, and then pressed with a huge force by screws, during which the fluid was seen to ooze out through the pores of the metal, and to stand, like a dew, upon its surface.

As water is thus penetrating, and its parts thus minute, it may easily be supposed that they enter into the composition of all bodies, vegetable, animal, and fossil. This every chymist's experience convinces him of; and the mixture is the more obvious, as it can always be separated, by a gentle heat, from those substances with which it had been united. Fire, as was said, will penetrate where water cannot pass; but then it is not so easily to be separated. But there is scarce any substance from which its water cannot be divorced. The parings or filings of lead, tin, and antimony, by distillation, yield water plentifully: the hardest stones, sea-salt, nitre, vitriol, and sulphur, are found to consist chiefly of water; into which they resolve by force of fire. "All birds, beasts, and fishes," says Newton, "insects, trees, and vegetables, with their parts, grow from water; and, by putrefaction, return to water again." In short, almost every substance that we see, owes its texture and firmness to the parts of wa ter that mix with its earth; and, deprived of this fluid, it becomes a mass of shapeless dust and ashes.

From hence we see, as was above hinted, that this most fluid body, when mixed with others, gives them consistence and form. Water, by being mixed with earth and ashes, and formed into a vessel, when baked before the fire, becomes a coppel, remarkable for this, that it will bear the utmost force of the hottest furnace that art can contrive. So the Chinese earth, of which porcelain is made, is nothing more than an artificial composition of earth and water, united by heat; and which a greater degree of heat could easily separate. Thus we see a body, extremely fluid of itself, in some measure assuming a new nature, by being united with others: we see a body, whose fluid and dissolving qualities are so obvious, giving consistence and hardness to all the substances of the earth.

From considerations of this kind, Thales, and many of the ancient philosophers, held that all things were made of water. In order to confirm this opinion, Helmont made an experiment, by divesting a quantity of earth of all its oils and salts, and then putting this earth, so prepared, into an earthen pot, which nothing but rain-water could enter, and planting a willow therein; this vegetable, so planted, grew up to a considerable height and bulk, merely from the accidental aspersion of rain-water; while the earth, in which it was planted, received no sensible diminution. From this experiment, he concluded, that water was the only nourishment of the vegetable tribe; and that vegetables, being the nourishment of animals, all organized substances, therefore, owed their support and being only to water. But this has been said by Woodward to be all a mistake: for he shows, that water being impregnated with earthy particles, is only the conveyer of such substances into the pores of vegetables, rather than an increaser of them, by its own bulk: he shows that water is ever found to afford much less nourishment, in proportion as it is purified by distillation. A plant in distilled water, will not grow so fast as in water not distilled: and if the same be distilled three or four times over, the plant will scarce grow at all, or receive any nourishment from it. So that water, as such, does not seem the proper nourishment of vegetables, but only the vehicle thereof, which contains the nutritious particles, and carries them through all parts of the plant. Water, in its pure state, may suffice to extend or swell the parts of a plant, but affords vegetable matter in a moderate proportion.

However this be, it is agreed on all sides, that water, such as we find it, is far from being a pure simple substance. The most genuine we know is mixed with exhalations and dissolutions of various kinds; and no expedient that has been hitherto discovered, is capable of purifying it entirely. If we filter and distil it a thousand times, according to Boerhaave, it will still depose a sediment: and by repeating the process we may evaporate it entirely away, but can never totally remove its impurities. Some, however, assert, that water, properly distilled, will have no sediment;* and that the little white speck which is found at the bottom of the still, is a substance that enters

---

* Hill's History of Fossils.

from without. Kircher used to show in his museum, a phial of water, that had been kept for fifty years, hermetically sealed ;* during which it had deposed no sediment, but continued as transparent as when first it was put in. How far, therefore, it may be brought to a state of purity by distillation, is unknown ; but we very well know, that all such water as we every where see, is a bed in which plants, minerals, and animals, are all found confusedly floating together.

Rain-water, which is a fluid of Nature's own distilling, and which has been raised so high by evaporation, is, nevertheless, a very mixed and impure substance. Exhalations of all kinds, whether salts, sulphurs, or metals, make a part of its substance, and tend to increase its weight. If we gather the water that falls, after a thunder-clap, in a sultry summer's day, and let it settle, we shall find a real salt sticking at the bottom. In winter, however, its impure mixtures are fewer, but still may be separated by distillation. But as to that which is generally caught pouring from the tops of houses, it is particularly foul, being impregnated with the smoke of the chimnies, the vapour of the slate or tiles, and with other impurities that birds and animals may have deposited there. Besides, though it should be supposed free from all these, it is mixed with a quantity of air, which, after being kept for some time, will be seen to separate.

Spring-water is next in point of purity. This, according to Dr. Halley, is collected from the air itself; which being sated with water, and coming to be condensed by the evening's cold, is driven against the tops of the mountains, where being condensed, and collected, it trickles down by the sides, into the cavities of the earth ; and running for a while underground, bubbles up in fountains upon the plain. This having made but a short circulation, has generally had no long time to dissolve or imbibe any foreign substances by the way.

River-water is generally more foul than the former.—Wherever the stream flows, it receives a tincture from its channel. Plants, minerals, and animals, all contribute to add to its impurities : so that such as live at the mouths of great rivers, generally are subject to all those disorders which contaminated and unwholesome waters are known to produce. Of all the river-water in the world, that of the Indus and the Thames is said to be the most light and wholesome.

The most impure fresh-water that we know, is that of stagnated pools and lakes, which, in summer, may be more properly considered as a jelly of floating insects, than a collection of water. In this, millions of little reptiles, undisturbed by any current, which might crush their frames to pieces, breed and engender. The whole teems with shapeless life, and only grows more fruitful by increasing putrefaction.

Of the purity of all these waters, the lightness, and not the transparency, ought to be the test. Water may be extremely clear and beautiful to the eye, and yet very much impregnated with mineral particles. In fact, sea-water is the most transparent of any, and yet

---

* Hermetically sealing a glass vessel, means no more than heating the mouth of the phial red hot ; and thus when the glass is become pliant, squeezing the mouth together with a pair of pincers, and then twisting it six or seven times round, which effectually closes it up.

Phalanger, or Surinam Rat, p. 272.

Babyroussa, p. 97.

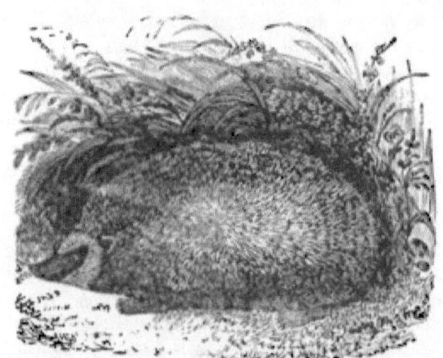

Hedgehog, p. 214.

it is well known to contain a large mixture of salt and bitumen. On the contrary, those waters which are lightest, have the fewest dissolutions floating in them; and may, therefore, be the most useful for all the purposes of life. But, after all, though much has been said upon this subject, and although waters have been weighed with great assiduity, to determine their degree of salubrity, yet neither this, nor their curdling with soap, nor any other philosophical standard whatsoever, will answer the purposes of true information. Experience alone ought to determine the useful or noxious qualities of every spring; and experience assures us, that different kinds of water are adapted to different constitutions. An incontestible proof of this, are the many medicinal springs throughout the world, whose peculiar benefits are known to the natives of their respective countries. These are of various kinds, according to the different minerals with which they are impregnated; hot, saline, sulphureous, bituminous, and oily. But the account of these will come most properly under that of the several minerals by which they are produced.

After all, therefore, we must be contented with but an impure mixture for our daily beverage. And yet, perhaps, this very mixture may often be more serviceable to our health than that of a purer kind. We know that it is so with regard to vegetables: and why not, also, in general, to man? Be this as it will, if we are desirous of having water in its greatest purity, we are ordered, by the curious in this particular, to distil it from snow, gathered upon the tops of the highest mountains, and to take none but the outer and superficial part thereof. This we must be satisfied to call pure water; but even this is far short of the pure unmixed philosophical element; which, in reality, is no where to be found.

As water is thus mixed with foreign matter, and often the repository of minute animals, or vegetable seeds, we need not be surprised that, when carried to sea, it is always found to putrefy. But we must not suppose that it is the element itself which thus grows putrid and offensive, but the substances with which it is impregnated. It is true, the utmost precautions are taken to destroy all vegetable and animal substances that may have previously been lodged in it, by boiling; but, notwithstanding this, there are some that will still survive the operation, and others that find their way during the time of its stowage. Seamen, therefore, assure us, that their water is generally found to putrefy twice, at least, and sometimes three times in a long voyage. In about a month after it has been at sea, when the bung is taken out of the cask, it sends up a noisome and dangerous vapour, which would take fire upon the application of a candle.* The whole body of the water then is found replete with little worm-like insects, that float, with great briskness, through all its parts. These generally live for about a couple of days; and then dying, by depositing their spoils, for a while increase the putrefaction. After a time, the heavier parts of these sinking to the bottom, the lighter float, in a scum, at the top; and this is what mariners call, the water's purging itself. There is still, however, another race of insects, which are bred, very proba-

* Phil. Trans. vol. v. part ii. p. 71.

bly, from the spoils of the former; and produce, after some time, similar appearances; these dying, the water is then thought to change no more. However, it very often happens, especially in hot climates, that nothing can drive these nauseous insects from the ship's store of water. They often increase to a very disagreeable and frightful size, so as to deter the mariner, though parching with thirst, from tasting that cup which they have contaminated.

This water, as thus described, therefore, is a very different fluid from that simple elementary substance upon which philosophical theories have been founded; and concerning the nature of which there have been so many disputes. Elementary water is no way compounded; but is without taste, smell, or colour; and incapable of being discerned by any of the senses, except the touch. This is the famous dissolvent of the chymists, into which, as they have boasted, they can reduce all bodies; and which makes up all other substances, only by putting on a different disguise. In some forms, it is fluid, transparent, and evasive of the touch; in others, hard, firm, and elastic. In some, it is stiffened by cold; in others, dissolved by fire. According to them, it only assumes external shapes from accidental causes; but the mountain is as much a body of water, as the cake of ice that melts on its brow; and even the philosopher himself is composed of the same materials with the cloud or meteor which he contemplates.

Speculation seldom rests where it begins. Others disallowing the universality of this substance, will not allow that in a state of nature there is any such thing as water at all. " What assumes the appearance," say they, " is nothing more than melted ice. Ice is the real element of Nature's making; and when found in a state of fluidity, it is then in a state of violence. All substances are naturally hard; but some more readily melt with heat than others. It requires a great heat to melt iron; a smaller heat will melt copper: silver, gold, tin, and lead, melt with smaller still: ice, which is a body like the rest, melts with a very moderate warmth; and quicksilver melts with the smallest warmth of all. Water, therefore, is but ice kept in continual fusion; and still returning to its former state, when the heat is taken away." Between these opposite opinions, the controversy has been carried on with great ardour, and much has been written on both sides; and yet, when we come to examine the debate, it will probably terminate in this question, whether cold or heat first began their operations upon water? This is a fact of very little importance, if known; and, what is more, it is a fact we can never know.

Indeed, if we examine into the operations of cold and heat upon water, we shall find that they produce somewhat similar effects. Water dilates in its bulk, by heat, to a very considerable degree; and, what is more extraordinary, it is likewise dilated by cold, in the same manner.

If water be placed over a fire, it grows gradually larger in bulk, as it becomes hot, until it begins to boil: after which no art can either increase its bulk or its heat. By increasing the fire, indeed, it may be more quickly evaporated away; but its heat and its bulk still continue the same. By the expanding of this fluid, by heat, philosophers have found a way to determine the warmth or the coldness of other bodies: for if put into a glass tube, by its swelling and rising it shows the quantity of heat in the body to which it is applied; by its contracting and sinking, it shows the absence of the same. Instead of using water in this instrument, which is called a thermometer, they now make use of spirit of wine, which is not apt to freeze, and which is endued even with a greater expansion by heat than water. The instrument consists of nothing more than a hollow ball of glass, with a long tube growing out of it. This being partly filled with spirits of wine tinctured red, so as to be seen when it rises, the ball is plunged into boiling water, which making the spirit within expand and rise in the tube, the water marks the greatest height to which it ascends; at this point the tube is to be broken off, and then hermetically sealed, by melting the glass with a blow-pipe: a scale being placed by the side, completes the thermometer. Now as the fluid expands or condenses with heat or cold, it will rise and fall in the tube in proportion; and the degree or quantity of ascent or descent will be seen in the scale.

No fire, as was said, can make water hotter, after it begins to boil. We can, therefore, at any time be sure of an equable certain heat; which is that of boiling water, which is invariably the same. The certainty of such a heat is not less useful than the instrument that measures it. It affords a standard, fixed, degree of heat over the whole world; boiling water being as hot in Greenland, as upon the coasts of Guinea. One fire is more intense than another; of heat there are various degrees; but boiling water is a heat every where the same, and easily procurable.

As heat thus expands water, so cold, when it is violent enough to freeze the same, produces exactly the same effect, and expands it likewise. Thus water is acted upon in the same manner by two opposite qualities; being dilated by both. As a proof that it is dilated by cold, we have only to observe the ice floating on the surface of a pond, which it would not do were it not dilated, and grown more bulky, by freezing, than the water, which remains unfroze. Mr. Boyle, however, put the matter past a doubt, by a variety of experiments.* Having poured a proper quantity of water into a strong earthen vessel, he exposed it, uncovered, to the open air, in frosty nights; and observed, that continually the ice reached higher than the water, before it was frozen. He filled also a tube with water, and stopped both ends with wax: the water, when frozen, was found to push out the stopples from both ends; and a rod of ice appeared at each end of the tube, which showed how much it was swollen by the cold within.

From hence, therefore, we may be very certain of the cold's dilating of the water; and experience also shows, that the force of this expansion has been found as great as any which heat has been found to produce. The touch-hole of a strong gun-barrel being stopped, and a plug of iron forcibly driven into the muzzle, after the barrel had been filled with water, it was placed in a mixture of ice and salt; the plug, though soldered to the barrel, at first gave way, but being fixed in more firmly, within a quarter of an hour the gun-barrel burst

* Boyle, vol. i. p. 610

with a loud noise, and blew up the cover of the box wherein it lay Such is its force in an ordinary experiment. But it has been known to burst cannons, filled with water, and then left to freeze; for the cold congealing the water, and the ice swelling, it became irresistible. The bursting of rocks, by frost, which is frequent in the northern climates, and is sometimes seen in our own, is an equal proof of the expansion of congealed water. For having by some means insinuated itself into the body of the rock, it has remained there till the cold was sufficient to affect it by congelation. But when once frozen, no obstacle is able to confine it from dilating; and, if it cannot otherwise find room, the rock must burst asunder.

This alteration in the bulk of water might have served as a proof that it was capable of being compressed into a narrower space than it occupied before; but, till of late, water was held to be incompressible. The general opinion was, that no art whatsoever could squeeze it into a narrower compass; that no power on earth, for instance, could force a pint of water into a vessel that held a hair's-breadth less than a pint. And this, said they, appears from the famous Florentine experiment; where the water, rather than suffer compressure, was seen to ooze through the pores of the solid metal; and, at length, making a cleft in the side, spur out with great vehemence. But later trials have proved that water is very compressible, and partakes of that elasticity which every other body possesses in some degree. Indeed, had not mankind been dazzled by the brilliancy of one inconclusive experiment, there were numerous reasons to convince them of its having the same properties with other substances. Ice, which is water in another state, is very elastic. A stone, flung slantingly along the surface of a pond, bounds from the water several times; which shows it to be elastic also. But the trials of Mr. Canton have put this past all doubt; which being somewhat similar to those of the great Boyle, who pressed it with weights properly applied, carry sufficient conviction.

What has been hitherto related, is chiefly applicable to the element of water alone; but its fluidity is a property that it possesses in common with several other substances, in other respects greatly differing from it. That quality which gives rise to the definition of a fluid, namely, that its parts are in a continual intestine motion, seems extremely applicable to water. What the shapes of those parts are, it would be vain to attempt to discover. Every trial only shows the futility of the attempt; all we find is, that they are extremely minute; and that they roll over each other with the greatest ease. Some, indeed, from this property alone, have not hesitated to pronounce them globular; and we have, in all our hydrostatical books, pictures of these little globes in a state of sliding and rolling over each other. But all this is merely the work of imagination; we know that substances of any kind, reduced very small, assume a fluid appearance, somewhat resembling that of water. Mr. Boyle, after finely powdering and sifting a little dry powder of Plaster of Paris, put it in a vessel over the fire, where it soon began to boil like water, exhibiting all the motions and appearances of a boiling liquor. Although but a powder, the parts of which we know are very different from each other, and just as accident has formed them, yet it heaved in great waves like water. Upon agitation, a heavy body will sink to the bottom, and a light one emerge to the top. There is no reason, then, to suppose the figure of the parts of water round, since we see their fluidity very well imitated by a composition, the parts of which are of various forms and sizes. The shape of the parts of water, therefore, we must be content to continue ignorant of. All we know is, that earth, air, and fire, conduce to separate the parts from each other.

Earthy substances divide the parts from each other, and keep them asunder. This division may be so great, that the water will entirely lose its fluidity thereby. Mud, potter's clay, and dried bricks, are but so many different combinations of earth and water, each substance in which the parts of water are most separated from each other, appearing to be the most dry. In some substances, indeed, where the parts of water are greatly divided, as in porcelain, for instance, it is no easy matter to recover and bring them together again; but they continue in a manner fixed and united to the manufactured clay. This circumstance led Doctor Cheney into a very peculiar strain of thinking. He suspected that the quantity of water, on the surface of the earth, was daily decreasing. For, says he, some parts of it are continually joined to vegetable, animal, and mineral substances, which no art can again recover. United with these, the water loses its fluidity; for if, continues he, we separate a few particles of any fluid, and fasten them to a solid body, or keep them asunder, they will be fluid no longer To produce fluidity, a considerable number of such particles are required; but here they are close, and destitute of their natural properties. Thus, according to him, the world is growing every day harder and harder, and the earth firmer and firmer; and there may come a time when every object around us may be stiffened in universal frigidity! However, we have causes enough of anxiety in this world already, not to add this preposterous concern to the number.

That air also contributes to divide the parts of water, we can have no manner of doubt; some have even disputed whether water be not capable of being turned into air. However, though this cannot be allowed, it must be granted, that it may be turned into a substance which greatly resembles air (as we have seen in the experiment of the œlipile) with all its properties; except that, by cold, this new-made air may be condensed again into water.

But of all the substances which tend to divide the parts of water, fire is the most powerful. Water, when heated into steam, acquires such force, and the parts of it tend to fly off from each other with such violence, that no earthly substance we know of is strong enough to confine them. A single drop of water, converted into steam, has been found capable of raising a weight of twenty tons; and would have raised twenty thousand, were the vessel confining it sufficiently strong, and the fire below increased in proportion.

From this easy yielding of its parts to external pressure, arises the art of determining the specific gravity of bodies by plunging them in water; with many other useful discoveries in that part of natural philosophy, called *hydrostatics*. The laws of this science, which Archimedes began, and Pascal, with some other of the moderns, have much improved, rather belongs to experimental than to natural history. However, I will take leave to mention some of the most striking paradoxes in this branch of science, which are as well confirmed by experiment, as rendered universal by theory. It would, indeed, be unpardonable, while discoursing on the properties of water, to omit giving some account of the manner in which it sustains such immense bulks as we see floating upon its soft and yielding surface: how some bodies, that are known to sink at one time, swim with ease, if their surface be enlarged: how the heaviest body, even gold itself, may be made to swim upon water; and how the lightest, such as cork, shall remain sunk at the bottom: how the pouring in of a single quart of water, will burst a hogshead hooped with iron: and how it ascends, in pipes, from the valley, to travel over the mountain: these are circumstances that are at first surprising; but, upon a slight consideration, lose their wonder.

* In order to conceive the manner in which all these wonders are effected, we must begin by observing that water is possessed of an invariable property, which has not hitherto been mentioned; that of always keeping its surface level and even. Winds, indeed, may raise it into waves: or art spurt it up in fountains; but ever, when left to itself, it sinks into a smooth even surface, of which no part is higher than another. If I should pour water, for instance, into the arm of a pipe of the shape of the letter U, the fluid would rise in the other arm just to the same height; because, otherwise, it would not find its level, which it invariably maintains. A pipe bending from one hill down into the valley, and rising by another, may be considered as a tube of this kind, in which the water, sinking in one arm, rises to maintain its level in the other. Upon this principle all water-pipes depend; which can never raise the water higher than the fountain from which they proceed.

Again, let us suppose for a moment, that the arms of the pipe already mentioned, may be made long or short at pleasure; and let us still further suppose, that there is some obstacle at the bottom of it, which prevents the water poured into one arm, from rising in the other. Now it is evident, that this obstacle at the bottom will sustain a pressure from the water in one arm, equal to what would make it rise in the other; and this pressure will be great, in proportion as the arm filled with water is tall. We may, therefore, generally conclude, that the bottom of every vessel is pressed by a force, in proportion to the height of the water in that vessel. For instance, if the vessel filled with water be forty feet high, the bottom of that vessel will sustain such a pressure as would raise the same water forty feet high, which is very great. From hence we see how extremely apt our pipes, that convey water to the city, are to burst; for descending from a hill of more than forty feet high, they are pressed by the water contained in them, with a force equal to what would raise it to more than forty feet high; and that this is sometimes able to burst a wooden pipe, we can have no room to doubt of.

* In the above sketch, the manner of demonstrating used by Monsieur D'Alembert is made use of, as the most obvious, and the most satisfactory. Vide Essai sur, &c.

Still recurring to our pipe, let us suppose one of its arms ten times as thick as the other; this will produce no effect whatsoever upon the obstacle below, which we supposed hindered its rise in the other arm; because, how thick soever the pipe may be, its contents would only rise to its own level; and it will, therefore, press the obstacle with a force equal thereto. We may, therefore, universally conclude, that the bottom of any vessel is pressed by its water, not as it is broad or narrow, but in proportion as it is high. Thus the water contained in a vessel not thicker than my finger, presses its bottom as forcibly as the water contained in a hogshead of an equal height; and, if we made holes in the bottoms of both, the water would burst out as forceful from the one as the other. Hence we may, with great ease, burst a hogshead with a single quart of water; and it has been often done. We have only,* for this, to place a hogshead on one end, filled with water; we then bore a hole in its top, into which we plant a narrow tin pipe, of about thirty feet high: by pouring a quart of water into this, at the top, as it continues to rise higher in the pipe, it will press more forcibly on the bottom and sides of the hogshead below, and at last burst it.

Still returning to our simple instrument of demonstration. If we suppose the obstacle at the bottom of the pipe to be moveable, so as that the force of the water can push it up into the other arm; such a body as quicksilver, for instance. Now, it is evident, that the weight of water weighing down upon this quicksilver in one arm, will at last press it up in the other arm; and will continue to press it upward, until the fluid in both arms be upon a par. So that here we actually see quicksilver, the heaviest substance in the world, except gold and platina, floating upon water, which is but a very light substance.

When we see water thus capable of sustaining quicksilver, we need not be surprised that it is capable of floating much lighter substances, ships, animals, or timber. When any thing floats upon water, we always see that a part of it sinks in the same. A cork, a ship, a buoy, each buries itself in a bed on the surface of the water; this bed may be considered as so much water displaced; the water will, therefore, lose so much of its own weight, as is equal to the weight of that bed of water which it displaces. If the body be heavier than a similar bulk of water, it will sink; if lighter, it will swim. Universally, therefore, a body plunged in water, loses as much of its weight as is equal to the weight of a body of water of its own bulk. Some light bodies, therefore, such as cork, lose much of their weight, and therefore swim; other more ponderous bodies sink, because they are heavier than their bulk of water.

Upon this simple theorem entirely depends the art of weighing metals hydrostatically. I have a guinea, for instance, and desire to know whether it be pure gold; I have weighed it in the usual way with another guinea, and find it exactly of the same weight, but still I have some suspicion, from its greater bulk, that it is not pure. In order to determine this, I have nothing more to do than to weigh it in water with that same guinea that I know to be good, and

* Nollet's Lectures.

of the same weight; and this will instantly show the difference; for the true ponderous metal will sink, and the false bulky one will be sustained in proportion to the greatness of its surface. Those whose business it is to examine the purity of metals, have a balance made for this purpose, by which they can precisely determine which is most ponderous, or, as it is expressed, which has the greatest specific gravity. Seventy-one pounds and a half of quicksilver is found to be equal in bulk to a hundred poundsweight of gold. In the same proportion sixty of lead, fifty-four of silver, forty-seven of copper, forty-five of brass, forty-two of iron, and thirty-nine of tin, are each equal to a hundred pounds of the same most ponderous of all metals.

This method of precisely determining the purity of gold, by weighing in water, was first discovered by Archimedes, to whom mankind have been indebted for many useful discoveries. Hiero, king of Sicily, having sent a certain quantity of gold to be made into a crown, the workman, it seems, kept a part for his own use, and supplied the deficiency with a baser metal. His fraud was suspected by the king, but could not be detected; till applying to Archimedes, he weighed the crown in water; and, by this method, informed the king of the quantity of gold which was taken away.

It has been said, that all fluids endeavour to preserve their level; and, likewise, that a body pressing on the surface, tended to destroy that level. From hence, therefore, it will easily be inferred, that the deeper any body sinks, the greater will be the resistance of the depressed fluid beneath. It will be asked, therefore, as the resistance increases in proportion as the body descends, how comes the body, after it is got a certain way, to sink at all? The answer is obvious. From the fluid above pressing it down with almost as great a force as the fluid beneath presses it up. Take away, by any art, the pressure of the fluid from above, and let only the resistance of the fluid from below be suffered to act, and after the body is gone down very deep, the resistance will be insuperable. To give an instance: a small hole opens in the bottom of a ship at sea, forty feet we will suppose below the surface of the water; through this the water bursts up with great violence; I attempt to stop it with my hand, but it pushes the hand violently away. Here the hand is, in fact, a body attempting to sink upon water, at a depth of forty feet, with the pressure from above taken away. The water, therefore, will overcome my strength; and will continue to burst in till it has got to its level: if I should then dive into the hold, and clap my hand upon the opening, as before, I should perceive no force acting against my hand at all: for the water above presses the hand as much down against the hole, as the water without presses it upward. For this reason, also, when we dive to the bottom of the water, we sustain a very great pressure from above, it is true, but it is counteracted by the pressure from below; and the whole acting uniformly on the surface of the body, wraps us close round without injury.

As I have deviated thus far, I will just mention one or two properties more, which water, and all such like fluids, is found to possess. And, first, their ascending in vessels which are emptied of air, as in our common pumps for instance. The air, however, being the agent in this case, we must previously examine its properties, before we undertake the explanation. The other property to be mentioned is, that of their ascending in small capillary tubes. This is one of the most extraordinary and inscrutable appearances in nature. Glass tubes may be drawn, by means of a lamp, as fine as a hair; still preserving their hollow within. If one of these be planted in a vessel of water, or spirit of wine, the liquor will immediately be seen to ascend; and it will rise higher, in proportion as the tube is smaller; a foot, two feet, and more. How does this come to pass? Is the air the cause? No: the liquor rises, although the air be taken away. Is attraction the cause? No: for quicksilver does not ascend, which it otherwise would. Many have been the theories of experimental philosophers to explain this property. Such as are fond of travelling in the regions of conjecture, may consult Hawksbee, Morgan, Jurin, or Watson, who have examined the subject with great minuteness. Hitherto, however, nothing but doubts, instead of knowledge, have been the result of their inquiries. It will not, therefore, become us to enter into the minuteness of the inquiry, when we have so many great wonders to call our attention away.

CHAPTER XIV.

OF THE ORIGIN OF RIVERS.

"The sun ariseth, and the sun goeth down, and pants for the "place from whence he arose. All things are filled with labour, and "man cannot utter it. All rivers run into the sea, yet the sea is not "full. Unto the place whence the rivers come, thither they return "again. The eye is not satisfied with seeing, nor the ear with hear-"ing."* Thus speaks the wisest of the Jews. And at so early a period was the curiosity of man employed in observing these great circulations of nature. Every eye attempted to explain those appearances; and every philosopher who has long thought upon the subject, seems to give a peculiar solution. The inquiry whence rivers are produced; whence they derive those unceasing stores of water, which continually enrich the world with fertility and verdure, has been variously considered, and divided the opinions of mankind more than any other topic in natural history.

In this contest the various champions may be classed under two leaders. M. De La Hire, who contends that rivers must be supplied from the sea, strained through the pores of the earth; and Dr. Halley, who has endeavoured to demonstrate that the clouds alone are sufficient for the supply.—Both sides have brought in mathematics to their aid; and have shown that long and laborious calculations can at any time be made to obscure both sides of a question.

De La Hire† begins his proofs, that rain-water, evaporated from the sea, is sufficient for the production of rivers; by showing, that

* Ecclesiastes, chap. l. 5, 7, 8. † Hist. de l'Acad. 1713, p. 56.

rain never penetrates the surface of the earth above sixteen inches. From thence he infers, that it is impossible for it, in many cases, to sink so as to be found at such considerable depths below. Rain-water, he grants, is often seen to mix with rivers, and to swell their currents; but a much greater part of it evaporates. "In fact," continues he, "if we suppose the earth every where covered with water, evaporation alone would be sufficient to carry off two feet nine inches of it in a year: and yet we very well know, that scarce nineteen inches of rain-water falls in that time; so that evaporation would carry off a much greater quantity than is ever known to descend. The small quantity of rain-water that falls is, therefore, but barely sufficient for the purposes of vegetation. Two leaves of a fig-tree have been found, by experiment, to imbibe from the earth, in five hours and a half, two ounces of water. This implies the great quantity of fluid that must be exhausted in the maintenance of one single plant. Add to this, that the waters of the river Rungis will, by calculation, rise to fifty inches; and the whole country from whence they are supplied never receives fifty inches in the year by rain. Besides this, there are many salt springs, which are known to proceed immediately from the sea, and are subject to its flux and reflux. In short, wherever we dig beneath the surface of the earth, except in a very few instances, water is to be found: and it is by this subterraneous water, that springs and rivers, nay, a great part of vegetation itself, is supported. It is this subterraneous water which is raised into steam, by the internal heat of the earth, that feeds plants. It is this subterraneous water that distils through its interstices; and there, cooling, forms fountains. It is this, that, by the addition of rains, is increased into rivers, and pours plenty over the whole earth."

On the other side of the question,* it is asserted, that the vapours which are exhaled from the sea, and driven by the winds upon land, are more than sufficient to supply not only plants with moisture, but also to furnish a sufficiency of water to the greatest rivers. For this purpose, an estimate has been made of the quantity of water emptied at the mouths of the greatest rivers; and of the quantity also raised from the sea by evaporation; and it has been found, that the latter by far exceeds the former. This calculation was made by Mr. Marriotte. By him it was found, upon receiving such rain as fell in a year, in a proper vessel fitted for that purpose, that, one year with another, there might fall about twenty inches of water upon the surface of the earth, throughout Europe. It was also computed that the river Seine, from its source to the city of Paris, might cover an extent of ground, that would supply it annually with above seven billions of cubic feet of this water, formed by evaporation. But upon computing the quantity which passed through the arches of one of its bridges in a year, it was found to amount only to two hundred and eighty millions of cubic feet, which is not above the sixth part of the former number. Hence, therefore, it appears, that this river may receive a supply, brought to it by the evaporated waters of the sea, six times greater than what it gives back to the sea by its current; and therefore, evaporation is more

* Phil. Trans. vol. ii. p. 123.

than sufficient for maintaining the greatest rivers, and supplying the purposes also of vegetation.

In this manner the sea supplies sufficient humidity to the air for furnishing the earth with all necessary moisture. One part of its vapours fall upon its own bosom, before they arrive upon land. Another part is arrested by the sides of mountains, and is compelled, by the rising stream of air, to mount upward towards the summits. Here it is presently precipitated, dripping down by the crannies of the stone. In some places, entering into the caverns of the mountain, it gathers in those receptacles, which being once filled, all the rest overflows; and breaking out by the sides of the hills, forms single springs. Many of these run down by the valleys, or guts, between the ridges of the mountain, and, coming to unite, form little rivulets or brooks; many of these meeting in one common valley, and gaining the plain ground, being grown less rapid, become a river: and many of these uniting, make such vast bodies of water, as the Rhine, the Rhone, and the Danube.

There is still a third part, which falls upon the lower grounds, and furnishes plants with their wonted supply. But the circulation does not rest even here; for it is again exhaled into vapour by the action of the sun; and afterwards returned to that great mass of waters whence it first arose. "This," adds Dr. Halley, "seems the most reasonable hypothesis; and much more likely to be true, than that of those who derive all springs from the filtering of the sea-waters, through certain imaginary tubes or passages within the earth; since it is well known that the greatest rivers have their most copious fountains the most remote from the sea."*

This seems the most general opinion; and yet, after all, it is still pressed with great difficulties; and there is still room to look out for a better theory. The perpetuity of many springs, which always yield the same quantity when the least rain or vapour is afforded, as well as when the greatest, is a strong objection. Derham† mentions a spring at Upminster, which he could never perceive by his eye to be diminished, in the greatest droughts, even when all the ponds in the country, as well as an adjoining brook, have been dry for several months together. In the rainy seasons, also, it was never overflowed; except sometimes, perhaps, for an hour or so, upon the immission of the external rains. He, therefore, justly enough concludes, that had this spring its origin from rain or vapour, there would be found an increase or decrease of its water, corresponding to the causes of its production.

Thus the reader, after, having been tossed from one hypothesis to another, must at last be content to settle in conscious ignorance. All that has been written upon this subject, affords him rather something to say, than something to think; something rather for others than for himself. Varenius, indeed, although he is at a loss for the origin of rivers, is by no means so as to their formation. He is pretty positive that all rivers are artificial. He boldly asserts, that their channels have been originally formed by the industry of man. His reasons are, that when a new spring breaks forth, the water does not make itself a

---

* Phil. Trans. vol. ii. p. 128.   † Derham Physico Theol.

new channel, but spreads over the adjacent land. "Thus," says he, "men are obliged to direct its course; or, otherwise, Nature would never have found one." He enumerates many rivers that are certainly known, from history, to have been dug by men. He alleges, that no salt-water rivers are found, because men did not want salt-water; and as for salt, that was procurable at a less expense than digging a river for it. However, it costs a speculative man but a small expense of thinking to form such a hypothesis. It may, perhaps, engross the reader's patience to detain him longer upon it.

Nevertheless, though Philosophy be thus ignorant, as to the production of rivers, yet the laws of their motion, and the nature of their currents, have been very well explained. The Italians have particularly distinguished themselves in this respect; and it is chiefly to them that we are indebted for the improvement.*

All rivers have their source either in mountains, or elevated lakes; and it is in their descent from these that they acquire that velocity which maintains their future current. At first their course is generally rapid and headlong; but it is retarded in its journey, by the continual friction against its banks, by the many obstacles it meets to divert its stream, and by the plains generally becoming more level as it approaches towards the sea.

If this acquired velocity be quite spent, and the plain through which the river passes is entirely level, it will, notwithstanding, still continue to run from the perpendicular pressure of the water, which is always in exact proportion to the depth. This perpendicular pressure is nothing more than the weight of the upper waters pressing the lower out of their places, and, consequently, driving them forward, as they cannot recede against the stream. As this pressure is greatest in the deepest parts of the river, so we generally find the middle of the stream most rapid; both because it has the greatest motion thus communicated by the pressure, and the fewest obstructions from the banks on either side.

Rivers thus set into motion are almost always found to make their own beds. Where they find the bed elevated, they wear its substance away, and deposit the sediment in the next hollow, so as in time to make the bottom of their channels even. On the other hand, the water is continually gnawing and eating away the banks on each side; and this with more force as the current happens to strike more directly against them. By these means it always has a tendency to render them more straight and parallel to its own course. Thus it continues to rectify its banks, and enlarge its bed; and, consequently to diminish the force of its stream, till there becomes an equilibrium between the force of the water, and the resistance of its banks, upon which both will remain without any further mutation. And it is happy for man that bounds are thus put to the erosion of the earth by water; and that we find all rivers only dig and widen themselves but to a certain degree.†

In those plains‡ and large valleys where great rivers flow, the bed of the river is usually lower than any part of the valley. But t often

---

* S. Guglielmini della Natura de Fiumi, passim.   † Ibid
‡ Buffon, de Fleuves, passim. vol. ii.

happens, that the surface of the water is higher than many of the grounds that are adjacent to the banks of the stream. If, after inundations, we take a view of some rivers, we shall find their banks appear above water, at a time that all the adjacent valley is overflowed. This proceeds from the frequent deposition of mud, and such like substances, upon the banks, by the rivers frequently overflowing; and thus, by degrees, they become elevated above the plain; and the water is often seen higher also.

Rivers, as every body has seen, are always broadest at the mouth, and grow narrower towards their source. But what is less known, and probably more deserving curiosity, is, that they run in a more direct channel as they immediately leave their sources; and that their sinuosities and turnings become more numerous as they proceed. It is a certain sign among the savages of North America, that they are near the sea when they find the rivers winding, and every now and then changing their direction. And this is even now become an indication to the Europeans themselves, in their journeys through those trackless forests. As those sinuosities, therefore, increase as the river approaches the sea, it is not to be wondered at that they sometimes divide, and thus disembogue by different channels. The Danube disembogues into the Euxine by seven mouths; the Nile by the same number; and the Wolga by seventy.

The currents* of rivers are to be estimated very differently from the manner in which those writers who have given us mathematical theories on this subject, represent them. They found their calculations upon the surface being a perfect plain from one bank to the other: but this is not the actual state of nature; for rivers, in general, rise in the middle; and this convexity is greatest in proportion as the rapidity of the stream is greater. Any person to be convinced of this, need only lay his eye, as nearly as he can, on a level with the stream, and looking across to the opposite bank, he will perceive the river in the midst to be elevated considerably above what it is at the edges. This rising, in some rivers, is often found to be three feet high; and is ever increased in proportion to the rapidity of the stream. In this case, the water in the midst of the current loses a part of its weight, from the velocity of its motion; while that at the sides, for the contrary reason, sinks lower. It sometimes, however, happens, that this appearance is reversed; for when tides are found to flow up with violence against the natural current of the water, the greatest rapidity is then found at the sides of the river, as the water there least resists the influx from the sea. On those occasions, therefore, the river presents a concave rather than a convex surface; and, as in the former case, the middle waters rose in a ridge, in this case they sink in a furrow.

The stream of all rivers is more rapid in proportion as its channel is diminished. For instance, it will be much swifter where it is ten yards broad, than where it is twenty; for the force behind still pushing the water forward, when it comes to the narrow part, it must make up by velocity what it wants in room.

* Buffon, de Fleuves, passim, vol. ii.

It often happens that the stream of a river is opposed by one of its jutting banks, by an island in the midst, the arches of a bridge, or some such obstacle. This produces not unfrequently a back current; and the water having passed the arch with great velocity, pushes the water on each side of its direct current. This produces a side current, tending to the bank; and not unfrequently a whirlpool; in which a large body of waters are circulated in a kind of cavity, sinking down in the middle. The central point of the whirlpool is always lowest, because it has the least motion: the other parts are supported, in some measure, by the violence of theirs, and consequently rise higher, as their motion is greater; so that towards the extremity of the whirlpool must be higher, than towards the centre.

If the stream of a river be stopped at the surface, and yet be free below; for instance, if it be laid over by a bridge of boats, there will then be a double current; the water at the surface will flow back, while that at the bottom will proceed with increased velocity. It often happens that the current at the bottom is swifter than at the top, when upon violent land-floods, the weight of waters toward the source, presses the waters at the bottom, before it has had time to communicate its motion to the surface. However, in all other cases, the surface of the stream is swifter than the bottom, as it is not retarded by rubbing over the bed of the river.

It might be supposed that bridges, dams, and other obstacles in the current of a river, would retard its velocity. But the difference they make is very inconsiderable. The water, by these stoppages, gets an elevation above the object; which, when it has surmounted, it gives a velocity that recompenses the former delay. Islands and turnings also retard the course of the stream but very inconsiderably; any cause which diminishes the quantity of the water, most sensibly diminishes the force and the velocity of the stream.

An increase* of water in the bed of the river always increases its rapidity; except in cases of inundation. The instant the river has overflowed its banks, the velocity of its current is always turned that way, and the inundation is perceived to continue for some days; which it would not otherwise do, if, as soon as the cause was discontinued, it acquired its former rapidity.

A violent storm, that sets directly up against the course of the stream, will always retard, and sometimes entirely stop its course. I have seen an instance of this, when the bed of a large river was left entirely dry for some hours, and fish were caught among the stones at the bottom.

Inundations are generally greater towards the source of rivers than farther down; because the current is generally swifter below than above; and that for the reasons already assigned.

A little river† may be received into a large one, without augmenting either its width or depth. This, which at first view seems a paradox, is yet very easily accounted for. The little river, in this case, only goes towards increasing the swiftness of the larger, and putting

* Buffon, vol. i. p. 62. † Guglielmini

its dormant waters into motion. In this manner the Venetian branch of the Po was pushed on by the Ferarese branch and that of Panaro, without any enlargement of its breadth or depth from these accessions.

A river tending to enter another, either perpendicularly, or in an opposite direction, will be diverted by degrees from that direction; and be obliged to make itself a more favourable entrance downward, and more conspiring with the stream of the former.

The union of two rivers into one, makes it flow the swifter: since the same quantity of water, instead of rubbing against four shores, now only rubs against two.—And, besides, the current being deeper, becomes, of consequence, more fitted for motion.

With respect to the places from whence rivers proceed, it may be taken for a general rule, that the largest* and highest mountains supply the greatest and most extensive rivers. It may be also remarked, in whatever direction the ridge of the mountain runs, the river takes an opposite course. If the mountain, for instance, stretches from north to south, the river runs from east to west; and so contrariwise. These are some of the most generally received opinions with regard to the course of rivers; however, they are liable to many exceptions; and nothing but an actual knowledge of each particular river can furnish us with an exact theory of its current.

The largest rivers of Europe are, first, the Wolga, which is about six hundred and fifty leagues in length, extending from Reschow to Astrachan. It is remarkable of this river, that it abounds with water during the summer months of May and June; but all the rest of the year is so shallow as scarce to cover its bottom, or allow a passage for loaded vessels that trade up its stream. It was up this river that the English attempted to trade into Persia, in which they were so unhappily disappointed, in the year 1741. The next in order is the Danube. The course of this is but about four hundred and fifty leagues, from the mountains of Switzerland to the Black Sea. It is so deep between Buda and Belgrade, that the Turks and Christians have fleets of men of war upon it; which frequently engaged, during the last war between the Ottomans and the Austrians; however, it is unnavigable further down, by reason of its cataracts, which prevent its commerce into the Black Sea. The Don, or Tanais, which is four hundred leagues from the source of that branch of it called the *Softna*, to its mouth in the Euxine Sea. In one part of its course, it approaches near the Wolga; and Peter the Great had actually begun a canal, by which he intended joining those two rivers; but this he did not live to finish. The Nieper, or Boristhenes, which rises in the middle of Muscovy, and runs a course of three hundred and fifty leagues, to empty itself into the Black Sea. The Old Cossacks inhabit the banks and islands of this river; and frequently cross the Black Sea, to plunder the maritime places on the coasts of Turkey. The Dwina; which takes its rise in a province of the same name in Russia, that runs a course of three hundred leagues, and disembogues into the White Sea, a little below Archangel.

* Dr. Halley.

The largest rivers of Asia are, the Hoanho, in China, which is eight hundred and fifty leagues in length, computing from its source at Raju Ribron, to its mouth in the Gulf of Changi. The Jenisca of Tartary, about eight hundred leagues in length, from the lake Selinga, to the Icy Sea. This river is, by some, supposed to supply most of that great quantity of drift wood which is seen floating in the seas, near the Arctic circle. The Oby, of five hundred leagues, running from the lake of Kila into the Northern Sea. The Amour, in Eastern Tartary, whose course is about five hundred and seventy-five leagues, from its source to its entrance into the sea of Kamtschatka. The Kiam, in China, five hundred and fifty leagues in length. The Ganges, one of the most noted rivers in the world, and about us long as the former. It rises in the mountains which separate India from Tartary; and running through the dominions of the Great Mogul, discharges itself by several mouths into the bay of Bengal. It is not only esteemed by the Indians for the depth and pureness of its stream, but for a supposed sanctity which they believe to be in its waters. It is visited annually by several hundred thousand pilgrims, who pay their devotions to the river as to a god: for savage simplicity is always known to mistake the blessings of the Deity, for the Deity himself. They carry their dying friends from distant countries, to expire on its banks; and to be buried in its stream. The water is lowest in April or May; but the rains beginning to fall soon after, the flat country is overflowed for several miles, till about the end of September; the waters then begin to retire, leaving a prolific sediment behind, that enriches the soil, and, in a few days' time, gives a luxuriance to vegetation, beyond what can be conceived by a European. Next to this may be reckoned the still more celebrated river Euphrates. This rises from two sources, northward of the city Erzerum, in Turcomania, and unites about three days' journey below the same, from whence, after performing a course of five hundred leagues, it falls into the gulf of Persia, fifty miles below the city of Bassora in Arabia. The river Indus is extended from its source to its discharge into the Arabian Sea, four hundred leagues.

The largest rivers of Africa are, the Senegal, which runs a course of not less than eleven hundred leagues, comprehending the Niger, which some have supposed to fall into it. However, later accounts seem to affirm that the Niger is lost in the sands, about three hundred miles up from the western coasts of Africa. Be this as it may, the Senegal is well known to be navigable for more than three hundred leagues up the country; and how much higher it may reach is not yet discovered, as the dreadful fatality of the inland parts of Africa, not only deters curiosity, but even avarice, which is a much stronger passion. At the end of last war, of fifty Englishmen that were sent to the factory at Galam, a place taken from the French, and nine hundred miles up the river, only one returned to tell the fate of his companions, who were destroyed by the climate. The celebrated river Nile is said to be nine hundred and seventy leagues, from its source among the Mountains of the Moon, in Upper Æthiopia, to its opening into the Mediterranean Sea. The sources of this river were considered as inscrutable by the ancients; and the causes of its periodi ca. inundation were equally unknown. They have both been ascertained by the missionaries who have travelled into the interior parts of Æthiopia. The Nile takes its rise in the kingdom of Gojam,* from a small aperture on the top of a mountain, which, though not above a foot and a half over, yet was unfathomable. This fountain, when arrived at the foot of the mountain, expands into a river; and being joined by others, forms a lake thirty leagues long, and as many broad; from this, its channel, in some measure, winds back to the country where it first began; from thence, precipitating by frightful cataracts, it travels through a variety of desert regions, equally formidable, such as Amhara, Olaca, Damot, and Xaoa. Upon its arrival in the kingdom of Upper Egypt, it runs through a rocky channel, which some late travellers have mistaken for its cataracts. In the beginning of its course, it receives many lesser rivers into it; and Pliny was mistaken, in saying that it received none. In the beginning also of its course, it has many windings; but, for above three hundred leagues from the sea, it runs in a direct line. Its annual overflowings arise from a very obvious cause, which is almost universal with the great rivers that take their source near the line. The rainy season, which is periodical in those climates, floods the rivers; and as this always happens in our summer, so the Nile is at that time overflown. From these inundations, the inhabitants of Egypt derive happiness and plenty; and, when the river does not arise to its accustomed heights, they prepare for an indifferent harvest. It begins to overflow about the seventeenth of June; it generally continues to augment for forty days, and decreases in about as many more. The time of increase and decrease, however, is much more inconsiderable now than it was among the ancients. Herodotus informs us, that it was a hundred days rising, and as many falling; which shows that the inundation was much greater at that time than at present. Mr. Buffon† has ascribed the present diminution, as well to the lessening of the Mountains of the Moon, by their substance having so long been washed down with the stream, as to the rising of the earth in Egypt, that has for so many ages received this extraneous supply. But we do not find, by the buildings that have remained since the times of the ancients, that the earth is much raised since then. Besides the Nile in Africa, we may reckon the Zara, and the Coanza, from the greatness of whose openings into the sea, and the rapidity of whose streams, we form an estimate of the great distance from whence they come. Their courses, however, are spent in watering deserts and savage countries, whose poverty or fierceness have kept strangers away.

But of all parts of the world, America, as it exhibits the most lofty mountains, so also it supplies the largest rivers. The foremost of these is the great river Amazon, which, from its source in the lake of Lauricocha, to its discharge into the Western Ocean, performs a course of more than twelve hundred leagues.‡ The breadth and depth of this river are answerable to its vast length; and, where its width is most contracted, its depth is augmented in proportion. So great is the body of its waters, that other rivers, though before the objects of

* Kircher Mundi. Subt. vol. ii. p. 72.   † Buffon, vol. ii. p. 82.   ‡ Ulloa, vol. i. p. 389

admiration, are lost in its bosom. It proceeds, after their junction, with its usual appearance, without any visible change in its breadth or rapidity; and, if we may so express it, remains great without ostentation. In some places it displays its whole magnificence, dividing into several large branches, and encompassing a multitude of islands; and, at length, discharges itself into the ocean, by a channel of a hundred and fifty miles broad. Another river, that may almost rival the former, is the St. Lawrence, in Canada, which rising in the lake Assiniboils, passes from one lake to another, from Christineaux to Alempigo; from thence to lake Superior; thence to the lake Hurons; to lake Erie; to lake Ontario; and, at last, after a course of nine hundred leagues, pours their collected waters into the Atlantic Ocean. The river Mississippi is of more than seven hundred leagues in length, beginning at its source near the lake Assiniboils, and ending at its opening into the gulf of Mexico. The river Plate runs a length of more than eight hundred leagues from its source in the river Parana, to its mouth. The river Oroonoko is seven hundred and fifty-five leagues in length, from its source near Pasto, to its discharge into the Atlantic Ocean.

Such is the amazing length of the greatest rivers; and even in some of these, the most remote sources very probably yet continue unknown. In fact, if we consider the number of rivers which they receive, and the little acquaintance we have with the regions through which they run, it is not to be wondered at that geographers are divided concerning the sources of most of them. As among a number of roots by which nourishment is conveyed to a stately tree, it is difficult to determine precisely that by which the tree is chiefly supplied; so among the many branches of a great river, it is equally difficult to tell which is the original. Hence it may easily happen, that a smaller branch is taken for the capital stream; and its runnings are pursued, and delineated, in prejudice of some other branch that better deserved the name and the description. In this manner* in Europe, the Danube is known to receive thirty lesser rivers: the Wolga, thirty-two or thirty-three. In Asia, the Hohano receives thirty-five; the Jenisea above sixty; the Oby as many; the Amour about forty; the Nanquin receives thirty rivers; the Ganges twenty; and the Euphrates about eleven. In Africa, the Senegal receives more than twenty rivers; the Nile receives not one for five hundred leagues upwards, and then only twelve or thirteen. In America, the river Amazon receives above sixty, and those very considerable; the river St. Lawrence about forty, counting those which fall into its lakes; the Mississippi receives forty; and the river Plate above fifty.

I mentioned the inundations of the Ganges and the Nile, but almost every other great river whose source lies within the tropics, have their stated inundations also. The river Pegu has been called, by travellers, the Indian Nile, because of the similar overflowings of its stream: this it does to an extent of thirty leagues on each side: and so fertilizes the soil, that the inhabitants send great quantities of rice into other countries, and have still abundance for their own consump-

* Buffon, vol. ii. p. 74.

tion. The river Senegal has likewise its inundations, which cover the whole flat country of Negroland, beginning and ending much about the same time with those of the Nile: as, in fact, both rivers rise from the same mountains. But the difference between the effects of the inundations in each river is remarkable: in the one, it distributes health and plenty; in the other, diseases, famine, and death. The inhabitants along the torrid coasts of the Senegal, can receive no benefit from any additional manure the river may carry down to their soil, which is by nature more than sufficiently luxuriant; or, even if they could they have not industry to turn it to any advantage. The banks, therefore, of the river, lie uncultivated, overgrown with rank and noxious herbage, and infested with thousands of animals of various malignity. Every new flood only tends to increase the rankness of the soil, and to provide fresh shelter for the creatures that infest it. If the flood continues but a few days longer than usual, the improvident inhabitants, who are driven up in the higher grounds, want provisions, and a famine ensues. When the river begins to return into its channel, the humidity and heat of the air are equally fatal; and the carcasses of infinite numbers of animals, swept away by the inundation, putrefying in the sun, produce a stench that is almost insupportable. But even the luxuriance of the vegetation becomes a nuisance. I have been assured, by persons of veracity who have been up the river Senegal, that there are some plants growing along the coast, the smell of which is so powerful, that it is hardly to be endured. It is certain, that all the sailors and soldiers who have been at any of our factories there, ascribe the unwholesomeness of the voyage up the stream, to the vegetable vapour. However this be, the inundations of the rivers in this wretched part of the globe, contribute scarce any advantage, if we except the beauty of the prospects which they afford. These, indeed, are finished beyond the utmost reach of art: a spacious glassy river, with its banks here and there fringed to the very surface by the mangrove-tree, that grows down into the water, presents itself to view; lofty forests of various colours, with openings between, carpeted with green plants, and the most gaudy flowers; beasts and animals, of various kinds, that stand upon the banks of the river, and, with a sort of wild curiosity, survey the mariners as they pass, contribute to heighten the scene. This is the sketch of an African prospect; which delights the eye, even while it destroys the constitution.

Besides these annually periodical inundations, there are many rivers that overflow at much shorter intervals. Thus most of those in Peru and Chili have scarce any motion by night; but upon the appearance of the morning sun, they resume their former rapidity; this proceeds from the mountain snows, which melting with the heat, increase the stream, and continue to drive on the current while the sun continues to dissolve them. Some rivers also flow with an even steady current, from their source to the sea; others flow with greater rapidity, their stream being poured down in a cataract, or swallowed by the sands, before they reach the sea.

The rivers of those countries that have been least inhabited, are usually more rocky, uneven, and broken into water-falls or cataracts.

than those where the industry of man has been more prevalent. Wherever man comes, nature puts on a milder appearance: the terrible and the sublime, are exchanged for the gentle and the useful; the cataract is sloped away into a placid stream; and the banks become more smooth and even.* It must have required ages to render the Rhone or the Loire navigable: their beds must have been cleaned and directed; their inequalities removed; and by a long course of industry, Nature must have been taught to conspire with the desires of her controller. Every one's experience must have supplied instances of rivers thus being made to flow more evenly, and more beneficially to mankind; but there are some whose currents are so rapid, and falls so precipitate, that no art can obviate; and that must for ever remain as amazing instances of incorrigible Nature.

Of this kind are the cataracts of the Rhine; one of which I have seen exhibit a very strange appearance; it was that at Schaffhausen, which was frozen quite across, and the water stood in columns where the cataract had formerly fallen. The Nile, as was said, has its cataracts. The river Vologda, in Russia, has two. The river Zara, in Africa, has one near its source. The river Velino, in Italy, has a cataract of above a hundred and fifty feet perpendicular. Near the city of Gottenburgh,† in Sweden, the river rushes down from a prodigious high precipice, into a deep pit, with a terrible noise, and such dreadful force, that those trees designed for the masts of ships, which are floated down the river, are usually turned upside down in their fall, and often are shattered to pieces, by being dashed against the surface of the water in the pit; this occurs if the masts fall sideways upon the water; but if they fall endways, they dive so far under water, that they disappear for a quarter of an hour, or more: the pit, into which they are thus plunged, has been often sounded with a line of some hundred fathoms long, but no ground has been found hitherto. There is also a cataract at Powerscourt, in Ireland, in which, if I am rightly informed, the water falls three hundred feet perpendicular; which is a greater descent than that of any other cataract in any part of the world. There is a cataract at Albany, in the province of New-York, which pours its stream fifty feet perpendicular. But of all the cataracts in the world, that of Niagara, in Canada, if we consider the great body of water that falls, must be allowed to be the greatest, and the most astonishing.

This amazing fall of water is made by the river St. Lawrence, in its passage from the lake Erie into the lake Ontario. We have already said that the St. Lawrence was one of the largest rivers in the world; and yet the whole of its waters are here poured down, by a fall of a hundred and fifty feet perpendicular. It is not easy to bring the imagination to correspond with the greatness of the scene; a river, extremely deep and rapid, and that serves to drain the waters of almost all North America into the Atlantic ocean, is here poured precipitately down a ledge of rocks, that rise, like a wall, across the whole bed of its stream. The width of the river a little above, is near three quarters of a mile broad; and the rocks, where it grows narrower,

* Buffon, vol. ii. p. 90. † Phil. Trans. vol. li. p. 375.

are four hundred yards over. Their direction is not straight across, but hollowing inwards like a horse-shoe; so that the cataract, which bends to the shape of the obstacle, rounding inwards, presents a kind of theatre the most tremendous in nature. Just in the middle of this circular wall of waters, a little island, that has braved the fury of the current, presents one of its points, and divides the stream at top into two; but it unites again long before it has got to the bottom. The noise of the fall is heard at several leagues distance; and the fury of the waters at the bottom of their fall is inconceivable. The dashing produces a mist that rises to the very clouds; and that produces a most beautiful rainbow, when the sun shines. It may easily be conceived, that such a cataract quite destroys the navigation of the stream; and yet some Indian canoes, as it is said, have been known to venture down it with safety.

Of those rivers that lose themselves in the sands, or are swallowed up by chasms in the earth, we have various information. What we are told by the ancients, of the Alpheus, in Arcadia, that sinks into the ground, and rises again near Syracuse in Sicily, where it takes the name of Arethusa, is rather more known than credited. But we have better information with respect to the river Tigris being lost in this manner under mount Taurus; of the Guadalquiver in Spain, being buried in the sands; of the river Greatah, in Yorkshire, running under ground, and rising again; and even of the great Rhine itself, a part of which is no doubt lost in the sands, a little above Leyden. But it ought to be observed of this river, that by much the greatest part arrives at the ocean; for, although the ancient channel which fell into the sea, a little to the west of that city, be now entirely choaked up, yet there are still a number of small canals, that carry a great body of waters to the sea; and, besides, it has also two very large openings, the Lech, and the Waal, below Rotterdam, by which it empties itself abundantly.

Be this as it will, nothing is more common in sultry and sandy deserts, than rivers being thus either lost in the sands, or entirely dried up by the sun. And hence we see, that under the line, the small rivers are but few; for such little streams as are common in Europe, and which with us receive the name of rivers, would quickly evaporate, in those parching and extensive deserts. It is even confidently asserted, that the great river Niger is thus lost before it reaches the ocean; and that its supposed mouths, the Gambia and the Senegal, are distinct rivers, that come a vast way from the interior parts of the country. It appears, therefore, that the rivers under the line are large; but it is otherwise at the poles,* where they must necessarily be small. In that desolate region, as the mountains are covered with perpetual ice, which melts but little, or not at all, the springs and rivulets are furnished with a very small supply. Here, therefore, men and beasts would perish, and die for thirst, if Providence had not ordered, that in the hardest winter, thaws should intervene, which deposit a small quantity of snow-water in pools under the ice; and from this source the wretched inhabitants drain a scanty beverage.

* Crantz's History of Greenland, vol. i. p. 41

Thus, whatever quarter of the globe we turn to, we shall find new reasons to be satisfied with that part of it in which we reside. Our rivers furnish all the plenty of the African stream, without its inundation; they have all the coolness of the polar rivulet, with a more constant supply; they may want the terrible magnificence of huge cataracts, or extensive lakes, but they are more navigable, and more transparent; though less deep and rapid than the rivers of the torrid zone, they are more manageable, and only wait the will of man to take their direction. The rivers of the torrid zone, like the monarchs of the country, rule with despotic tyranny; profuse in their bounties, and ungovernable in their rage. The rivers of Europe, like their kings, are the friends, and not the oppressors of the people; bounded by known limits, abridged in the power of doing ill, directed by human sagacity, and only at freedom to distribute happiness and plenty

---

CHAPTER XV.

OF THE OCEAN IN GENERAL; AND OF ITS SALTNESS.

If we look upon a map of the world, we shall find that the ocean occupies considerably more of the globe, than the land is found to do. This immense body of water is diffused round both the Old and New Continent, to the south; and may surround them also to the north, for what we know, but the ice in those regions has stopped our inquiries. Although the ocean, properly speaking, is but one extensive sheet of waters, continued over every part of the globe, without interruption, and although no part of it is divided from the rest, yet geographers have distinguished it by different names; as the Atlantic or Western Ocean, the Northern Ocean, the Southern Ocean, the Pacific Ocean, and the Indian Ocean. Others have divided it differently, and given other names; as the Frozen Ocean, the Inferior Ocean, or the American Ocean. But all these being arbitrary distinctions, and not of Nature's making, the naturalist may consider them with indifference.

In this vast receptacle, almost all the rivers of the earth ultimately terminate; nor do such great supplies seem to increase its stores; for it is neither apparently swollen by their tribute, nor diminished by their failure; it still continues the same. Indeed, what is the quantity of water of all the rivers and lakes in the world, compared to that contained in this great receptacle?* If we should offer to make a rude estimate, we shall find that all the rivers in the world, flowing into the bed of the sea, with a continuance of their present stores, would take up at least eight hundred years to fill it to its present height. For, supposing the sea to be eighty-five millions of square miles in extent, and a quarter of a mile, upon an average, in depth, this, upon calculation, will give about twenty-one millions of cubic miles of water, as the contents of the whole ocean. Now, to estimate the quantity of water which all the rivers supply, take any one of them; the Po, for

* Buffon, vol. ii. p. 70.

instance, the quantity of whose discharge into the sea, is known to be one cubic mile of water in twenty-six days. Now it will be found, upon a rude computation, from the quantity of ground the Po, with its influent streams, covers, that all the rivers of the world furnish about two thousand times that quantity of water. In the space of a year, therefore, they will have discharged into the sea about twenty-six thousand cubic miles of water; and not till eight hundred years, will they have discharged as much water as is contained in the sea at present. I have not troubled the reader with the odd numbers, lest he should imagine I was giving precision to a subject that is incapable of it.

Thus great is the assemblage of waters diffused round our habitable globe; and yet immeasurable as they seem, they are mostly rendered subservient to the necessities and the conveniences of so little a being as man. Nevertheless, if it should be asked whether they be made for him alone, the question is not easily resolved. Some philosophers have perceived so much analogy to man in the formation of the ocean, that they have not hesitated to assert its being made for him alone. The distribution of land and water,* say they, is admirable: the one being laid against the other so skilfully, that there is a just equipoise of the whole globe. Thus the Northern Ocean balances against the Southern; and the New Continent is an exact counterweight to the Old. As to any objection from the ocean's occupying too large a share of the globe, they contend, that there could not have been a smaller surface employed to supply the earth with a due share of evaporation. On the other hand, some take the gloomy side of the question; they either magnify† its apparent defects; or assert, that‡ what seem defects to us, may be real beauties to some wiser order of beings. They observe, that multitudes of animals are concealed in the ocean, and but a small part of them are known; the rest therefore, they fail not to say, were certainly made for their own benefit, and not for ours. How far either of these opinions be just, I will not presume to determine; but of this we are certain, that God has endowed us with abilities to turn this great extent of waters to our own advantage. He has made these things, perhaps, for other uses; but he has given us faculties to convert them to our own. The much agitated question, therefore, seems to terminate here. We shall never know whether the things of this world have been made for our use; but we very well know that we have been made to enjoy them. Let us then boldly affirm, that the earth, and all its wonders, are ours; since we are furnished with powers to force them into our service. Man is the lord of all the sublunary creation; the howling savage, the winding serpent, with all the untameable and rebellious offspring of Nature, are destroyed in the contest, or driven at a distance from his habitations. The extensive and tempestuous ocean, instead of limiting or dividing his power, only serves to assist his industry, and enlarge the sphere of his enjoyments. Its billows, and its monsters, instead of presenting a scene of terror, only call up the courage of this little intrepid being; and the greatest danger that man now fears on the deep, is

---

* Derham Physico-Theol. † Burnet's Theory, passim. ‡ Pope's Ethic Epistles, passim.

from his fellow creatures. Indeed, when I consider the human race as Nature has formed them, there is but very little of the habitable globe that seems made for them. But when I consider them as accumulating the experience of ages, in commanding the earth, there is nothing so great or so terrible. What a poor contemptible being is the naked savage, standing on the beach of the ocean, and trembling at its tumults! How little capable is he of converting its terrors into benefits; or of saying, Behold an element made wholly for my enjoyments! He considers it as an angry deity, and pays it the homage of submission. But it is very different when he has exercised his mental powers; when he has learned to find his own superiority, and to make it subservient to his commands. It is then that his dignity begins to appear, and that the true Deity is justly praised for having been mindful of man; for having given him the earth for his habitation, and the sea for an inheritance.

This power which man has obtained over the ocean, was at first enjoyed in common; and none pretended to a right in that element where all seemed intruders. The sea, therefore, was open to all till the time of the emperor Justinian. His successor Leo granted such as were in possession of the shore, the sole right of fishing before their respective territories. The Thracian Bosphorus was the first that was thus appropriated; and from that time it has been the struggle of most of the powers of Europe to obtain an exclusive right in this element. The republic of Venice claims the Adriatic. The Danes are in the possession of the Baltic. But the English have a more extensive claim to the empire of all the seas, encompassing the kingdoms of England, Scotland, and Ireland; and although these have been long contested, yet they are now considered as their indisputable property. Every one knows that the great power of the nation is exerted on this element; and that the instant England ceases to be superior upon the ocean, its safety begins to be precarious.

It is in some measure owing to our dependence upon the sea, and to our commerce there, that we are so well acquainted with its extent and figure. The bays, gulfs, currents, and shallows of the ocean, are much better known and examined, than the provinces and kingdoms of the earth itself. The hopes of acquiring wealth by commerce, has carried man to much greater length than the desire of gaining information could have done. In consequence of this, there is scarce a strait or a harbour, scarce a rock or a quicksand, scarce an inflection of the shore, or the jutting of a promontory, that has not been minutely described. But as these present very little entertainment to the imagination, or delight to any but those whose pursuits are lucrative, they need not be dwelt upon here. While the merchant and the mariner are solicitous in describing currents and soundings, the naturalist is employed in observing wonders, though not so beneficial, yet to him of a much more important nature. The saltness of the sea seems to be the foremost.

Whence the sea has derived that peculiar bitterish saltness which we find in it, appears, by Aristotle, to have exercised the curiosity of naturalists in all ages. He supposed (and mankind were for ages content with the solution) that the sun continually raised dry saline ex

Panther, p. 119.

Leopard, p. 119.

Ounce, p. 121.

Jaguar, p. 120.

THE EARTH.

halations from the earth, and deposited them upon the sea; and hence, say his followers, the waters of the sea are more salt at top than at bottom. But, unfortunately for this opinion, neither of the facts is true. Sea-salt is not to be raised by the vapours of the sun; and sea-water is not salter at the top than at the bottom. Father Bohours is of opinion, that the Creator gave the waters of the ocean their saltness at the beginning; not only to prevent their corruption, but to enable them to bear greater burthens. But their saltness does not prevent their corruption; for stagnant sea-water, like fresh, soon grows putrid: and, as for their bearing greater burthens, fresh-water answers all the purposes of navigation quite as well. The established opinion, therefore, is that of Boyle,* who supposes, "That the sea's saltness is supplied not only from rocks or masses of salt at the bottom of the sea, but also from the salt which the rains and rivers, and other waters, dissolve in their passage through many parts of the earth, and at length carry with them to the sea." But as there is a difference in the taste of rock-salt found at land, and that dissolved in the waters of the ocean, this may be produced by the plenty of nitrous and bituminous bodies that, with the salts, are likewise washed into that great receptacle. These substances being thus once carried to the sea, must for ever remain there; for they do not rise by evaporation, so as to be returned back from whence they came. Nothing but the fresh waters of the sea rise in vapours; and all the saltness remains behind. From hence it follows, that every year the sea must become more and more salt; and this speculation, Dr. Halley carries so far, as to lay down a method of finding out the age of the world by the saltness of its waters. "For if it be observed,"† says he, "what quantity of salt is at present contained in a certain weight of water taken up from the Caspian Sea, for example, and, after some centuries, what greater quantity of salt is contained in the same weight of water, taken from the same place; we may conclude, that in proportion as the saltness has increased in a certain time, so much must it have increased before that time; and we may thus, by the rule of proportion, make an estimate of the whole time wherein the water would acquire the degree of saltness it should be then possessed of." All this may be fine; however, an experiment, begun in this century, which is not to be completed till some centuries hence, is rather a little mortifying to modern curiosity; and, I am induced to think, the inhabitants round the Caspian Sea, will not be apt to undertake the inquiry.

This saltness is found to prevail in every part of the ocean; and as much at the surface as at the bottom. It is also found in all those seas that communicate with the ocean; but rather in a less degree.

The great lakes, likewise, that have no outlets nor communication with the ocean, are found to be salt; but some of them in less proportion. On the contrary, all those lakes through which rivers run into the sea, however extensive they be, are, notwithstanding, very fresh: for the rivers do not deposit their salts in the bed of the lake, but carry them with their currents into the ocean. Thus the lakes Ontario and Erie, in North America, although for magnitude they

---

* Boyle, vol. iii. p. 221.   † Phil. Trans. vol. v. p. 218.

may be considered as inland seas, are, nevertheless, fresh-water lakes; and kept so by the river St. Lawrence, which passes through them. But those lakes that have no communication with the sea, nor any rivers going out, although they be less than the former, are, however, always salt. Thus, that which goes by the name of the Dead Sea, though very small, when compared to those already mentioned, is so exceedingly salt, that its waters seem scarcely capable of dissolving any more. The lakes of Mexico and of Titicaca, in Peru, though of no great extent, are, nevertheless, salt; and both for the same reason.

Those who are willing to turn all things to the best, have not failed to consider this saltness of the sea, as a peculiar blessing from Providence, in order to keep so great an element sweet and wholesome. What foundation there may be in the remark, I will not pretend to determine; but we shall shortly find a much better cause for its being kept sweet, namely, its motion.

On the other hand, there have been many who have considered the subject in a different light, and have tried every endeavour to make salt-water fresh, so as to supply the wants of mariners in long voyages, or when exhausted of their ordinary stores. At first it was supposed simple distillation would do; but it was soon found that the bitter part of the water still kept mixed. It was then tried by uniting salt of tartar with sea-water, and distilling both; but here the expense was greater than the advantage. Calcined bones were next thought of; but a hogshead of calcined bones, carried to sea, would take up as much room as a hogshead of water, and was more hard to be obtained. In this state, therefore, have the attempts to sweeten sea-water rested; the chymist satisfied with the reality of his invention; and the mariner convinced of its being useless. I cannot, therefore, avoid mentioning a kind of succedaneum which has been lately conceived to answer the purposes of fresh-water, when mariners are quite exhausted. It is well known, that persons who go into a warm bath, come out several ounces heavier than they went in: their bodies having imbibed a correspondent quantity of water. This more particularly happens, if they have been previously debarred from drinking, or go in with a violent thirst; which they quickly find quenched, and their spirits restored. It was supposed, that in case of a total failure of fresh-water at sea, a warm bath might be made of sea-water, for the use of mariners; and that their pores would thus imbibe the fluid, without any of its salts, which would be seen to crystallize on the surface of their bodies. In this manner, it is supposed, a sufficient quantity of moisture may be procured to sustain life, till time or accident furnish a more copious supply.

But, however this be, the saltness of the sea can by no means be considered as a principal cause in preserving its waters from putrefaction. The ocean has its currents, like rivers, which circulate its contents round the globe; and these may be said to be the great agents that keep it sweet and wholesome. Its saltness alone would by no means answer this purpose: and some have even imagined, that the various substances with which it is mixed, rather tend to promote putrescence than impede it. Sir Robert Hawkins, one of our most enlightened navigators, gives the following account of a calm, in which the sea, continuing for some time without motion, began to assume a very formidable appearance. "Were it not," says he, "for the moving of the sea, by the force of winds, tides, and currents, it would corrupt all the world. The experiment of this I saw in the year 1590, lying with a fleet about the islands of Azores, almost six months; the greatest part of which time we were becalmed. Upon which all the sea became so replenished with several sorts of jellies, and forms of serpents, adders, and snakes, as seemed wonderful some green, some black, some yellow, some white, some of divers colours; and many of them had life; and some there were a yard and a half, and two yards long; which, had I not seen, I could hardly have believed. And hereof are witnesses all the company of the ships which were then present; so that hardly a man could draw a bucket of water clear of some corruption. In which voyage, towards the end thereof, many of every ship fell sick, and began to die apace. But the speedy passage into our country was a remedy to the crazed, and a preservative for those that were not touched."

This shows, abundantly, how little the sea's saltness was capable of preserving it from putrefaction: but to put the matter beyond all doubt, Mr. Boyle* kept a quantity of sea-water, taken up in the English Channel, for some time barrelled up; and, in the space of a few weeks, it began to acquire a fetid smell: he was also assured by one of his acquaintance, who was becalmed for twelve or fourteen days in the Indian Sea, that the water, for want of motion, began to stink; and that had it continued much longer, the stench would probably have poisoned him. It is the motion, therefore, and not the saltness of the sea, that preserves it in its present state of salubrity; and this, very probably, by dashing and breaking in pieces the rudiments, if I may so call them, of the various animals that would otherwise breed there, and putrefy.

There are some advantages, however, which are derived from the saltness of the sea. Its waters being evaporated, furnish that salt which is used for domestic purposes; and although in some places it is made from springs, and in others dug out of mines, yet the greatest quantity is made only from the sea. That which is called *bay-salt*, (from its coming to us by the bay of Biscay) is a stronger kind, made by evaporation in the sun; that called *common-salt*, is evaporated in pans over the fire, and is of a much inferior quality to the former.

Another benefit arising from the quantity of salt dissolved in the sea, is, that it thus becomes heavier, and, consequently, more buoyant. Mr. Boyle, who examined the difference between sea-water and fresh, found that the former appeared to be about a forty-fifth part heavier than the latter. Those, also, who have had opportunities of bathing in the sea, pretend to have experienced a much greater ease in swimming there than in fresh water. However, as we see they have only a forty-fifth part more of their weight sustained by it, I am apt to doubt whether so minute a difference can be practically perceivable. Be this as it may, as sea-water alters in its weight from fresh, so it is found also to differ from itself in different parts of the ocean. In

* Boyle, vol. iii. p. 222.

general, it is perceivable to be heavier, and consequently salter, the nearer we approach the line.*

But there is an advantage arising from the saltness of the waters of the sea, much greater than what has been yet mentioned; which is, that their congelation is thus retarded. Some, indeed, have gone so far as to say, that sea-water never freezes :† but this is an assertion contradicted by experience. However, it is certain that it requires a much greater degree of cold to freeze it than fresh-water; so that, while rivers and springs are seen converted into one solid body of ice, the sea is always fit for navigation, and no way affected by the coldness of the severest winter. It is, therefore, one of the greatest blessings we derive from this element, that, when at land all the stores of nature are locked up from us, we find the sea ever open to our necessities, and patient of the hand of industry.

But it must not be supposed, because in our temperate climate we never see the sea frozen, that it is in the same manner open in every part of it. A very little acquaintance with the accounts of mariners, must have informed us, that at the polar regions it is embarrassed with mountains and moving sheets of ice, that often render it impassable. These tremendous floats are of different magnitudes; sometimes rising more than a thousand feet above the surface of the water :‡ sometimes diffused into plains of above two hundred leagues in length; and, in many parts, sixty or eighty broad. They are usually divided by fissures; one piece following another so close, that a person may step from one to the other. Sometimes mountains are seen rising amidst these plains, and presenting the appearance of a variegated landscape, with hills and valleys, houses, churches, and towers. These are appearances in which all naturalists are agreed; but the great contest is respecting their formation. Mr. Buffon asserts,§ that they are formed from fresh-water alone, which congealing at the mouths of great rivers, accumulate those huge masses that disturb navigation. However, this great naturalist seems not to have been aware that there are two sorts of ice floating in these seas; the flat ice, and the mountain ice: the one formed of sea-water only; the other of fresh.‖

The flat or driving ice, is entirely composed of sea-water; which, upon dissolution, is found to be salt; and is readily distinguished from the mountain, or fresh-water ice, by its whiteness, and want of transparency. This ice is much more terrible to mariners than that which rises up in lumps: a ship can avoid the one, as it is seen at a distance; but it often gets in among the other, which, sometimes closing, crushes it to pieces. This, which manifestly has a different origin from the fresh-water ice, may perhaps have been produced in the Icy Sea, beneath the pole; or along the coasts of Spitzbergen or Nova Zembla.

The mountain ice, as was said, is different in every respect, being formed of fresh-water, and appearing hard and transparent; it is generally of a pale green colour, though some pieces are of a beautiful sky-blue; many large masses also appear gray, and some black

---

* Phil. Trans. vol. ii. p. 207. † Macrobius. ‡ Crantz's History of Greenland, vol. i p. 31.
§ Buffon, vol. ii. p. 91. ‖ Crantz.

If examined more nearly, they are found to be incorporated with earth, stones, and brush-wood, washed from the shore. On these also are sometimes found, not only earth, but nests with birds' eggs, at several hundred miles from land. The generality of these, though almost totally fresh, have nevertheless a thick crust of salt-water frozen upon them, probably from the power that ice has sometimes to produce ice. Such mountains as are here described, are most usually seen at spring-time, and after a violent storm, driving out to sea, where they at first terrify the mariner, and are soon after dashed to pieces by the continual washing of the waves; or driven into the warmer regions of the south, there to be melted away. They sometimes, however, strike back upon their native shores, where they seem to take root at the feet of mountains; and, as Martius tells us, are sometimes higher than the mountains themselves. Those seen by him were blue, full of clefts and cavities made by the rain, and crowned with snow, which, alternately thawing and freezing every year, augmented their size. These, composed of materials more solid than that driving at sea, presented a variety of agreeable figures to the eye, that with a little help from fancy assumed the appearance of trees in blossom; the inside of churches, with arches, pillars, and windows; and the blue-coloured rays, darting from within, presented the resemblance of a glory.

If we inquire into the origin and formation of these, which, as we see, are very different from the former, I think we have a very satisfactory account of them in Crantz's History of Greenland; and I will take leave to give the passage with a very few alterations. "These mountains of ice," says he, "are not salt, like the sea-water, but sweet; and therefore, can be formed no where except on the mountains, in rivers, in caverns, and against the hills near the sea-shore. The mountains of Greenland are so high, that the snow which falls upon them, particularly on the north-side, is in one night's time wholly converted into ice: they also contain clefts and cavities, where the sun seldom or never injects his rays: besides these, are projections, or landing-places, on the declivities of the steepest hills, where the rain and snow-water lodge, and quickly congeal. When now the accumulated flakes of snow slide down, or fall with the rain from the eminences above, on these prominences; or, when here and there a mountain-spring comes rolling down to such a lodging-place, where the ice has already seated itself, they all freeze, and add their tribute to it. This, by degrees, waxes to a body of ice, that can no more be overpowered by the sun; and which, though it may indeed, at certain seasons, diminish by a thaw, yet, upon the whole, through annual acquisitions, it assumes an annual growth. Such a body of ice is often prominent far over the rocks. It does not melt on the upper surface, but underneath; and often cracks into many larger or smaller clefts, from whence the thawed water trickles out. By this, it becomes at last so weak, that being overloaded with its own ponderous bulk, it breaks loose, and tumbles down the rocks with a terrible crash. Where it happens to overhang a precipice on the shore, it plunges into the deep with a shock like thunder; and with such an agitation of the water, as will overset a boat at some distance, as many a poor Greenlander has fatally experienced." Thus are these amazing ice-mountains launched forth to sea, and found floating in the waters 'round both the poles. It is these that have hindered mariners from discovering the extensive countries that lie round the South Pole; and that probably block up the passage to China by the North.

I will conclude this chapter, with one effect more, produced by the saltness of the sea; which is the luminous appearance of its waves in the night. All who have been spectators of a sea by night, a little ruffled with winds, seldom fail of observing its fiery brightness. In some places it shines as far as the eye can reach;[*] at other times, only when the waves boom against the side of the vessel, or the oar dashes into the water. Some seas shine often; others more seldom; some, ever when particular winds blow; and others within a narrow compass; a long tract of light being seen along the surface, whilst all the rest is hid in total darkness. It is not easy to account for these extraordinary appearances: some have supposed that a number of luminous insects produced the effect, and this is in reality sometimes the case; in general, however, they have every resemblance to that light produced by electricity; and, probably, arise from the agitation and dashing of the saline particles of the fluid against each other. But the manner in which this is done, for we can produce nothing similar by any experiments hitherto made, remains for some happier accident to discover. Our progress in the knowledge of nature is slow: and it is a mortifying consideration, that we are hitherto more indebted for success to chance than industry.

CHAPTER XVI.

OF THE TIDES, MOTION, AND CURRENTS OF THE SEA; WITH THEIR EFFECTS.

It was said in the former chapter, that the waters of the sea were kept sweet by their motion; without which they would soon putrefy, and spread universal infection. If we look for final causes, here indeed we have a great and an obvious one that presents itself before us. Had the sea been made without motion, and resembling a pool of stagnant water, the nobler races of animated nature would shortly be at an end. Nothing would then be left alive but swarms of ill-formed creatures, with scarcely more than vegetable life; and subsisting by putrefaction. Were this extensive bed of waters entirely quiescent, millions of the smaller reptile kinds would there find a proper retreat to breed and multiply in: they would find there no agitation, no concussion in the parts of the fluid to crush their feeble frames, or to force them from the places where they were bred: there they would multiply in security and ease, enjoy a short life, and putrefying thus again, give nourishment to numberless others, as little worthy of existence as themselves. But the motion of this great element, effects

[*] Boyle, vol. i. p. 234.

ally destroys the number of these viler creatures; its currents and its tides produce continual agitations, the shock of which they are not able to endure; the parts of the fluid rubbing against each other, destroy all the viscidities; and the ocean, if I may so express it, acquires health by exercise.

The most obvious motion of the sea, and the most generally acknowledged, is that of its tides. This element is observed to flow for certain hours, from south towards the north; in which motion or flux, which lasts about six hours, the sea gradually swells; so that entering the mouths of rivers, it drives back the river-waters to their heads. After a continual flux of six hours, the sea seems to rest for a quarter of an hour; and then begins to ebb, or retire back again, from north to south, for six hours more; in which time the waters sinking, the rivers resume their natural course. After a seeming pause of a quarter of an hour, the sea again begins to flow as before: and thus it has alternately risen and fallen, twice a-day, since the creation.

This amazing appearance did not fail to excite the curiosity, as it did the wonder of the ancients. After some wild conjectures of the earliest philosophers, it became well known in the time of Pliny, that the tides were entirely under the influence, in a small degree, of the sun; but in a much greater of the moon. It was found that there was a flux and reflux of the sea, in the space of twelve hours, fifty minutes, which is exactly the time of a lunar day. It was observed, that whenever the moon was in the meridian, or, in other words, as nearly as possible over any part of the sea, that the sea flowed to that part, and made a tide there; on the contrary, it was found, that when the moon left the meridian, the sea began to flow back again from whence it came; and there might be said to ebb. Thus far the waters of the sea seemed very regularly to attend the motions of the moon. But as it appeared, likewise, that when the moon was in the opposite meridian, as far off on the other side of the globe, that there was a tide on this side also; so that the moon produced two tides, one by her greatest approach to us, and another by her greatest distance from us: in other words, the moon, in once going round the earth, produced two tides, always at the same time; one on the other part of the globe directly under her; and the other, on the part of the globe directly opposite.

Mankind continued for several ages content with knowing the general cause of these wonders, hopeless of discovering the particular manner of the moon's operation. Kepler was the first who conjectured that attraction was the principal cause, asserting, that the sphere of the moon's operation extended to the earth, and drew up its waters. The precise manner in which this is done, was discovered by Newton.

The moon has been found, like all the rest of the planets, to attract and to be attracted by the earth. This attraction prevails throughout our whole planetary system. The more matter there is contained in any body, the more it attracts; and its influence decreases in proportion as the distance, when squared, increases. This being premised, let us see what must ensue upon supposing the moon in the meridian of any tract of the sea. The surface of the water immediately under the moon, is nearer the moon than any other part of the globe is; and, therefore, must be more subject to its attraction than the waters any where else. The waters will, therefore, be attracted by the moon, and rise in a heap; whose eminence will be the highest where the attraction is greatest. In order to form this eminence, it is obvious that the surface, as well as the depths, will be agitated; and that wherever the waters run from one part, succeeding waters must run to fill up the space it has left. Thus the waters of the sea, running from all parts, to attend the motion of the moon, produce the flowing of the tide; and it is high tide at that part wherever the moon comes over it, or to its meridian.

But when the moon travels onward, and ceases to point over the place where the waters were just risen, the cause here of their rising ceasing to operate, they will flow back by their natural gravity into the lower parts from whence they had travelled; and this retiring of the waters will form the ebbing of the sea.

Thus the first part of the demonstration is obvious; since, in general, it requires no great sagacity to conceive that the waters nearest the moon are most attracted, or raised highest by the moon. But the other part of the demonstration, namely, how there come to be high tides at the same time, on the opposite side of the globe, and where the waters are farthest from the moon, is not so easy to conceive. To comprehend this, it must be observed, that the part of the earth and its waters that are farthest from the moon, are the parts of all others that are least attracted by the moon: it must also be observed, that all the waters, when the moon is on the opposite side of the earth, must be attracted by it in the same direction that the earth itself attracts them; that is, if I may so say, quite through the body of the earth, towards the moon itself. This, therefore, being conceived, it is plain that those waters which are farthest from the moon, will have less weight than those of any other part, on the same side of the globe; because the moon's attraction, which conspires with the earth's attraction, is there least. Now, therefore, the waters farthest from the moon, having less weight, and being lightest, will be pressed on all sides, by those that, having more attraction, are heavier: they will be pressed, I say, on all sides; and the heavier waters flowing in, will make them swell and rise, in an eminence directly opposite to that on the other side of the globe, caused by the more immediate influence of the moon.

In this manner, the moon, in one diurnal revolution, produces two tides; one raised immediately under the sphere of its influence, and the other directly opposite to it. As the moon travels, this vast body of waters rears upward, as if to watch its motions; and pursues the same constant rotation. However, in this great work of raising the tides, the sun has no small share; it produces its own tides constantly every day, just as the moon does, but in a much less degree, because the sun is at an immensely greater distance. Thus there are solar tides, and lunar tides. When the forces of these two great luminaries concur, which they always do when they are either in the same, or in opposite parts of the heavens, they jointly produce a much greater tide, than when they are so situated in the heavens, as much to make peculiar tides of their own. To express the very same thing technically; in the conjunctions and oppositions of the sun and moon, the attraction of the sun conspires with the attraction of the moon; by which means the high spring-tides are formed. But in the quadratures of the sun and moon, the water raised by the one is depressed by the other; and hence the lower neap-tides have their production. In a word, the tides are greatest in the syzigies, and least in the quadratures.

This theory well understood, and the astronomical terms previously known, it may readily be brought to explain the various appearances of the tides, if the earth were covered with a deep sea, and the waters uninfluenced by shoals, currents, straits, or tempests. But in every part of the sea, near the shores, the geographer must come in to correct the calculations of the astronomer. For, by reason of the shallowness of some places, and the narrowness of the straits in others, there arises a great diversity in the effect, not to be accounted for without an exact knowledge of all the circumstances of the place. In the great depths of the ocean, for instance, a very slow and imperceptible motion of the whole body of water will suffice to raise its surface several feet high; but if the same increase of water is to be conveyed through a narrow channel, it must rush through it with the most impetuous rapidity. Thus, in the English Channel, and the German Ocean, the tide is found to flow strongest in those places that are narrowest; the same quantity of water being, in this case, driven through a smaller passage. It is often seen, therefore, pouring through a strait with great force; and by its rapidity, considerably raised above the surface of that part of the ocean into which it runs.

This shallowness and narrowness in many parts of the sea give also rise to a peculiarity in the tides of some parts of the world. For in many places, and in our own seas in particular, the greatest swell of the tide is not while the moon is in its meridian height, and directly over the place, but some time after it has declined from thence. The sea, in this case, being obstructed, pursues the moon with what dispatch it can, but does not arrive with all its waters till long after the moon has ceased to operate. Lastly, from this shallowness of the sea, and from its being obstructed by shoals and straits, we may account for the Mediterranean, the Baltic, and the Black Sea, having no sensible tides. These, though to us they seem very extensive, are not however large enough to be affected by the influence of the moon: and as to their communication with the ocean, through such narrow inlets it is impossible in a few hour's time that they should receive and return water enough to raise or depress them in any considerable degree.

In general, therefore, we may observe, that all tides are much higher, and more considerable, in the torrid zone, than in the rest of the ocean; the sea in those parts being generally deeper, and less affected by changeable winds, or winding shores.* The greatest tide we know of, is that at the mouth of the river Indus, where the water rises thirty feet in height. How great, therefore, must have been the amazement of Alexander's soldiers at so strange an appearance! They

* Buffon, vol. ii. p. 187

who always before had been accustomed only to the scarcely perceptible risings of the Mediterranean, or the minute intumescence of the Black Sea, when made at once spectators of a river rising and falling thirty feet in a few hours, must, no doubt, have felt the most extreme awe, and, as we are told,* a mixture of curiosity and apprehension. The tides are also remarkably high on the coasts of Malay, in the straits of Sunda, in the Red Sea, at the mouth of the river St. Lawrence, along the coasts of China and Japan, at Panama, and in the gulf of Bengal. The tides at Tonquin, however, are the most remarkable in the world. In this part there is but one tide, and one ebb, in twenty-four hours; whereas, as we have said before, in other places there are two. Besides, there, twice in each month, there is no tide at all, when the moon is near the equinoctial, the water being for some time quite stagnant. These, with some other odd appearances attending the same phænomena, were considered by many as inscrutable; but Sir Isaac Newton, with peculiar sagacity, adjudged them to arise from the concurrence of two tides, one from the South Sea, and the other from the Indian Ocean. Of each of these tides there come successively two every day; two at one time greater, and two at another that are less. The time between the arrival of the two greater, is considered by him as high tide; the time between the two lesser, as ebb. In short, with this clue, that great mathematician solved every appearance, and so established his theory as to silence every opposer.

This fluctuation of the sea from the tides, produces another, and more constant rotation of its waters, from the east to the west, in this respect following the course of the moon. This may be considered as one great and general current of the waters of the sea; and although it be not every where distinguishable, it is nevertheless every where existent, except when opposed by some particular current or eddy produced by partial and local causes. This tendency of the sea towards the west, is plainly perceivable in all the great straits of the ocean; as, for instance, in those of Magellan, where the tide running in from the east, rises twenty feet high, and continues flowing six hours; whereas the ebb continues but two hours, and the current is directed to the west. This proves that the flux is not equal to the reflux; and that from both results a motion of the sea westward, which is more powerful during the time of the flux than the reflux.

But this motion westward has been sensibly observed by navigators, in their passage back from India to Madagascar, and so on to Africa. In the great Pacific Ocean also it is very perceivable; but the places where it is most obvious, are, as was said, in those straits which join one ocean to another. In the straits between the Maldivia islands, in the gulf of Mexico, between Cuba and Jucatan. In the straits of the gulf of Paria, the motion is so violent, that it hath received the appellation of the Dragon's Mouth. Northward, in the sea of Canada, in Waigat's straits, in the straits of Java, and in short, in every strait where the ocean on one part pours into the ocean on the other. In this manner, therefore, is the sea carried with an unceasing circulation round the globe; and, at the same time that its waters are pushed

* Quintus Curtius.

back or forward with the tide, they have thus a progressive current to the west, which though less observable, is not the less real.

Besides these two general motions of the sea, there are others which are particular to many parts of it, and are called *currents*. These are found to run in all directions, east, west, north, and south; being formed, as was said above, by various causes; the prominence of the shores, the narrowness of the straits, the variation of the wind, and the inequalities at the bottom. These, though no great object to the philosopher, as their causes are generally local and obvious, are nevertheless of the most material consequence to the mariner; and, without a knowledge of which, he could never succeed. It often has happened, that when a ship has unknowingly got into one of these, every thing seems to go forward with success, the mariners suppose themselves every hour approaching their wished-for port, the wind fills their sails, and the ship's prow seems to divide the water; but, at last, by miserable experience they find, that instead of going forward, they have been all the time receding. The business of currents, therefore, makes a considerable article in navigation; and the direction of their stream, and their rapidity, has been carefully set down. This some do by the observation of the surface of the current; or by the driving of the froth along the shore; or by throwing out what is called the *log-line*, with a buoy made for that purpose, and by the direction and motion of this, they judge of the setting and the rapidity of the current.

These currents are generally found to be most violent under the equator, where indeed all the motions of the ocean are most perceivable. Along the coasts of Guinea, if a ship happens to overshoot the mouth of any river it is bound to, the current prevents its return; so that it is obliged to steer out to sea, and take a very large compass, in order to correct the former mistake. These set in a contrary direction to the general motion of the sea westward; and that so strongly, that a passage which, with the current, is made in two days, is with difficulty performed in six weeks against it. However, they do not extend above twenty leagues from the coast; and ships going to the East Indies, take care not to come within the sphere of their action. At Sumatra, the currents, which are extremely rapid, run from south to north: there are also strong currents between Madagascar and the Cape of Good Hope. On the western coasts of America, the current always runs from the south to the north, where a south wind, continually blowing, most probably occasions this phænomenon. But the currents that are most remarkable, are those continually flowing in the Mediterranean sea, both from the ocean by the straits of Gibraltar, and at its other extremity, from the Euxine sea by the Archipelago. This is one of the most extraordinary appearances in nature, this large sea receiving not only the numerous rivers that fall into it, such as the Nile, the Rhone, and the Po, but also a very great influx from the Euxine sea on one part, and the ocean on the other. At the same time, it is seen to return none of those waters it is thus known to receive. Outlets running from it there are none; no rivers but such as bring it fresh supplies; no straits but what are constantly pouring their waters into it: it has, therefore, been the wonder of mankind in every age, how, and by what means, this vast concourse of waters are disposed of; or how this sea, which is always receiving, and never returning, is no way fuller than before. In order to account for this, some have said, that the water was reconveyed by subterraneous passages into the Red Sea.* There is a story told of an Arabian califf, who caught a dolphin in this sea, admiring the beauty of which, he let it go again, having previously marked it by a ring of iron. Some time after a dolphin was caught in the Red Sea, and quickly known by the ring to be the same that had been taken in the Mediterranean before. Such, however, as have not been willing to found their opinions upon a story, have attempted to account for the disposal of the waters of the Mediterranean by evaporation. For this purpose they have entered into long calculations upon the extent of its surface, and the quantity of water that would be raised from such a surface in a year. They then compute how much water runs in by its rivers and straits in that time; and find, that the quantity exhausted by evaporation, greatly exceeds the quantity supplied by rivers and seas. This solution, no doubt, would be satisfactory, did not the ocean, and the Euxine, evaporate as well as the Mediterranean: and as these are subject to the same drain, it must follow, that all the seas will in this respect be upon a par; and, therefore, there must be some other cause for this unperceived drain, and continual supply. This seems to be satisfactorily enough accounted for by Dr. Smith, who supposes an under current running through the straits of Gibraltar to carry out as much water into the ocean, as the upper current continually carries in from it. To confirm this, he observes, that nearer home, between the north and the south foreland, the tide is known to run one way at top, and the ebb another way at bottom. This double current he also confirms by an experiment communicated to him by an able seaman, who being with one of the king's frigates in the Baltic, found he went with his boat into the mid-stream, and was carried violently by the current; upon which a basket was sunk, with a large cannon-ball, to a certain depth of water, which gave a check to the boat's motion; as the basket sunk still lower, the boat was driven, by the force of the water below, against the upper current; and the lower the basket was let down, the stronger the under current was found, and the quicker was the boat's motion against the upper stream, which seemed not to be above four fathom deep. From hence we readily infer, that the same cause may operate at the straits of Gibraltar; and that while the Mediterranean seems replenishing at top, it may be emptying at bottom.

The number of the currents at sea are impossible to be recounted, nor indeed are they always known; new ones are daily produced by a variety of causes, and as quickly disappear. When a regular current is opposed by another in a narrow strait, or where the bottom of the sea is very uneven, a whirlpool is often formed. These were formerly considered as the most formidable obstructions to navigation; and the ancient poets and historians speak of them with terror; they are described as swallowing up ships, and dashing them against the rocks at the bottom: apprehension did not fail to add imaginary

* Kircher Mundi. Subt. vol. 1.

terrors to the description, and placed at the centre of the whirlpool a dreadful den, fraught with monsters whose howlings served to add new horrors to the dashings of the deep. Mankind at present, however, view these eddies of the sea with very little apprehension; and some have wondered how the ancients could have so much overcharged their descriptions. But all this is very naturally accounted for. In those times when navigation was in its infancy, and the slightest concussion of the waves generally sent the poor adventurer to the bottom, it is not to be wondered at that he was terrified at the violent agitations in one of these. When his little ship, but ill fitted for opposing the fury of the sea, was got within the vortex, there was then no possibility of ever returning. To add to the fatality, they were always near the shore; and along the shore was the only place where this ill-provided mariner durst venture to sail. These were, therefore, dreadful impediments to his navigation; for if he attempted to pass between them and the shore, he was sometimes sucked in by the eddy; and if he attempted to avoid them out at sea, he was often sunk by the storm. But in our time, and in our present improved state of navigation, Charybdis, and the Euripus, with all the other irregular currents of the Mediterranean, are no longer formidable. Mr. Addison, not attending to this train of thinking, upon passing through the straits of Sicily, was surprised at the little there was of terror in the present appearance of Scylla and Charybdis; and seems to be of opinion, that their agitations are much diminished since the times of antiquity. In fact, from the reasons above, all the wonders of the Mediterranean sea are described in much higher colours than they merit, to us who are acquainted with the more magnificent terrors of the ocean. The Mediterranean is one of the smoothest and most gentle seas in the world: its tides are scarce perceivable, except in the gulf of Venice, and shipwrecks are less known there than in any other part of the world.

It is in the ocean, therefore, that these whirlpools are particularly dangerous, where the tides are violent, and the tempests fierce. To mention only one, that called *the Maelstroom*, upon the coasts of Norway, which is considered as the most dreadful and voracious in the world. The name it has received from the natives, signifies *the navel of the sea*; since they suppose that a great share of the water of the sea is sucked up and discharged by its vortex. A minute description of the internal parts is not to be expected, since none who were there ever returned to bring back information. The body of the waters that form this whirlpool, are extended in a circle above thirteen miles in circumference.* In the midst of this stands a rock, against which the tide in its ebb is dashed with inconceivable fury. At this time it instantly swallows up all things that come within the sphere of its violence, trees, timber, and shipping. No skill in the mariner, nor strength of rowing, can work an escape: the sailor at the helm finds the ship at first go in a current opposite to his intentions; his vessel's motion, though slow in the beginning, becomes every moment more rapid; it goes round in circles still narrower and narrower, till at last

* Kircher, Mund. Subt. vol. i. p. 156.

it is dashed against the rocks, and instantly disappears: nor is it seen again for six hours; till the tide flowing, it is vomited forth with the same violence with which it was drawn in. The noise of this dreadful vortex still farther contributes to increase its terror, which, with the dashing of the waters, and the dreadful valley, if it may be so called, caused by their circulation, makes one of the most tremendous objects in nature.

CHAPTER XVII.

OF THE CHANGES PRODUCED BY THE SEA UPON THE EARTH.

FROM what has been said, as well of the earth as of the sea, they both appear to be in continual fluctuation. The earth, the common promptuary that supplies subsistence to men, animals, and vegetables, is continually furnishing its stores to their support. But the matter which is thus derived from it, is soon restored and laid down again to be prepared for fresh mutations. The transmigration of souls is, no doubt, false and whimsical; but nothing can be more certain than the transmigration of bodies: the spoils of the meanest reptile may go to the formation of a prince; and, on the contrary, as the poet has it, the body of Cæsar may be employed in stopping a beer-barrel. From this, and other causes, therefore, the earth is in continual change. Its internal fires, the deviation of its rivers, and the falling of its mountains, are daily altering its surface; and geography can scarce recollect the lakes and the valleys that history once described.

But these changes are nothing to the instability of the ocean. It would seem that inquietude was as natural to it as its fluidity. It is first seen with a constant and equable motion going towards the west; the tides then interrupt this progression, and for a time drive the waters in a contrary direction; besides these agitations, the currents act their part in a smaller sphere, being generally greatest where the other motions of the sea are least; namely, nearest the shore: the winds also contribute their share in this universal fluctuation; so that scarcely any part of the sea is wholly seen to stagnate.

> Nil enim quiescit, undis impellitur unda,
> Et spiritus et calor toto se corpore miscent.

As this great element is thus changed, and continually labouring internally, it may be readily supposed that it produces correspondent changes upon its shores, and those parts of the earth subject to its influence. In fact it is every day making considerable alterations, either by overflowing its shores in one place, or deserting them in others; by covering over whole tracts of country that were cultivated and peopled, at one time; or by leaving its bed to be appropriated to the purposes of vegetation, and to supply a new theatre for human industry, at another.

In this struggle between the earth and the sea for dominion, the greatest number of our shores seem to defy the whole rage of the waves, both by their height, and the rocky materials of which they are composed. The coasts of Italy, for instance,* are bordered with rocks of marble of different kinds, the quarries of which may easily be distinguished at a distance from sea, and appear like perpendicular columns of the most beautiful kinds of marble, ranged along the shore. In general, the coasts of France, from Brest to Bourdeaux, are composed of rocks; as are also those of Spain and England, which defend the land, and only are interrupted, here and there, to give an egress to rivers, and to grant the conveniences of bays and harbours to our shipping. It may in general be remarked, that wherever the sea is most violent and furious, there the boldest shores, and of the most compact materials, are found to oppose it There are many shores several hundred feet perpendicular, against which the sea, when swollen with tides or storms, rises and beats with inconceivable fury. In the Orkneys,† where the shores are thus formed, it sometimes, when agitated by a storm, rises two hundred feet perpendicular, and dashes up its spray, together with sand and other substances that compose its bottom, upon land, like showers of rain.

From hence, therefore, we may conceive how the violence of the sea, and the boldness of the shore, may be said to have made each other. Where the sea meets no obstacles, it spreads its waters with a gentle intumescence, till all its power is destroyed, by wanting depth to aid the motion. But when its progress is checked in the midst, by the prominence of rocks, or the abrupt elevation of the land, it dashes with all the force of its depth against the obstacle, and forms, by its repeated violence, that abruptness of the shore which confines its impetuosity. Where the sea is extremely deep, or very much vexed by tempests, it is no small obstacle that can confine its rage; and for this reason we see the boldest shores projected against the deepest waters; all less impediments having long before been surmounted and washed away. Perhaps of all the shores in the world, there is not one so high as that on the west of St. Kilda, which, upon a late admeasurement,‡ was found to be six hundred fathoms perpendicular above the surface of the sea. Here also, the sea is deep, turbulent, and stormy; so that it requires great force in the shore to oppose its violence. In many parts of the world, and particularly upon the coasts of the East Indies, the shores, though not high above water, are generally very deep, and consequently the waves roll against the land with great weight and irregularity. The rising of the waves against the shore, is called by mariners *the surf of the sea:* and in shipwrecks is generally fatal to such as attempt to swim on shore. In this case no dexterity in the swimmer, no float he can use, neither swimming-girdle nor cork-jacket will save him; the weight of the superincumbent waves breaks upon him at once, and crushes him with certain ruin. Some few of the natives, however, have the art of swimming and of navigating their little boats near those shores where a European is sure of instant destruction.

In places where the force of the sea is less violent, or its tides less rapid, the shores are generally seen to descend with a more gradual declivity. Over these, the waters of the tide steal by almost imper

---

* Buffon, vol. ii. p. 190. † Ibid. vol. ii. p. 191. ‡ Description of St. Kilda

ceptible degrees, covering them for a large extent, and leaving them bare on its recess. Upon these shores, as was said, the sea seldom beats with any great violence, as a large wave has not depth sufficient to float it onwards, so that here only are to be seen gentle surges making calmly towards land, and lessening as they approach. As the sea, in the former description, is generally seen to present prospects of tumult and uproar, here it more usually exhibits a scene of repose and tranquil beauty. Its waters which, when surveyed from the precipice, afforded a muddy greenish hue, arising from their depth and position to the eye*, when regarded from a shelving shore wear the colour of the sky, and seem rising to meet it. The deafening noise of the deep sea, is here converted into gentle murmurs; instead of the water dashing against the face of the rock, it advances and recedes, still going forward, but with just force enough to push its weeds and shells, by insensible approaches, to the shore.

There are other shores, besides those already described, which either have been raised by art, to oppose the sea's approaches, or, from the sea's gaining ground, are threatened with imminent destruction. The sea's being thus seen to give and take away lands at pleasure, is, without question, one of the most extraordinary considerations in all natural history. In some places it is seen to obtain the superiority by slow and certain approaches; or to burst in at once, and overwhelm all things in undistinguished destruction; in other places it departs from its shores, and where its waters have been known to rage, it leaves fields covered with the most beautiful verdure.

The formation of new lands by the sea's continually bringing its sediment to one place, and by the accumulation of its sands in another, is easily conceived. We have had many instances of this in England. The island of Oxney, which is adjacent to Romney-marsh, was produced in this manner. This had for a long time been a low level, continually in danger of being overflown by the river Rother; but the sea, by its depositions, has gradually raised the bottom of the river, while it has hollowed the mouth; so that the one is sufficiently secured from inundations, and the other is deep enough to admit ships of considerable burthen. The like also may be seen at that bank called the *Doggersands*, where two tides meet, and which thus receives new increase every day, so that in time the place seems to promise fair for being habitable earth. On many parts of the coasts of France, England, Holland, Germany, and Prussia, the sea has been sensibly known to retire.† Hubert Thomas asserts, in his Description of the Country of Liege, that the sea formerly encompassed the city of Tongres, which, however, is at present thirty-five leagues distant from it. this assertion he supports by many strong reasons; and among others, by the iron rings fixed in the walls of the town, for fastening the ships that came into the port. In Italy there is a considerable piece of ground gained at the mouth of the river Arno; and Ravenna, that once stood by the sea-side, is now considerably removed from it. But we need scarce mention these, when we find that the whole republic of Holland seems to be a conquest upon the

---

* Newton's Optic's, p. 133—167. † Buffon, vol. vi. p. 424

sea, and, in a manner, rescued from its bosom. The surface of the earth, in this country, is below the level of the bed of the sea; and I remember, upon approaching the coast, to have looked down upon it from the sea, as into a valley; however, it is every day rising higher by the depositions made upon it by the sea, the Rhine, and the Meuse; and those parts which formerly admitted large men of war, are now known to be too shallow to receive ships of very moderate burthen.* The province of Jucatan, a peninsula in the gulf of Mexico, was formerly a part of the sea. This tract, which stretches out into the ocean a hundred leagues, and which is above thirty broad, is every where, at a moderate depth below the surface, composed of shells, which evince that its land once formed the bed of the sea. In France, the town of Aigues Mortes was a port in the time of St. Louis, which is now removed more than four miles from the sea.—Psalmodi, in the same kingdom, was an island in the year 815, but is now more than six miles from the shore. All along the coasts of Norfolk, I am very well assured, that in the memory of man the sea has gained fifty yards in some places, and lost as much in others.

Thus numerous, therefore, are the instances of new lands having been produced from the sea, which, as we see, is brought about two different ways: first, by the waters raising banks of sand and mud where their sediment is deposited; and, secondly, by their relinquishing the shore entirely, and leaving it unoccupied to the industry of man.

But as the sea has been thus known to recede from some lands, so has it, by fatal experience, been found to encroach upon others; and probably these depredations on one part of the shore, may account for their dereliction from another; for the current which rested upon some certain bank, having got an egress in some other place, it no longer presses upon its former bed, but pours all its stream into the new entrance; so that every inundation of the sea may be attended with some correspondent dereliction of another shore.

However this be, we have numerous histories of the sea's inundations, and its burying whole provinces in its bosom. Many countries that have been thus destroyed, bear melancholy witness to the truth of history; and show the tops of their houses, and the spires of their steeples, still standing at the bottom of the water. One of the most considerable inundations we have in history, is that which happened in the reign of Henry I. which overflowed the estates of the Earl Godwin, and forms now that bank called the Goodwin Sands. In the year 1546, a similar eruption of the sea destroyed a hundred thousand persons in the territory of Dort; and yet a greater number round Dullart. In Friezland, and Zealand, there were more than three hundred villages overwhelmed; and their ruins continue still visible at the bottom of the water in a clear day. The Baltic sea has, by slow degrees, covered a large part of Pomerania; and, among others, destroyed and overwhelmed the famous port of Vineta. In the same manner, the Norwegian sea has formed several little islands from the main land, and still daily advances upon the continent. The German sea has advanced upon the shores of Holland, near Catt; so that the ruins of

* Buffon, vol. vi. p. 424.

an ancient citadel of the Romans, which was formerly built upon this coast, are now actually under water. To these accidents several more might be added; our own historians, and those of other countries, abound with them; almost every flat shore of any extent, being able to show something that it has lost, or something that it has gained from the sea.

There are some shores on which the sea has made temporary depredations; where it has overflowed, and after remaining perhaps some ages, it has again retired of its own accord, or been driven back by the industry of man.* There are many lands in Norway, Scotland, and the Maldivia Islands, that are at one time covered with water, and at another free. The country round the isle of Ely, in the time of Bede, about a thousand years ago, was one of the most delightful spots in the whole kingdom; it was not only richly cultivated, and produced all the necessaries of life, but grapes also, that afforded excellent wine. The accounts of that time are copious in the description of its verdure and fertility; its rich pastures covered with flowers and herbage; its beautiful shades, and wholesome air. But the sea, breaking in upon the land, overwhelmed the whole country, took possession of the soil, and totally destroyed one of the most fertile valleys in the world. Its air, from being dry and healthful, from that time became most unwholesome, and clogged with vapours; and the small part of the country that, by being higher than the rest, escaped the deluge, was soon rendered uninhabitable, from its noxious vapours. Thus this country continued under water for some centuries; till at last the sea, by the same caprice which had prompted its invasions, began to abandon the earth in like manner. It has continued for some ages to relinquish its former conquests; and although the inhabitants can neither boast the longevity, nor the luxuries of their former preoccupants, yet they find ample means of subsistence; and if they happen to survive the first years of their residence there, they are often known to arrive at a good old age.

But although history be silent as to many other inundations of the like kind, where the sea has overflowed the country, and afterwards retired, yet we have numberless testimonies of another nature, that prove it beyond the possibility of doubt: I mean those numerous trees that are found buried at considerable depths in places where either rivers or the sea has accidentally overflown.† At the mouth of the river Ness, near Bruges, in Flanders, at the depth of fifty feet, are found great quantities of trees lying as close to each other as they do in a wood: the trunks, the branches, and the leaves, are in such perfect preservation, that the particular kind of each tree may instantly be known. About five hundred years ago, this very ground was known to have been covered with the sea; nor is there any history or tradition of its having been dry ground, which we can have no doubt must have been the case. Thus we see a country flourishing in verdure, producing large forests, and trees of various kinds, overwhelmed by the sea. We see this element depositing its sediment to a height of fifty feet; and its waters must, therefore, have risen much higher. We see the

* Buffon, vol. ii. p. 425.   † Ibid, vol. ii. p. 403

same, after it has thus overwhelmed and sunk the land so deep beneath its slime, capriciously retiring from the same coasts, and leaving that habitable once more, which it had formerly destroyed. All this is wonderful; and perhaps, instead of attempting to inquire after the cause, which has hitherto been inscrutable, it will best become us to rest satisfied with admiration.

At the city of Modena in Italy, and about four miles round it, wherever it is dug, when the workmen arrive at the depth of sixty-three feet, they come to a bed of chalk, which they bore with an auger five feet deep: they then withdraw from the pit, before the auger is removed, and upon its extraction, the water bursts up through the aperture with great violence, and quickly fills this new-made well, which continues full, and is affected neither by rains nor droughts. But that which is most remarkable in this operation, is the layers of earth as we descend. At the depth of fourteen feet are found the ruins of an ancient city, paved streets, houses, floors, and different pieces of Mosaic. Under this is found a solid earth, that would induce one to think had never been removed; however, under it is found a soft oozy earth, made up of vegetables; and at twenty-six feet depth, large trees entire, such as walnut-trees, with the walnuts still sticking on the stem, and their leaves and branches in exact preservation. At twenty-eight feet deep, a soft chalk is found, mixed with a vast quantity of shells; and this bed is eleven feet thick. Under this, vegetables are found again, with leaves, and branches of trees as before; and thus alternately chalk and vegetable earth to the depth of sixty-three feet. These are the layers wherever the workmen attempt to bore; while in many of them they also find pieces of charcoal, bones, and bits of iron. From this description, therefore, it appears, that this country has been alternately overflowed and deserted by the sea, one age after another: nor were these overflowings and retirings of trifling depth, or of short continuance. When the sea burst in, it must have been a long time in overwhelming the branches of the fallen forest with its sediment; and still longer in forming a regular bed of shells eleven feet over them. It must have, therefore, taken an age, at least, to make any one of these layers; and we may conclude, that it must have been many ages employed in the production of them all. The land also, upon being deserted, must have had time to grow compact, to gather fresh fertility, and to be drained of its waters before it could be disposed to vegetation, or before its trees could have shot forth again to maturity.

We have instances nearer home of the same kind given us in the Philosophical Transactions; one of them by Mr. Derham. An inundation of the sea, at Dagenham, in Essex, laying bare a part of the adjacent pasture for about two hundred feet wide, and, in some places, twenty deep, it discovered a number of trees that had lain there for many ages before: these trees, by laying long under ground, were become black and hard, and their fibres so tough, that one might as easily break a wire, as any of them: they lay so thick in the place where they were found, that in many parts he could step from one to another: he conceived also, that not only all the adjacent marshes, for several hundred acres, were covered underneath with such timber, but also the marshes along the mouth of the Thames, for several miles. The meeting with these trees at such depths, he ascribes to the sediment of the river, and the tides, which constantly washing over them, have always left some part of their substance behind, so as, by repeated alluvions, to work a bed of vegetable earth over them, to the height at which he found it.

The levels of Hatfield-Chace, in Yorkshire, a tract above eighteen thousand acres, which was yearly overflown, was reduced to arable and pasture-land, by one sir Cornelius Vermusden, a Dutchman. At the bottom of this wide exient, are found millions of the roots and bodies of trees, of such as this island either formerly did, or does at present produce. The roots of all stand in their proper postures; and by them, as thick as ever they could grow, the respective trunks of each, some above thirty yards long. The oaks, some of which have been sold for fifteen pounds a piece, are as black as ebony, very lasting, and close grained. The ash-trees are as soft as earth, and are commonly cut in pieces by the workmen's spades, and as soon as flung up into the open air, turn to dust. But all the rest, even the willows themselves, which are softer than the ash, preserve their substance and texture to this very day. Some of the firs appear to have vegetated, even after they were fallen, and to have, from their branches, struck up large trees, as great as the parent trunk. It is observable, that many of these trees have been burnt, some quite through, some on one side, some have been found chopped and squared, others riven with great wooden wedges, all sufficiently manifesting, that the country which was deluged had formerly been inhabited. Near a great root of one tree, were found eight coins of the Roman emperors; and, in some places, the marks of the ridge and furrow were plainly perceivable, which testified that the ground had formerly been patient of cultivation.

The learned naturalist who has given this description,[*] has pretty plainly evinced, that this forest, in particular, must have been thus levelled by the Romans; and that the falling of the trees must have contributed to the accumulation of the waters. " The Romans," says he, " when the Britons fled, always pursued them into the fortresses of low woods, and miry forests: in these the wild natives found shelter; and, when opportunity offered, issued out, and fell upon their invaders without mercy. In this manner the Romans were at length so harassed, that orders were issued out for cutting down all the woods and forests in Britain. In order to effect this, and destroy the enemy the easier, they set fire to the woods, composed of pines, and other inflammable timber, which spreading, the conflagration destroyed not only the forest, but infinite numbers of the wretched inhabitants who had taken shelter therein. When the pine-trees had thus done what mischief they could, the Romans then brought their army nearer, and, with whole legions of the captive Britons, cut down most of the trees that were yet left standing; leaving only here and there some trees untouched, as monuments of their fury. These, unneedful of their labour, being destitute of the support of the underwood, and of their

---

[*] Phil. Trans. vol. iv. part ii. p. 214.

neighbouring trees, were easily overthrown by the winds, and, without interruption, remained on the places where they happened to fall. The forest, thus fallen, must necessarily have stopped up the currents, both from land and sea; and turned into great lakes, what were before but temporary streams. The working of the waters here, the consumption and decay of rotten boughs and branches, and the vast increase of water-moss which flourishes upon marshy grounds, soon formed a covering over the trunks of the fallen trees, and raised the earth several feet above its former level. The earth thus every day swelling, by a continual increase from the sediment of the waters, and by the lightness of the vegetable substances of which it was composed, soon overtopt the waters by which this intumescence was at first effected; so that it entirely got rid of its inundations, or only demanded a slight assistance from man for that purpose." This may be the origin of all bogs, which are performed by the putrefaction of vegetable substances, mixed with the mud and slime deposited by waters, and at length acquiring a sufficient consistency.

From this we see what powerful effects the sea is capable of producing upon its shores, either by overflowing some, or deserting others; by altering the direction of these, and rendering those craggy and precipitate, which before were shelving. But the influence it has upon these, is nothing to that which it has upon that great body of earth, which forms its bottom. It is at the bottom of the sea that the greatest wonders are performed, and the most rapid changes are produced; it is there that the motion of the tides and the currents have their whole force, and agitate the substances of which their bed is composed. But all these are almost wholly hid from human curiosity; the miracles of the deep are performed in secret; and we have but little information from its abysses, except what we receive by inspection at very shallow depths, or by the plummet, or from divers, who are kown to descend from twenty to thirty fathoms.*

The eye can reach but a very short way into the depths of the sea; and that only when its surface is glassy and serene. In many seas it perceives nothing but a bright sandy plain at bottom, extending for several hundred miles, without an intervening object. But in others, particularly in the Red Sea, it is very different: the whole bottom of this extensive bed of waters is, literally speaking, a forest of submarine plants and corals, formed by insects for their habitation, sometimes branching out to a great extent.—Here are seen the madrepores, the sponges, mosses, sea-mushrooms, and other marine productions, covering every part of the bottom; so that some have even supposed the sea to have taken its name from the colour of its plants below. However, these plants are by no means peculiar to this sea, as they are found in great quantities in the Persian Gulf, along the coast of Africa, and those of Provence and Catalonia.

The bottom of many parts of the sea, near America, presents a very different, though a very beautiful appearance. This is covered with vegetables, which make it look as green as a meadow, and beneath are seen thousands of turtles, and other sea animals, feeding thereon

* Phil Trans. vol. iv. part ii. p. 192.

In order to extend our knowledge of the sea to greater depths, recourse has been had to the plummet; which is generally made of a lump of lead of about forty pounds weight, fastened to a cord.* This, however, only answers in moderate depths; for when a deep sea is to be sounded, the matter of which the cord is composed, being lighter than the water, floats upon it, and when let down to a considerable depth, its length so increases its surface, that it is often sufficient to prevent the lead from sinking; so that this may be the reason that some parts of the sea are said to have no bottom.

In general, we learn from the plummet, that the bottom of the sea is tolerably even where it has been examined; and that the farther from the shore, the sea is in general the deeper. Notwithstanding, in the midst of a great and unfathomable ocean, we often find an island raising its head, and singly braving its fury. Such islands may be considered as the mountains of the deep; and, could we for a moment imagine the waters of the ocean removed or dried away, we should probably find the inequalities of its bed resembling those that are found at land. Here extensive plains, there valleys, and, in many places, mountains of amazing height. M. Buache has actually given us a map of that part of its bottom, which lies between Africa and America, taken from the several soundings of mariners: in it we find the same uneven surface that we do upon land, the same eminences, and the same depressions. In such an imaginary prospect, however, there would be this difference, that as the tops of land-mountains appear the most barren and rocky, the tops of sea-mountains would be found the most verdant and fruitful.

The plummet, which thus gives us some idea of the inequalities of the bottom, leaves us totally in the dark as to every other particular; recourse, therefore, has been had to divers: these, either being bred up in this dangerous way of life, and accustomed to remain some time under water without breathing, or assisted by means of a diving-bell, have been able to return some confused and uncertain accounts of the places below. In the great diving-bell improved by Dr. Halley, which was large enough to contain five men, and was supplied with fresh air by buckets, that alternately rose and fell, they descended fifty fathom. In this huge machine, which was let down from the mast of the ship, the doctor himself went down to the bottom, where, when the sea was clear, and especially when the sun shone, he could see perfectly well to write or read, and much more to take up any thing that was underneath: at other times, when the water was troubled and thick, it was as dark as night below, so that he was obliged to keep a candle lighted at the bottom. But there was one thing very remarkable, that the water which from above was usually seen of a green colour, when looked at from below, appeared to him of a very different one, casting a redness upon one of his hands, like that of damask roses:†—a proof of the sea's taking its colour not from any thing floating in it, but from the different reflections of the rays of light. Upon the whole, the accounts we have received from the bottom, by this contrivance, are but few. We learn from it, and

* Boyle, vol. ii. p. 5.   † Newton's Optics. p. 56

from divers in general, that while the surface of the sea may be deformed by tempests, it is usually calm and temperate below;* that some divers, who have gone down when the weather was calm, and came up when it was tempestuous, were surprised at their not perceiving the change at the bottom. This, however, must not be supposed to obtain with regard to the tides, and the currents, as they are seen constantly shifting their bottom; taking their bed with great violence from one place, and depositing it upon another. We are informed, also, by divers, that the sea grows colder in proportion as they descend to the bottom; that as far as the sun's rays pierce, it is influenced by their warmth; but lower, the cold becomes almost intolerable. A person of quality, who had been himself a diver, as Mr. Boyle informs us, declared, that though he seldom descended above three or four fathoms, yet he found it so much colder than near the top, that he could not well endure it; and that being let down in a great divingbell, although the water could not immediately touch him, he found the air extremely cold upon his first arrival at the bottom.

From divers also we learn, that the sea, in many places, is filled with rocks at bottom; and, that among their clefts, and upon their sides, various substances sprout forward, which are either really vegetables, or the nests of insects, increased to some magnitude. Some of these assume the shape of beautiful flowers; and, though soft when taken up, soon harden, and are kept in the cabinets of the curious.

But of all those divers who have brought us information from the bottom of the deep, the famous Nicola Pesce, whose performances are told us by Kircher, is the most celebrated. I will not so much as pretend to vouch for the veracity of Kircher's account, which he assures us he had from the archives of the kings of Sicily; but it may serve to enliven a heavy chapter. "In the times of Frederic, king of Sicily, there lived a celebrated diver, whose name was Nicholas, and who, from his amazing skill in swimming, and his perseverance under water, was surnamed the *Fish*. This man had, from his infancy, been used to the sea; and earned his scanty subsistence by diving for corals and oysters; which he sold to the villagers on shore. His long acquaintance with the sea, at last, brought it to be almost his natural element. He frequently was known to spend five days in the midst of the waves, without any other provisions than the fish which he caught there, and ate raw. He often swam over from Sicily to Calabria, a tempestuous and dangerous passage, carrying letters from the king. He was frequently known to swim among the gulfs of the Lipari islands, no way apprehensive of danger.

"Some mariners out at sea, one day observed something at some distance from them, which they regarded as a sea-monster; but, upon its approach, it was known to be Nicholas, whom they took into their ship. When they asked him whither he was going in so stormy and rough a sea, and at such a distance from land, he shewed them a packet of letters, which he was carrying to one of the towns of Italy, exactly done up in a leather bag, in such a manner as that they could not be wetted by the sea. He kept them thus company for some

---

* Boyle, vol. iii. p. 242.

time on their voyage, conversing and asking questions; and after eating a hearty meal with them, he took his leave, and, jumping into the sea, pursued his voyage alone.

"In order to aid these powers of enduring in the deep, Nature seemed to have assisted him in a very extraordinary manner; for the spaces between his fingers and toes were webbed, as in a goose; and his chest became so very capacious, that he could take in, at one inspiration, as much breath as would serve him for a whole day.

"The account of so extraordinary a person did not fail to reach the king himself; who, actuated by the general curiosity, ordered that Nicholas should be brought before him. It was no easy matter to find Nicholas, who generally spent his time in the solitudes of the deep; but, at last, however, after much searching, he was found, and brought before his Majesty. The curiosity of this monarch had been long excited by the accounts he had heard of the bottom of the gulf of Charybdis; he therefore conceived that it would be a proper opportunity to have more certain information; and commanded our poor diver to examine the bottom of this dreadful whirlpool: as an incitement to his obedience, he ordered a golden cup to be flung into it. Nicholas was not insensible of the danger to which he was exposed; dangers best known only to himself; and he therefore presumed to remonstrate; but the hopes of the reward, the desire of pleasing the king, and the pleasure of shewing his skill, at last prevailed. He instantly jumped into the gulf, and was swallowed as instantly up in its bosom. He continued for three quarters of an hour below; during which time the king and his attendants remained upon shore anxious for his fate; but he at last appeared, buffeting upon the surface, holding the cup in triumph in one hand, and making his way good among the waves with the other. It may be supposed he was received with applause, upon his arrival on shore; the cup was made the reward of his adventure; the king ordered him to be taken proper care of; and, as he was somewhat fatigued and debilitated by his labour, after a hearty meal he was put to bed, and permitted to refresh himself by sleeping.

"When his spirits were thus restored, he was again brought to satisfy the king's curiosity with a narrative of the wonders he had seen; and his account was to the following effect:—He would never, he said, have obeyed the king's commands, had he been apprized of half the dangers that were before him. There were four things, he said, that rendered the gulf dreadful, not only to men, but even to the fishes themselves: first, the force of the water bursting up from the bottom, which requires great strength to resist: secondly, the abruptness of the rocks, that on every side threatened destruction; thirdly, the force of the whirlpool, dashing against those rocks; and fourthly, the number and magnitude of the polypous fish, some of which appeared as large as a man, and which, every where sticking against the rocks, projected their fibrous arms to entangle him. Being asked how he was able so readily to find the cup that had been thrown in, he replied, that it happened to be flung by the waves into the cavity of a rock, against which he himself was urged in his descent. This account, however, did not satisfy the king's curiosity: being requested

Spanish Pointer, p. 125.

American Wild Dog, p. 126.

Esquimaux Dog, p. 126.

to venture once more into the gulf for further discoveries, he at first refused; but the king, desirous of having the most exact information possible of all things to be found in the gulf, repeated his solicitations; and, to give them still greater weight, produced a larger cup than the former, and added also a purse of gold. Upon these considerations, the unfortunate Pessacola once again plunged into the whirlpool, and was never heard of more."

CHAPTER XVIII.

A SUMMARY ACCOUNT OF THE MECHANICAL PROPERTIES OF AIR.

Having described the earth and the sea, we now ascend into that fluid which surrounds them both; and which, in some measure, supports and supplies all animated nature. As upon viewing the bottom of the ocean from its surface, we see an infinity of animals moving therein, and seeking food; so, were some superior being to regard the earth at a proper distance, he might consider us in the same light: he might, from his superior station, behold a number of busy little beings, immersed in the aerial fluid, that every where surrounds them, and sedulously employed in procuring the means of subsistence. This fluid, though too fine for the gross perception of its inhabitants, might, to his nicer organs of sight, be very visible; and, while he at once saw into its operations, he might smile at the varieties of human conjecture concerning it: he might readily discern, perhaps, the height above the surface of the earth to which this fluid atmosphere reaches; he might exactly determine that peculiar form of its parts which gives it the spring or elasticity with which it is endued; he might distinguish which of its parts were pure, incorruptible air, and which only made for a little time to assume the appearance, so as to be quickly returned back to the element from whence it came. But as for us, who are immersed at the bottom of this gulf, we must be contented with a more confined knowledge; and, wanting a proper point of prospect, remain satisfied with a combination of the effects.

One of the first things that our senses inform us of is, that although the air is too fine for our sight, it is very obvious to our touch. Although we cannot see the wind contained in a bladder, we can very readily feel its resistance; and though the hurricane may want colour, we often fatally experience that it does not want force. We have equal experience of the air's spring or elasticity: the bladder, when pressed, returns again upon the pressure being taken away; a bottle, when filled, often bursts, from the spring of air which is included.

So far the slightest experience reaches; but, by carrying experiment a little farther, we learn, that air also is heavy; a round glass vessel being emptied of its air, and accurately weighed, has been found lighter than when it was weighed with the air in it. Upon computing the superior weight of the full vessel, a cubic foot of air is found to weigh something more than an ounce.

From this experiment, therefore, we learn, that the earth, and all things upon its surface, are every where covered with a ponderous fluid, which rising very high over our heads, must be proportionably heavy. For instance, as in the sea, a man at the depth of twenty feet sustains a greater weight of water, than a man at the depth of but ten feet; so will a man at the bottom of a valley have a greater weight of air over him, than a man on the top of a mountain.

From hence we may conclude, that we sustain a very great weight of air; and although, like men walking at the bottom of the sea, we cannot feel the weight which presses equally round us, yet the pressure is not the less real. As in morals, we seldom know the blessings that surround us till we are deprived of them; so here we do not perceive the weight of the ambient fluid till a part of it is taken away. If, by any means, we contrive to take away the pressure of the air from any one part of our bodies, we are soon made sensible of the weight upon the other parts. Thus, if we clap our hand upon the mouth of a vessel from whence the air has been taken away, there will thus be air on one side, and none on the other; upon which, we shall instantly find the hand violently sucked inwards; which is nothing more than the weight of the air upon the back of the hand that forces it into the space which is empty below.

As, by this experiment, we perceive that the air presses with great weight upon every thing on the surface of the earth, so by other experiments we learn the exact weight with which it presses. First, if the air be exhausted out of any vessel, a drinking vessel for instance,* and this vessel be set with the mouth downwards in water, the water will rise up into the empty space, and fill the inverted glass; for the external air will, in this case, press up the water where there is no weight to resist; as, one part of a bed being pressed, makes the other parts, that have no weight upon them, rise. In this case, as was said, the water being pressed without, will rise in the glass; and would continue to rise (if the empty glass were tall enough) thirty-two feet high. In fact, there have been pipes made purposely for this experiment, of above thirty-two feet high; in which, upon being exhausted, the water has always risen to the height of thirty-two feet: there it has always rested, and never ascended higher. From this, therefore, we learn, that the weight of the air which presses up the water, is equal to a pillar or column of water, which is thirty-two feet high: as it is just able to raise such a column, and no more. In other words, the surface of the earth is every where covered with a weight of air, which is equivalent to a covering of thirty-two feet deep of water; or to a weight of twenty-nine inches and a half of quicksilver, which is known to be just as heavy as the former.

Thus we see that the air, at the surface of the earth, is just as heavy as thirty-two feet of water, or twenty-nine inches and a half of quicksilver; and it is easily found, by computation, that to raise water thirty-two feet, will require a weight of fifteen pounds upon every square inch. Now, if we are fond of computations, we have only to calcu-

---

* This may be done by burning a bit of paper in the same, and then quickly turning it down upon the water.

late how many square inches are in the surface of an ordinary human body, and allowing every inch to sustain fifteen pounds, we may amaze ourselves at the weight of air we sustain. It has been computed, and found, that our ordinary load of air amounts to within a little of forty thousand pounds: this is wonderful! but wondering is not the way to grow wise.

Notwithstanding this be our ordinary load, and our usual supply, there are, at different times, very great variations. The air is not, like water, equally heavy at all seasons; but sometimes is lighter, and sometimes more heavy. It is sometimes more compressed, and sometimes more elastic or springy, which produces the same effects as at increase of its weight. The air, which at one time raises water thirty-two feet in the tube, and quicksilver twenty-nine inches, will not at another raise the one to thirty feet, or the other to twenty-six inches. This makes, therefore, a very great difference in the weight we sustain; and we are actually known, by computation, to carry at one time four thousand pounds of air more than at another.

The reason of this surprising difference in the weight of air, is either owing to its pressure from above, or to an increase of vapour floating in it. Its increased pressure is the consequence of its spring or elasticity, which cold and heat sensibly affect, and are continually changing.

This elasticity of the air is one of its most amazing properties; and to which it should seem nothing can set bounds. A body of air that may be contained in a nutshell, may easily, with heat, be dilated into a sphere of unknown dimensions. On the contrary, the air contained in a house, may be compressed into a cavity not larger than the eye of a needle. In short, no bounds can be set to its confinement or expansion; at least, experiment has hitherto found its attempts indefinite. In every situation, it retains its elasticity; and the more closely we compress it, the more strongly does it resist the pressure. If to the increasing the elasticity on one side by compression, we increase it on the other side by heat, the force of both soon becomes irresistible; and a certain French philosopher* supposed, that air thus confined and expanding, was sufficient for the explosion of a world.

Many instruments have been formed to measure and determine these different properties of the air; and which serve several useful purposes. The barometer serves to measure its weight; to tell us when it is heavier, and when lighter. It is composed of a glass tube or pipe, of about thirty inches in length, closed up at one end; this tube is then filled with quicksilver; this done, the maker clapping his finger upon the open end, inverts the tube, and plunges the open end, finger and all, into a bason of quicksilver, and then takes his finger away; now the quicksilver in the tube will, by its own weight, endeavour to descend into that in the bason; but the external air, pressing on the surface of the quicksilver in the bason without, and no air being in the tube at top, the quicksilver will continue in the tube, being pressed up, as was said, by the air, on the surface of the bason below. The height at which it is known to stand in the tube, is usually about twen-

---

* Monsieur Amontons.

ty-nine inches, when the air is heavy; but not above twenty-six, when the air is very light. Thus, by this instrument, we can, with some exactness, determine the weight of the air; and, of consequence, tell before-hand the changes of the weather. Before fine dry weather, the air is charged with a variety of vapours, which float in it unseen, and render it extremely heavy, so that it presses up the quicksilver; or, in other words, the barometer rises. In moist, rainy weather, the vapours are washed down, or there is not heat sufficient for them to rise, so that the air is then sensibly lighter, and presses up the quicksilver with less force; or, in other words, the barometer is seen to fall. Our constitutions seem also to correspond with the changes of the weather-glass; they are braced, strong, and vigorous, with a large body of air upon them; they are languid, relaxed, and feeble, when the air is light, and refuses to give our fibres their proper tone.

But although the barometer thus measures the weight of the air with exactness enough for the general purposes of life, yet it is often affected with a thousand irregularities, that no exactness in the instrument can remedy, nor no theory account for. When high winds blow, the quicksilver generally is low: it rises higher in cold weather than in warm; and is usually higher at morning and evening than at midday: it generally descends lower after rain than it was before it. There are also frequent changes in the air, without any sensible alteration in the barometer.

As the barometer is thus used in predicting the changes of the weather, so it is also serviceable in measuring the heights of mountains, which mathematicians cannot so readily do: for, as the higher we ascend from the surface of the earth, the air becomes lighter, so the quicksilver in the baromter will descend in proportion. It is found to sink at the rate of the tenth part of an inch for every ninety feet we ascend; so that in going up a mountain, if I find the quicksilver fallen an inch, I conclude, that I am got upon an ascent of near nine hundred feet high. In this there has been found some variation; into a detail of which, it is not the business of a natural historian to enter.

In order to determine the elasticity of air, the wind-gun has been invented, which is an instrument variously made; but in all upon the principle of compressing a large quantity of air into a tube, in which there is an ivory ball, and then giving the compressed elastic air free power to act, and drive the ball as directed. The ball, thus driven, will pierce a thick board; and will be as fatal, at small distances, as if driven with gunpowder. I do not know whether ever the force of this instrument has been assisted by means of heat; certain I am, that this, which could be very easily contrived by means of phosphorus, or any other hot substance applied to the barrel, would give such a force as I doubt whether gunpowder itself could produce.

The air-pump is an instrument contrived to exhaust the air from round a vessel adapted to that purpose, called a receiver. This method of exhausting, is contrived in the simple instrument, by a piston, like that of a syringe, going down into the vessel, and thus pushing out its air; which, by means of a valve, is prevented from returning into the vessel again. But this, like all other complicated instruments, will be better understood by a minute inspection, than an hour's description: it may suffice here to observe, that by depriving animals, and other substances, of all air, it shews us what the benefits and effects of air are in sustaining life, or promoting vegetation.

The digester is an instrument of still more extraordinary effects than any of the former; and sufficiently discovers the amazing force of air, when its elasticity is augmented by fire. A common tea-kettle, if the spout were closed up, and the lid put firmly down, would serve to become a digester, if strong enough. But the instrument used for this purpose is a strong metal pot, with a lid to screw close on, so that, when down, no air can get in or return: into this pot meat and bones are put, with a small quantity of water, and then the lid screwed close: a lighted lamp is put underneath, and, what is very extraordinary, (yet equally true) in six or eight minutes the whole mass, bones and all, are dissolved into a jelly; so great is the force and elasticity of the air contained within, struggling to escape, and breaking in pieces all the substances with which it is mixed. Care, however, must be taken not to heat this instrument too violently; for then the inclosed air would become irresistible, and burst the whole, with, perhaps, a fatal explosion.

There are numberless other useful instruments made to depend on the weight, the elasticity, or the fluidity of the air, which do not come within the plan of the present work; the design of which is not to give an account of the inventions that have been made for determining the nature and properties of air, but a mere narrative of its effects. The description of the pump, the forcing-pump, the fire-engine, the steam-engine, the syphon, and many others, belong not to the naturalist, but the experimental philosopher: the one gives a history of Nature, as he finds she presents herself to him; and he draws the obvious picture: the other pursues her with close investigation, tortures her by experiment to give up her secrets, and measures her latent qualities with laborious precision. Much more, therefore, might be said of the mechanical effects of air, and of the conjectures that have been made respecting the form of its parts; how some have supposed them to resemble little hoops, coiled up in a spring; others, like fleeces of wool; others, that the parts are endued with a repulsive quality, by which, when squeezed together, they endeavour to fly off, and recede from each other. We might have given the disputes relative to the height to which this body of air extends above us, and concerning which there is no agreement. We might have inquired how much of the air we breathe is elementary, and not reducible to any other substance; and of what density it would become, if it were supposed to be continued down to the centre of the earth. At that place we might, with the help of figures, and a bold imagination, have shown it twenty thousand times heavier than its bulk of gold. We might also prove it millions of times purer than upon earth, when raised to the surface of the atmosphere. But these speculations do not belong to natural history; and they have hitherto produced no great advantages in that branch of science to which they more properly appertain.

CHAPTER XIX.

AN ESSAY TOWARDS A NATURAL HISTORY OF THE AIR.

A LATE eminent philosopher has considered our atmosphere as one large chymical vessel, in which an infinite number of various operations are constantly performing. In it all the bodies of the earth are continually sending up a part of their substance by evaporation, to mix in this great alembic, and to float awhile in common. Here minerals, from their lowest depths, ascend in noxious, or in warm vapours, to make a part of the general mass; seas, rivers, and subterraneous springs, furnish their copious supplies; plants receive and return their share; and animals, that by living upon, consume this general store, are found to give it back in greater quantities when they die.* The air, therefore, that we breathe, and upon which we subsist, bears very little resemblance to that pure elementary body which was described in the last chapter; and which is rather a substance that may be conceived, than experienced to exist. Air, such as we find it, is one of the most compounded bodies in all nature. Water may be reduced to a fluid every way resembling air, by heat; which, by cold, becomes water again. Every thing we see gives off its parts to the air, and has a little floating atmosphere of its own round it. The rose is encompassed with a sphere of its own odorous particles; while the night-shade infects the air with a scent of a more ungrateful nature. The perfume of musk flies off in such abundance, that the quantity remaining becomes sensibly lighter by the loss. A thousand substances that escape all our senses, we know to be there; the powerful emanations of the load-stone, the effluvia of electricity, the rays of light, and the insinuations of fire. Such are the various substances through which we move, and which we are constantly taking in at every pore, and returning again with imperceptible discharge.

This great solution, or mixture of all earthly bodies, is continually operating upon itself; which, perhaps, may be the cause of its unceasing motion: but it operates still more visibly upon such grosser substances as are exposed to its influence; for scarcely any substance is found capable of resisting the corroding qualities of the air. The air, say the chymists, is a chaos, furnished with all kinds of salts and menstruums; and, therefore, it is capable of dissolving all kinds of bodies. It is well known, that copper and iron are quickly covered, and eaten with rust; and that in the climates near the equator, no art can keep them clean. In those dreary countries, the instruments, knives, and keys, that are kept in the pocket, are nevertheless quickly encrusted; and the great guns, with every precaution, after some years, become useless. Stones, as being less hard, may be readily supposed to be more easily soluble. The marble of which the noble monuments of Italian antiquity are composed, although in one of the finest climates in the world, shew the impressions which have been made upon them by the air. In many places they seem worm-eaten by time; and, in others, they appear crumbling into dust. Gold alone seems to be ex

* Boyle, vol. ii. p. 593.

empted from this general state of dissolution; it is never found to contract rust, though exposed never so long; the reason of this seems to be, that sea-salt, which is the only menstruum capable of acting upon, and dissolving gold, is but very little mixed with the air; for salt being a very fixed body, and not apt to volatilize, and rise with heat, there is but a small proportion of it in the atmosphere. In the elaboratories, and shops, however, where salt is much used, and the air is impregnated with it, gold is found to rust as well as other metals.

Bodies of a softer nature are obviously destroyed by the air.* Mr. Boyle says, that silks brought to Jamaica, will, if there exposed to the air, rot, even while they preserve their colour; but if kept therefrom, they both retain their strength and gloss. The same happens in Brasil, where their clothes, which are black, soon turn of an iron colour; though, in the shops, they preserve their proper hue.† In these tropical climates also, such are the putrescent qualities of the air, that white sugar will sometimes be full of maggots. Drugs and plasters lose their virtue, and become verminous. In some places they are obliged to expose the sweetmeats by day in the sun, otherwise the night air would quickly cause them to putrefy. On the contrary, in the cold arctic regions, animal substances, during their winter, are never known to putrefy; and meat may be kept for months without any salt whatsoever. This experiment happily succeeded with the eight Englishmen that were accidentally left upon the inhospitable coasts of Greenland, at a place where seven Dutchmen had perished but a few years before; for killing some rein-deer for their subsistence and having no salt to preserve the flesh, to their great surprise they soon found it did not want any, as it remained sweet during their eight months' continuance upon that shore.

These powers with which air is endued over unorganized substances, are exerted in a still stronger manner over plants, animals of an inferior nature, and, lastly, over man himself. Most of the beauty, and the luxuriance of vegetation, is well known to be derived from the benign influence of the air: and every plant seems to have its favourite climate, not less than its proper soil. The lower ranks of animals also, seem formed for their respective climates, in which only they can live. Man alone seems the child of every climate, and capable of existing in all. However, this peculiar privilege does not exempt him from the influences of the air; he is as much subject to its malignity, as the meanest insect or vegetable.

With regard to plants, air is so absolutely necessary for their life and preservation, that they will not vegetate in an exhausted receiver. All plants have within them a quantity of air, which supports and agitates their juices. They are continually imbibing fresh nutriment from the air, to increase this store, and to supply the wants which they sustain from evaporation. When, therefore, the external air is drawn from them, they are no longer able to subsist. Even that quantity of air which they before were possessed of, escapes through their pores, into the exhausted receiver; and as this continues to be pumped away, they become languid, grow flaccid, and die. However

* Buffon, vol. iii. p. 62.   † Ibid, vol. iii. p. 68.

the plant or flower thus ceasing to vegetate, is kept, by being secured from the external air, a much longer time sweet than it would have continued, had it been openly exposed.

That air which is so necessary to the life of vegetables, is still more so to that of animals; there are none found, how seemingly torpid soever, that do not require their needful supply. Fishes themselves will not live in water from whence the air is exhausted; and it is generally supposed that they die in frozen ponds, from the want of this necessary to animal existence. Many have been the animals that idle curiosity has tortured in the prison of a receiver, merely to observe the manner of their dying. We shall, from a thousand instances, produce that of the viper, as it is known to be one of the most vivacious reptiles in the world; and as we shall feel but little compassion for its tortures. Mr. Boyle took a new-caught viper, and shutting it up into a small receiver, began to pump away the air.* "At first, upon the air's being drawn away, it began to swell; some time after he had done pumping, it began to gape, and open its jaws; being thus compelled to open its jaws, it once more resumed its former lankness; it then began to move up and down within, as if to seek for air, and after a while foamed a little, leaving the foam sticking to the inside of the glass; soon after the body and neck grew prodigiously tumid, and a blister appeared upon its back; an hour and a half after the receiver was exhausted, the distended viper moved, and gave manifest signs of life; the jaws remained quite distended; as it were from beneath the epiglottis, came the black tongue, and reached beyond it; but the animal seemed, by its posture, not to have any life; the mouth also was grown blackish within; and in this situation it continued for twenty-three hours. But upon the air's being re-admitted, the viper's mouth was presently closed, and soon after opened again; and for some time those motions continued, which argued the remains of life." Such is the fate of the most insignificant or minute reptile that can be thus included. Mites, fleas, and even the little eels that are found swimming in vinegar, die for want of air. Not only these, but the eggs of these animals, will not produce in vacuo, but require air to bring them to perfection.

As in this manner air is necessary to their subsistence, so also it must be of a proper kind, and not impregnated with foreign mixtures. That factitious air which is pumped from plants or fluids, is generally, in a short time, fatal to them. Mr. Boyle has given us many experiments to this purpose. After having shewn that all vegetable, and most mineral substances, properly prepared, may afford air, by being placed in an exhausted receiver, and this in such quantities, that some have thought it a new substance, made by the alteration which the mineral or plant has undergone by the texture of its parts being loosened in the operation—having shewn, I say, that this air may be drawn in great quantities from vegetable, animal, or mineral substances, such as apples, cherries, amber burnt, or hartshorn†—he included a frog in artificial air, produced from paste; in seven minutes' space it suf-

---

* Boyle's Physico-Mechan. Exper. passim
† Boyle's Physico-Mechan. vol. ii. p. 598.

fered convulsions, and at last lay still, and being taken out, recovered no motion at all, but was dead. A bird inclosed in artificial air, from raisins, died in a quarter of a minute, and never stirred more. A snail was put into the receiver, with air of paste; in four minutes it ceased to move, and was dead, although it had survived in vacuo for several hours: so that factitious air proved a greater enemy to animals than even a vacuum itself.

Air also may be impregnated with fumes that are instantly fatal to animals. The fumes of hot iron, copper, or any other heated metal, blown into the place where an animal is confined, instantly destroy it. We have already mentioned the vapours in the grotto Del Cane suffocating a dog. The ancients even supposed, that these animals, as they always ran with their noses to the ground, were the first that felt any infection. In short, it should seem that the predominance of any one vapour, from any body, how wholesome soever in itself, becomes infectious; and that we owe the salubrity of the air to the variety of its mixture.

But there is no animal whose frame is more sensibly affected by the changes of the air than man. It is true, he can endure a greater variety of climates than the lower orders generally are able to do; but it is rather by the means which he has discovered of obviating their effects, than by the apparent strength of his constitution. Most other animals can bear cold or hunger better, endure greater fatigues in proportion, and are satisfied with shorter repose. The variations of the climate, therefore, would probably affect them less, if they had the same means or skill in providing against the severities of the change. However this be, the body of man is an instrument much more nicely sensible of the variations of the air, than any of those which his own art has produced; for his frame alone seems to unite all their properties, being invigorated by the weight of the air, relaxed by its moisture, enfeebled by its heat, and stiffened by its frigidity.

But it is chiefly by the predominance of some peculiar vapour, that the air becomes unfit for human support. It is often found, by dreadful experience, to enter into the constitution, to mix with its juices, and to putrefy the whole mass of blood. The nervous system is not less affected by its operations; palsies and vertigoes are caused by its damps; and a still more fatal train of distempers by its exhalations. In order that the air should be wholesome, it is necessary, as we have seen, that it should not be of one kind, but the compound of several substances; and the more various the composition, to all appearance the more salubrious. A man, therefore, who continues in one place, is not so likely to enjoy this wholesome variety, as he who changes his situation; and, if I may so express it, instead of waiting for a renovation of air, walks forward to meet its arrival. This mere motion, independent even of the benefits of exercise, becomes wholesome, by thus supplying a great variety of that healthful fluid by which we are sustained.

A thousand accidents are found to increase these bodies of vapour, that make one place more or less wholesome than another. Heat may raise them in too great quantities; and cold may stagnate them. Minerals may give off their effluvia in such proportion as to keep away all other kind of air ; vegetables may render the air unwholesome by their supply ; and animal putrefation seems to furnish a quantity of vapour, at least as noxious as any of the former. All these united, generally make up the mass of respiration, and are, when mixed together, harmless; but any one of them, for a long time singly predominant, becomes at length fatal.

The effects of heat in producing a noxious quality in the air, are well known. Those torrid regions under the Line are always unwholesome. At Senegal, I am told, the natives consider forty as a very advanced time of life, and generally die of old age at fifty. At Carthagena,* in America, where the heat of the hottest day ever known in Europe is continual, where, during their winter season, these dreadful heats are united with a continual succession of thunder, rain, and tempests, arising from their intenseness, the wan and livid complexions of the inhabitants might make strangers suspect that they were just recovered from some dreadful distemper ; the actions of the natives are conformable to their colour ; in all their motions there is somewhat relaxed and languid ; the heat of the climate even affects their speech, which is soft and slow, and their words generally broken. Travellers from Europe retain their strength and ruddy colour in that climate, possibly for three or four months ; but afterwards suffer such decays in both, that they are no longer to be distinguished from the inhabitants by their complexion. However, this languid and spiritless existence is frequently drawled on sometimes even to eighty. Young persons are generally most affected by the heat of the climate, which spares the more aged ; but all, upon their arrival on the coasts, are subject to the same train of fatal disorders. Few nations have experienced the mortality of these coasts, so much as our own : in our unsuccessful attack upon Carthagena, more than three parts of our army were destroyed by the climate alone ; and those that returned from that fatal expedition, found their former vigour irretrievably gone. In our more fortunate expedition, which gave us the Havana, we had little reason to boast of our success; instead of a third, not a fifth part of the army were left survivors of their victory, the climate being an enemy that even heroes cannot conquer.

The distempers that thus proceed from the cruel malignity of those climates, are many : that, for instance, called the *Chapotonadas*, carries off a multitude of people ; and extremely thins the crews of European ships, whom gain tempts into those inhospitable regions. The nature of this distemper is but little known, being caused in some persons by cold, in others by indigestion. But its effects are far from being obscure ; it is generally fatal in three or four days : upon its seizing the patient, it brings on what is there called the *black vomit*, which is the sad symptom after which none are ever found to recover. Some, when the vomit attacks them, are seized with a delirium, that, were they not tied down, they would tear themselves to pieces, and thus expire in the midst of this furious paroxysm. This disorder, in milder climates, takes the name of the *bilious fever*, and is attended with milder symptoms, but very dangerous in all.

* Ulloa, vol. i. p. 42.

There are many other disorders incident to the human body, that seem the offspring of heat; but to mention no other, tnat very lassitude which prevails in all the tropical climates, may be considered as a disease. The inhabitants of India,* says a modern philosopher, sustain an unceasing languor, from the heats of their climate, and are torpid in the midst of profusion. For this reason, the great Disposer of nature has clothed their country with trees of an amazing height, whose shade might defend them from the beams of the sun; and whose continual freshness might, in some measure, temperate their fierceness. From these shades, therefore, the air receives refreshing moisture, and animals a cooling protection. The whole race of savage animals retire, in the midst of the day, to the very centre of the forest, not so much to avoid their enemy man, as to find a defence against the raging heats of the season. This advantage which arises from shade in torrid climates, may probably afford a solution for that extraordinary circumstance related by Boyle, which he imputes to a different cause. In the island of Ternate, belonging to the Dutch, a place that had been long celebrated for its beauty and healthfulness, the clove-trees grew in such plenty, that they in some measure lessened their own value: for this reason, the Dutch resolved to cut down the forests, and thus to raise the price of the commodity: but they had soon reason to repent of their avarice; for such a change ensued, by cutting down the trees, that the whole island, from being healthy and delightful, having lost its charming shades, became extremely sickly, and has actually continued so to this day. Boerhaave considered heat so prejudicial to health, that he was never seen to go near a fire.

An opposite set of calamities are the consequence, in climates where the air is condensed by cold. In such places, all that train of distempers which are known to arise from obstructed perspiration, are very common;† eruptions, boils, scurvy, and a loathsome leprosy, that covers the whole body with a scurf, and white putrid ulcers. These disorders also are infectious; and, while they thus banish the patient from society, they generally accompany him to the grave. The men of those climates seldom attain to the age of fifty; but the women, who do not lead such laborious lives, are found to live longer.

The autumnal complaints which attend a wet summer, indicate the dangers of a moist air. The long continuance of an east wind also, shews the prejudice of a dry one. Mineral exhalations, when copious, are every where known to be fatal; and although we probably owe the increase and luxuriance of vegetation to a moderate degree of their warmth, yet the natives of those countries where there are mines in plenty, but too often experience the noxious effects of their vicinity. Those trades also that deal in the preparations of metals of all kinds, are always unwholesome; and the workmen, after some time, are generally seen to labour under palsies, and other nervous complaints. The vapours from some vegetable substances are well known to be attended with dangerous effects. The shade of the machinel tree, in America, is said to be fatal, as was that of the juniper, if we may credit the ancients. Those who walk through fields of poppies, or in any

* Linne's Amœnitates, vol. v. p. 444.   † Crantz's History of Greenland. vol. i. p. 235.

manner prepare those flowers for making opium, are very sensibly affected with the drowsiness they occasion. A physician of Mr. Boyle's acquaintance, causing a large quantity of black hellebore to be pounded in a mortar, most of the persons who were in the room, and especially the person who pounded it, were purged by it, and some of them strongly. He also gathered a certain plant in Ireland, which the person who beat in a mortar, and the physician who was standing near, were so strongly affected by, that their hands and faces swelled to an enormous size, and continued tumid for a long time after.

But neither mineral nor vegetable steams are so dangerous to the constitution, as those proceeding from animal substances, putrefying either by disease or death. The effluvia that comes from diseased bodies, propagate that frightful catalogue of disorders which are called *infectious*. The parts which compose vegetable vapours and mineral exhalations, seem gross and heavy, in comparison of these volatile vapours, that go to great distances, and have been described as spreading desolation over the whole earth. They fly every where; penetrate every where; and the vapours that fly from a single disease, render it soon epidemic.

The plague is the first upon the list in this class of human calamities. From whence this scourge of man's presumption may have its beginning, is not well known: but we well know that it is propagated by infection. Whatever be the general state of the atmosphere, we learn, from experience, that the noxious vapours, though but singly introduced at first, taint the air by degrees; every person infected tends to add to the growing malignity; and, as the disorder becomes more general, the putrescence of the air becomes more noxious, so that the symptoms are aggravated by continuance. When it is said that the origin of this disorder is unknown, it implies, that the air seems to be but little employed in first producing it. There are some countries, even in the midst of Africa, that we learn have never been infected with it; but continue, for centuries, unmolested. On the contrary, there are others, that are generally visited once a year, as in Egypt, which, nevertheless, seems peculiarly blessed with the serenity and temperature of its climate. In the former countries, which are of vast extent, and many of them very populous, every thing should seem to dispose the air to make the plague continual among them. The great heats of the climate, the unwholesomeness of the food, the sloth and dirt of the inhabitants, but, above all, the bloody battles which are continually fought among them, after which heaps of dead bodies are left unburied, and exposed to putrefaction. All these one might think would be apt to bring the plague among them, and yet, nevertheless, we are assured by Leo Africanus, that in Numidia the plague is not known once in a hundred years; and that in Negroland, it is not known at all. This dreadful disorder, therefore, must have its rise, not from any previous disposition of the air, but from some particular cause, beginning with one individual, and extending the malignity, by communication, till at last the air becomes actually tainted by the generality of the infection.

The plague which spread itself over the whole world, in the year 1346, as we are told by Mezeray, was so contagious, that scarce a viliage, or even a house, escaped being infected by it. Before it had reached Europe, it had been for two years travelling from the great kingdom of Cathay, where it began by a vapour most horridly fœtid; this broke out of the earth like a subterranean fire, and upon the first instant of its eruption, consumed and desolated above two hundred leagues of that country, even to the trees and stones.

In that great plague which desolated the city of London, in the year 1665, a pious and learned schoolmaster of Mr. Boyle's acquaintance, who ventured to stay in the city, and took upon him the humane office of visiting the sick and the dying, who had been deserted by better physicians, averred, that being once called to a poor woman who had buried her children of the plague, he found the room where she lay so little that it scarce could hold any more than the bed whereon she was stretched. However, in this wretched abode, beside her, in an open coffin, her husband lay, who had some time before died of the same disease; and whom she, poor creature, soon followed. But what shewed the peculiar malignity of the air, thus suffering from animal putrefaction, was, that the contagious steams had produced spots on the very wall of their wretched apartment: and Mr. Boyle's own study, which was contiguous to a pesthouse, was also spotted in the same frightful manner. Happily for mankind, this disorder, for more than a century has not been known in our island; and, for this last age, has abated much of its violence, even in those countries where it is most common. Diseases, like empires, have their revolutions; and those which for a while were the scourge of mankind, sink unheard of, to give place to new ones, more dreadful, as being less understood.

For this revolution in disorders, which has employed the speculation of many, Mr. Boyle accounts in the following manner: "Since," says he, "there want not causes in the bowels of the earth, to make considerable changes amongst the materials that Nature has plentifully treasured up in those magazines, and as those noxious steams are abundantly supplied to the surface, it may not seem improbable, that in this great variety, some may be found capable of affecting the human frame in a particular manner, and thus of producing new diseases. The duration of these may be greater or less, according to the lastingness of those subterraneous causes that produced them. On which account, it need be no wonder that some diseases have but a short duration, and vanish not long after they appear; whilst others may continue longer, as having under ground more settled and durable causes to maintain them."

From the recital of this train of mischiefs produced by the air, upon minerals, plants, animals, and man himself, a gloomy mind may be apt to dread this indulgent nurse of nature as a cruel and an inexorable step-mother: but it is far otherwise; and, although we are sometimes injured, yet almost all the comforts and blessings of life spring from its propitious influence. It would be needless to observe that it is absolutely necessary for the support of our lives; for of this every moment's experience assures us. But how it contributes to this support, is not so readily comprehended. All allow it to be a friend, to whose benefits we are constantly obliged; and yet, to this hour, philosophers are divided as to the nature of the obligation. The dispute is whether the air is only useful by its weight to force our juices into circulation:* or, whether, by containing a peculiar spirit, it mixes with the blood in our vessels, and acts like a spur to their industry.† Perhaps it may exert both these useful offices at the same time. Its weight may give the blood its progressive motion, through the larger vessels of the body; and its admixture with it, cause those contractions of all the vessels, which serve to force it still more strongly forward, through the minutest channels of the circulation. Be this as it may, it is well known, that that part of our blood which has just received the influx of the air in our bodies, is of a very different colour from that which has almost performed its circuit. It has been found, that the arterial blood which has been immediately mixed with the air in the lungs, and, if I may so express it, is just beginning its journey through the body, is of a fine florid scarlet colour; while, on the contrary, the blood of the veins that is returning from having performed its duty, is of a blackish crimson hue. Whence this difference of colour should proceed, is not well understood; we only know the fact, that this florid colour is communicated by the air; and we are well convinced, that this air has been admitted into the blood for very useful purposes.

Besides this vital principle in animals, the air also gives life and body to flame. A candle quickly goes out in an exhausted receiver; for having soon consumed the quantity of air, it then expires for want of a fresh supply. There has been a flame contrived that will burn under water; but none has yet been found, that will continue to burn without air. Gunpowder, which is the most catching and powerful fire we know, will not go off in an exhausted receiver: nay, if a train of gunpowder be laid, so as that one part may be fired in the open air, yet the other part in vacuo will remain untouched, and unconsumed. Wood also set on fire, immediately goes out; and its flame ceases upon removing the air; for something is then wanting to press the body of the fire against that of the fuel, and to prevent the too speedy diffusion of the flame. We frequently see cooks, and others, whose business it is to keep up strong fires, take proper precautions to exclude the beams of the sun from shining upon them, which effectually puts them out. This they are apt to ascribe to a wrong cause; namely, the operation of the light; but the real fact is, that the warmth of the sun-beams lessens and dissipates the body of the air that goes to feed the flame; and the fire, of consequence, languishes for want of a necessary supply.

The air, while it thus kindles fire into flame, is, notwithstanding, found to moderate the rays of light, to dissipate their violence, and to spread a uniform lustre over every object. Were the beams of the sun to dart directly upon us, without passing through this protecting medium, they would either burn us up at once, or blind us with their effulgence. But by going through the air, they are reflected, refracted, and turned from their direct course, a thousand different ways; and thus are more evenly diffused over the face of nature.

Among the other necessary benefits the air is of to us, one of the

---

* Keil Robinson.   ᵃ Whytt upon vital and involuntary Motions.

principal is its conveyance of sound. Even the vibrations of a bell, which have the loudest effect that we know of, cease to be heard, when under the receiver of an air-pump. Thus all the pleasures we receive from conversation with each other, or from music, depend entirely upon the air.

Odours likewise are diffused only by the means of air; without this fluid to swim in, they would for ever remain torpid in their respective substances; and the rose would affect us with as little sensations of pleasure, as the thorn on which it grew.

Those who are willing to augment the catalogue of the benefits we receive from this element, assert also, that tastes themselves would be insipid, were it not that the air presses their parts upon the nerves of the tongue and palate, so as to produce their grateful effects. Thus, continue they, upon the tops of high mountains, as on the Peak of Teneriffe, the most poignant bodies, as pepper, ginger, salt, and spice, have no sensible taste, for want of their particles being thus sent home to the sensory. But we owe the air sufficient obligations, not to be studious of admitting this among the number: in fact, all substances have their taste, as well on the tops of the mountains, as in the bottom of the valley; and I have been one of many, who have ate a very savoury dinner on the Alps.

It is sufficient, therefore, that we regard the air as the parent of health and vegetation; as a kind dispenser of light and warmth; and as the conveyer of sounds and odours. This is an element of which avarice will not deprive us; and which power cannot monopolize. The treasures of the earth, the verdure of the fields, and even the refreshments of the stream, are too often seen going only to assist the luxuries of the great; while the less fortunate part of mankind stand humble spectators of their encroachments. But the air no limitations can bound, nor any landmarks restrain. In this benign element, all mankind can boast an equal possession; and for this we all have equal obligations to Heaven. We consume a part of it, for our own sustenance, while we live; and, when we die, our putrefying bodies give back the supply, which, during life, we had accumulated from the general mass.

CHAPTER XX.

OF WINDS, IRREGULAR AND REGULAR.

WIND is a current of air. Experimental philosophers produce an artificial wind, by an instrument called an *æolipile*. This is nothing more than a hollow copper ball, with a long pipe: a tea-kettle might be readily made into one, if it were entirely closed at the lid, and the spout left open; through this spout it is to be filled with water, and then set upon the fire, by which means it produces a violent blast, like wind, which continues while there is any water remaining in the instrument. In this manner water is converted into a rushing air; which, if caught as it goes out, and left to cool, is again quickly converted into its former element. Besides this, as was mentioned in the former chapter, almost every substance contains some portions of air. Vegetables, or the bodies of animals left to putrefy, produce it in a very copious manner. But it is not only seen thus escaping from bodies, but it may be very easily made to enter into them. A quantity of air may be compressed into water, so as to be intimately blended with it. It finds a much easier admission into wine, or any fermented liquor; and an easier still, into spirits of wine. Some salts suck up the air in such quantities, that they are made sensibly heavier thereby, and often are melted by its moisture. In this manner, most bodies, being found either capable of receiving or affording it, we are not to be surprised at those streams of air that are continually fleeting round the globe.—Minerals, vegetables, and animals, contribute to increase the current; and are sending off their constant supplies. These, as they are differently affected by cold or heat, by mixture or putrefaction, all yield different quantities of air at different times; and the loudest tempests, and most rapid whirlwinds, are formed from their united contributions.

The sun is the principal instrument in rarefying the juices of plants, so as to give an escape to their imprisoned air; it is also equally operative in promoting the putrefaction of animals. Mineral exhalations are more frequently raised by subterranean heat. The moon, the other planets, the seasons, are all combined in producing these effects in a smaller degree. Mountains give a direction to the courses of the air. Fires carry a current of air along their body. Night and day alternately chill and warm the earth, and produce an alternate current of its vapours. These, and many other causes, may be assigned for the variety, and the activity of the winds, their continual change, and uncertain duration.

With us on land, as the wind proceeds from so many causes, and meets such a variety of obstacles, there can be but little hopes of ever bringing its motions to conform to theory; or of foretelling how it may blow a minute to come. The great Bacon, indeed, was of opinion, that by a close and regular history of the winds, continued for a number of ages together, and the particulars of each observation reduced to general maxims, we might at last come to understand the variations of this capricious element; and that we could foretell the certainty of a wind, with as much ease as we now foretell the return of an eclipse. Indeed, his own beginnings in this arduous undertaking seem to speak the possibility of its success; but, unhappily for mankind, this investigation is the work of ages, and we want a Bacon to direct the process.

To be able, therefore, with any plausibility, to account for the variations of the wind upon land, is not to be at present expected; and to understand any thing of their nature, we must have recourse to those places where they are more permanent and steady. This uniformity and steadiness we are chiefly to expect upon the ocean. There, where there is no variety of substances to furnish the air with various and inconstant supplies, where there are no mountains to direct the course of its current, but where all is extensively uniform and even; in such a place, the wind arising from a simple cause, must have but one simple motion. In fact, we find it so. There are many parts of the world where the winds, that with us are so uncertain, pay their stated visits. In some places they are found to blow one way by day, and another by night; in others, for one half of the year they go in a direction contrary to their former course: but what is more extraordinary still, there are some places where the winds never change, but for ever blow the same way. This is particularly found to obtain between the tropics in the Atlantic and Æthiopic oceans; as well as in the great Pacific sea.

Few things can appear more extraordinary to a person who has never been out of our variable latitudes, than this steady wind, that for ever sits in the sail, sending the vessel forward; and as effectually preventing its return. He who has been taught to consider that nothing in the world is so variable as the winds, must certainly be surprised to find a place where there is nothing more uniform. With us their inconstancy has become a proverb; with the natives of those distant climates they may talk of a friend or a mistress as fixed and unchangeable as the winds, and mean a compliment by the comparison. When our ships are once arrived into the proper latitudes of the great Pacific ocean, the mariner forgets the helm, and his skill becomes almost useless: neither storms nor tempests are known to deform the glassy bosom of that immense sheet of waters: a gentle breeze, that for ever blows in the same direction, rests upon the canvass, and speeds the navigator. In the space of six weeks, ships are thus known to cross an immense ocean, that takes more than so many months to return. Upon returning, the trade-wind, which has been propitious, is then avoided; the mariner is generally obliged to steer into the northern latitudes, and to take the advantage of every casual wind that offers, to assist him into port. This wind, which blows with such constancy one way, is known to prevail not only in the Pacific ocean, but also in the Atlantic, between the coasts of Guinea and Brazil; and, likewise, in the Æthiopic ocean. This seems to be the great universal wind, blowing from the east to the west, that prevails in all the extensive oceans, where the land does not frequently break the general current. Were the whole surface of the globe an ocean, there would probably be but this one wind, for ever blowing from the east, and pursuing the motions of the sun westward. All the other winds seem subordinate to this; and many of them are made from the deviations of its current. To form, therefore, any conception relative to the variations of the wind in general, it is proper to begin with that which never varies.

There have been many theories to explain this invariable motion of the winds; among the rest, we cannot omit that of Dr. Lyster, for its strangeness. " The sea," says he, " in those latitudes, is generally covered over with green weeds, for a great extent; and the air produced from the vegetable perspiration of these, produces the trade-wind." The theory of Cartesius was not quite so absurd. He alleged, that the earth went round faster than its atmosphere at the equator; so that its motion, from west to east, gave the atmosphere an imaginary one from east to west; and thus an east wind was eternally seen to prevail. Rejecting those arbitrary opinions, conceived without force, and asserted without proof, Dr. Halley has given one more plausible; which seems to be the reigning system of the day.

To conceive his opinion clearly, let us for a moment suppose the whole surface of the earth to be an ocean, and the air encompassing it on every side, without motion. Now it is evident, that that part of the air which lies directly under the beams of the sun, will be rarefied; and if the sun remained for ever in the same place, there would be a great vacuity in the air, if I may so express it, beneath the place where the sun stood. The sun moving forward, from east to west, this vacuity will follow too, and still be made under it. But while it goes on to make new vacuities, the air will rush in to fill up those the sun has already made; in other words, as it is still travelling forward, the air will continually be rushing in behind, and pursue its motions from east to west. In this manner the air is put into motion by day; and by night the parts continue to impel each other, till the next return of the sun, that gives a new force to the circulation.

In this manner is explained the constant east wind that is found blowing round the globe, near the equator. But it is also known, that as we recede from the equator on either side, we come into a trade-wind, that continually blows from the poles, from the north on one side, or the south on the other, both directing towards the equator. This also proceeds from a similar cause with the former; for the air being more rarefied in those places over which the sun more directly darts its rays, the currents will come both from the north and the south, to fill up the intermediate vacuity.

These two motions, namely, the general one from east to west, and the more particular one from both the poles, will account for all the phænomena of trade-winds; which, if the whole surface of the globe were sea, would undoubtedly be constant, and for ever continue to blow in one direction. But there are a thousand circumstances to break these air-currents into smaller ones; to drive them back against their general course; to raise or depress them; to condense them into storms, or to whirl them in eddies. In consequence of this, regard must be often had to the nature of the soil, the position of the high mountains, the course of the rivers, and even to the luxuriance of vegetation.

If a country lying directly under the sun be very flat and sandy, and if the land be low and extensive, the heats occasioned by the reflection of the sun-beams produces a very great rarefaction of the air. The deserts of Africa, which are conformable to this description, are scarcely ever fanned by a breath of wind by day; but the burning sun is continually seen blazing in intolerable splendour above them. For this reason, all along the coasts of Guinea, the wind is always perceived blowing in upon land, in order to fill up the vacuity caused by the sun's operation. In those shores, therefore, the wind blows in a contrary direction to that of its general current; and is constantly found setting in from the west.

From the same cause it happens, that those constant calms, attended with deluges of rain, are found in the same part of the ocean. For this tract being placed in the middle, between the westerly winds blowing on the coast of Guinea, and the easterly trade-winds that move at some distance from shore, in a contrary direction, the tendency of that part of the air that lies between these two opposite currents is indifferent to either, and so rests between both in torpid serenity; and the weight of the incumbent atmosphere, being diminished by the continual contrary winds blowing from hence, it is unable to keep the vapours suspended that are copiously borne thither; so that they fall in continual rains.

But it is not to be supposed, that any theory can account for all the phænomena of even those winds that are known to be most regular. Instead of a complete system of the trade-winds, we must rather be content with an imperfect history. These,* as was said, being the result of a combination of effects, assume as great a variety as the causes producing them are various.

Besides the great general wind above-mentioned, in those parts of the Atlantic that lie under the temperate zone, a north wind prevails constantly during the months of October, November, December, and January. These, therefore, are the most favourable months for embarking for the East-Indies, in order to take the benefit of these winds, for crossing the Line: and it has been often found, by experience, that those who had set sail five months before, were not in the least farther advanced in their voyage, than those who waited for the favourable wind. During the winter, off Nova Zembla, and the other arctic countries, a north wind reigns almost continually. In the Cape de Verde Islands, a south wind prevails during the month of July. At the Cape of Good Hope, a northwest wind blows during the month of September. There are also regular winds, produced by various causes, upon land. The ancient Greeks were the first who observed a constant breeze, produced by the melting of the snows, in some high neighbouring countries. This was perceived in Greece, Thrace, Macedonia, and the Ægean sea. The same kind of winds are now remarked in the kingdom of Congo, and the most southern parts of Africa. The flux and reflux of the sea also produces some regular winds, that serve the purposes of trade; and, in general, it may be observed, that wherever there is a strong current of water, there is a current of air that seems to attend it.

Beside these winds that are found to blow in one direction, there are, as was said before, others that blow for certain months of the year one way, and the rest of the year the contrary way; these are called the *Monsoons*, from a famous pilot of that name, who first used them in navigation with success.† In all that part of the ocean that lies between Africa and India, the east winds begin at the month of January, and continue till about the commencement of June. In the month of August or September, the contrary direction takes place; and the west winds prevail for three or four months. The interval between these winds, that is to say, from the end of June to the beginning of August, there is no fixed wind; but the sea is usually tossed by violent tempests, proceeding from the north. These winds are always subject to their greatest variations, as they approach the land; so that on one side of the great peninsula of India, the coasts are, for near

* Buffon, vol. ii. p. 230.   † Varenii Geographia Generalis, cap. 21

half the year, harassed by violent hurricanes, and northern tempests; while, on the opposite side, and all along the coasts of C romandel, these dreadful tempests are wholly unknown. At Java, and Ceylon, a west wind begins to reign in the month of September; but at fifteen degrees of south latitude, this wind is found to be lost, and the great general trade-wind from the east is perceived to prevail. On the contrary, at Cochin, in China, the west wind begins in March; so that these monsoons prevail, at different seasons, throughout the Indies. So that the mariner takes one part of the year to go from Java to the Moluccas; another from Cochin to Molucca; another from Molucca to China; and still another to direct him from China to Japan.

There are winds also that may be considered as peculiar to certain coasts: for example, the south wind is almost constant upon the coasts of Chili and Peru; western winds almost constantly prevail on the coast of Terra Magellanica; and in the environs of the Straits le Maire. On the coasts of Malabar, north and north-west winds prevail continually; along the coast of Guinea, the north-west wind is also very frequent; and, at a distance from the coasts, the north-east is always found prevailing. From the beginning of November to the end of December, a west wind prevails on the coasts of Japan; and, during the whole winter, no ships can leave the port of Cochin, on account of the impetuosity of the winds that set upon the coast. These blow with such vehemence, that the ports are entirely choked up with sand, and even boats are not able to enter. However, the east winds that prevail for the other half of the year, clear the mouths of their harbours from the accumulations of the preceding winter, and set the confined ships at liberty. At the Straits of Babelmandel, there is a south wind that periodically returns, and which is always followed by a north-east.

Besides winds thus peculiar to certain coasts, there are others found to prevail on all the coasts, in warm climates, which, during one part of the day, blow from the shore, and, during another part of it, blow from the sea. The sea-breeze, in those countries, as Dampier observes, commonly rises in the morning about nine, proceeding slowly, in a fine small black curl, upon tne surface of the the water, and making its way to refresh the shore. It is gentle at first, but increases gradually till twelve, then insensibly sinks away, and is totally hushed at five. Upon its ceasing, the land-breeze begins to take its turn, which increases gradually till twelve at night, and is succeeded, in the morning, by the sea-breeze again. Without all doubt, nothing could be more fortunate, for the inhabitants of the warm countries, where those breezes blow, than this alternate refreshment, which they feel at those seasons when it is most wanted. The heat, on some coasts, would be insupportable, were it not for such a supply of air, when the sun has rarefied all that which lay more immediately under the coast. The sea-breeze temperates the heat of the sun by day; and the land-breeze corrects the malignity of the dews and vapours by night. Where these breezes, therefore, prevail, and they are very common, the inhabitants enjoy a share of health and happiness, unknown to those that live much farther up the country, or such as live in similar latitudes without this advantage. The cause of these obvi ously seems to arise from the rarefaction of the air by the sun, as their duration continues with its appearance, and alters when it goes down. The sun, it is observed, equally diffusing his beams upon land and sea the land, being a more solid body than the water, receives a greater quantity of heat, and reflects it more strongly. Being thus, therefore, heated to a greater degree than the waters, it, of consequence, drives the air from land out to sea; but, its influence being removed, the air returns to fill up the former vacuity. Such is the usual method of accounting for this phænomenon; but, unfortunately, these sea and land breezes are visitants that come at all hours. On the coasts of Malabar,* the land-breezes begin at midnight, and continue till noon; then the sea-breezes take their turn, and continue till midnight. While again, at Congo, the land-breezes begin at five, and continue till nine the next day.

But if the cause of these be so inscrutable, that are, as we see, tolerably regular in their visitations, what shall we say to the winds of our own climate, that are continually shifting, and incapable of rest? Some general causes may be assigned, which nothing but particular experience can apply. And, in the first place, it may be observed, that clouds, and heat, and in short, whatever either increases the density or the elasticity of the air, in any one place, will produce a wind there: for the increased activity of the air thus pressing more powerfully on the parts of it that are adjacent, will drive them forward, and thus go on, in a current, till the whole comes to an equality.

In this manner, as a denser air produces a wind on one hand; so will any accident, that contributes to lighten the air, produce it on the other: for, a lighter air may be considered as a vacuity, into which the neighbouring air will rush: and hence it happens, that when the barometer marks a peculiar lightness in the air, it is no wonder that it foretels a storm.

The winds upon large waters are generally more regular than those upon land. The wind at sea generally blows with an even steady gale; the wind at land puffs by intervals, increasing its strength, and remitting it, without any apparent cause. This, in a great measure, may be owing to the many mountains, towers, or trees, that it meets in its way, all contributing either to turn it from its course, or interrupt its passage.

The east wind blows more constantly than any other, and for an obvious reason: all other winds are, in some measure, deviations from it, and partly may owe their origin thereto. It is generally likewise the most powerful, and for the same reason.

There are often double currents of the air. While the wind blows one way, we frequently see the clouds move another. This is generally the case before thunder; for it is well known that the thunder-cloud always moves against the wind: the cause of this surprising appearance has hitherto remained a secret. From hence we may conclude, that weathercocks only inform us of that current of the air, which is near the surface of the earth; but are often erroneous with regard to the upper regions, and, in fact, Derham has often found them erroneous

* Button, vol. ii. p. 252

Winds are generally more powerful on elevated situations than on the plain, because their progress is interrupted by fewer obstacles. In proportion as we ascend the heights of a mountain, the violence of the weather seems to increase until we have got above the region of storms, where all is usually calm and serene. Sometimes, however, the storms rise even to the tops of the highest mountains; as we learn from those who have been on the Andes, and as we are convinced by the deep snows that crown even the highest.

Winds blowing from the sea are generally moister, and more attended with rains, than those which blow over extensive tracts of land; for the sea gives off more vapours to the air, and these are rolled forward upon land, by the wind's blowing from thence.* For this reason our easterly winds, that blow from the continent, are dry, in comparison of those that blow from the surface of the ocean, with which we are surrounded on every other quarter.

In general the winds are more boisterous in spring and autumn than at other seasons: for that being the time of high tides, the sea may communicate a part of its motions to the winds. The sun and moon, also, which then have a greater effect upon the waters, may also have some influence upon the winds: for, there being a great body of air surrounding the globe, which, if condensed into water, would cover it to the depth of thirty-two feet, it is evident that the sun and moon will, to a proportionable degree, affect the atmosphere, and make a tide of air. This tide will be scarcely perceivable, indeed; but, without doubt, it actually exists; and may contribute to increase the vernal and autumnal storms, which are then known to prevail.

Upon narrowing the passage through which the air is driven, both the density and the swiftness of the wind is increased. For, as currents of water flow with greater force and rapidity by narrowing their channels, so also will a current of air, driven through a contracted space, grow more violent and irresistible. Hence we find those dreadful storms that prevail in the defiles of mountains, where the wind, pushing from behind through a narrow channel, at once increases in speed and density, levelling, or tearing up, every obstacle that rises to obstruct its passage.

Winds reflected from the sides of mountains and towers, are often found to be more forceful than those in direct progression. This we frequently perceive near lofty buildings, such as churches or steeples, where winds are generally known to prevail, and are much more powerful than at some distance. The air, in this case, by striking against the side of the building, acquires additional density, and, therefore blows with more force.

These different degrees of density, which the air is found to possess, sufficiently shew that the force of the winds do not depend upon their velocity alone; so that those instruments called *anemometers*, which are made to measure the velocity of the wind, will by no means give us certain information of the force of the storm. In order to estimate this with exactness, we ought to know its density; which also these are not calculated to discover. For this reason we often see

* Derham's Physico-Theol.

storms, with very powerful effects, that do not seem to shew any great speed; and, on the contrary, we see these wind-measurers go round with great swiftness, when scarce any damage has followed from the storm.

Such is the nature and the inconstancy of the irregular winds with which we are best acquainted. But their effects are much more formidable in those climates near the tropics, where they are often found to break in upon the steady course of the trade-winds, and to mark their passage with destruction. With us, the tempest is but rarely known, and its ravages are registered as an uncommon calamity; but in the countries that lie between the tropics, and for a good space beyond them, its visits are frequent, and its effects are anticipated. In these regions the winds vary their terrors; sometimes involving all things in a suffocating heat; sometimes mixing all the elements of fire, air, earth, and water together; sometimes, with a momentary swiftness, passing over the face of the country, and destroying all things in their passage; and sometimes raising whole sandy deserts in one country, to deposit them upon some other. We have little reason, therefore, to envy these climates the luxuriance of their soil, or the brightness of their skies. Our own muddy atmosphere, that wraps us round in obscurity, though it fails to gild our prospects with sunshine, or our groves with fruitage, nevertheless answers the call of industry. They may boast of a plentiful, but precarious harvest; while, with us, the labourer toils in a certain expectation of a moderate, but a happy return.

In Egypt,[*] a kingdom so noted for its fertility, and the brightness of its atmosphere during summer, the south winds are so hot, that they almost stop respiration; besides which, they are charged with such quantities of sand, that they sometimes darken the air as with a thick cloud. These sands are so fine, and driven with such violence, that they penetrate every where, even into chests, be they shut ever so closely. If these winds happen to continue for any length of time, they produce epidemic diseases, and are often followed by a great mortality. It is also found to rain but very seldom in that country; however, the want of showers is richly compensated by the copiousness of their dews, which greatly tend to promote vegetation.

In Persia, the winter begins in November, and continues till March. The cold at that time is intense enough to congeal the water; and snow falls in abundance upon their mountains. During the months of March and April, winds arise, that blow with great force, and seem to usher in the heats of summer. These return again, in autumn, with some violence; without, however, producing any dreadful effects. But, during their summer, all along the coasts of the Persian Gulf, a very dangerous wind prevails, which the natives call the *Sameyel*, still more dreadful and burning than that of Egypt, and attended with instant and fatal effects. This terrible blast, which was perhaps, the pestilence of the ancients, instantly kills all those that it involves in its passage. What its malignity consists in, none can tell, as none have ever survived its effects, to give information. It frequently, as I am told, assumes a visible form, and darts, in a kind of bluish va

---

[*] Buffon, vol. ii. p. 258

pour, along the surface of the country. The natives, not only o Persia, but of Arabia, talk of its effects with terror; and their poets have not failed to heighten them with the assistance of imagination. They have described it as under the conduct of a minister of vengeance, who governs its terrors, and raises, or depresses it, as he thinks proper.* These deadly winds are also known along the coasts of India, at Negapatam, Masulipatam, and Petapoli. But, luckily for mankind, the shortness of their duration diminishes the injuries that might ensue from their malignity.

The Cape of Good Hope, as well as many islands in the West-Indies, are famous for their hurricanes, and that extraordinary kind of cloud which is said to produce them. This cloud, which is the forerunner of an approaching hurricane, appears, when first seen, like a small black spot, on the verge of the horizon; and is called, by sailors, *the bull's eye*, from being seen so minute at a vast distance. All this time, a perfect calm reigns over the sea and land, while the cloud grows gradually broader as it approaches. At length, coming to the place where its fury is to fall, it invests the whole horizon with darkness. During all the time of its approach, a hollow murmur is heard in the cavities of the mountains; and beasts and animals, sensible of its approach, are seen running over the fields, to seek for shelter. Nothing can be more terrible than its violence when it begins. The houses in those countries, which are made of timber, the better to resist its fury, bend to the blast like osiers, and again recover their rectitude. The sun, which, but a moment before, blazed with meridian splendour, is totally shut out; and a midnight darkness prevails, except that the air is incessantly illuminated with gleams of lightning, by which one can easily see to read. The rain falls, at the same time, in torrents; and its descent has been resembled to what pours from the spouts of our houses after a violent shower. These hurricanes are not less offensive to the sense of smelling also, and never come without leaving the most noisome stench behind them. If the seamen also lay by their wet clothes, for twenty-four hours, they are all found swarming with little white maggots, that were brought with the hurricane. Our first mariners, when they visited these regions, were ignorant of its effects, and the signs of its approach; their ships, therefore, were dashed to the bottom at the first onset; and numberless were the wrecks which the hurricane occasioned. But, at present, being forewarned of its approach, they strip their masts of all their sails, and thus patiently abide its fury. These hurricanes are common in all the tropical climates. On the coasts of Guinea they have frequently three or four in a day, that thus shut out the heavens, for a little space; and, when past, leave all again in former splendour. They chiefly prevail, on that coast, in the intervals of the trade-winds; the approach of which clears the air of its meteors, and gives these mortal showers that little degree of wholesomeness which they possess. They chiefly obtain there during the months of April and May; they are known, at Loango, from January to April; on the opposite coast of Africa, the hurricane season begins in May;

* Herbelot, Bibliotheque Orientale.

Mangabey, or White Eyelid Monkey, p. 267.

Red Monkey, p. 266.

White-nosed Monkey, p. 267.

and, in general, whenever a trade-wind begins to cease, these irregular tempests are found to exert their fury.

All this is terrible:—but there is a tempest, known in those climates, more formidable than any we have hitherto been describing, which is called, by the Spaniards, a *Tornado*. As the former was seen arriving from one part of the heavens, and making a line of destruction; so the winds in this seem to blow from every quarter, and settle upon one destined place, with such fury, that nothing can resist their vehemence. When they have all met, in their central spot, then the whirlwind begins with circular rapidity. The sphere every moment widens, as it continues to turn, and catches every object that lies within its attraction. This also, like the former, is preceded by a flattering calm; the air is every where hushed, and the sea is as smooth as polished glass: however, as its effects are more dreadful than those of the ordinary hurricane, the mariner tries all the power of his skill to avoid it; which, if he fails of doing, there is the greatest danger of his going to the bottom. All along the coasts of Guinea, beginning about two degrees north of the line, and so downward, lengthwise, for about a thousand miles, and as many broad, the ocean is unnavigable, upon account of these tornadoes. In this torrid region there reigns unceasing tornadoes, or continual calms; among which, whatever ship is so unhappy as to fall, is totally deprived of all power of escaping. In this dreadful repose of all the elements, the solitary vessel is obliged to continue, without a single breeze to assist the mariner's wishes, except those whirlwinds, which only serve to increase his calamity. At present, therefore, this part of the ocean is totally avoided; and, although there may be much gold along the coasts of that part of Africa, to tempt avarice, yet there is something, much more dreadful than the fabled dragon of antiquity, to guard the treasure. As the internal parts of that country are totally unknown to travellers, from their burning sands and extensive deserts, so here we find a vast tract of ocean, lying off its shores, equally unvisited by the mariner.

But of all these terrible tempests that deform the face of nature, and repress human presumption, the sandy tempests of Arabia and Africa are the most terrible, and strike the imagination most strongly. To conceive a proper idea of these, we are by no means to suppose them resembling those whirlwinds of dust that we sometimes see scattering in our air, and sprinkling their contents upon our roads or meadows. The sand-storm of Africa exhibits a very different appearance. As the sand of which the whirlwind is composed is excessively fine, and almost resembles the parts of water, its motion entirely resembles that of a fluid; and the whole plain seems to float onward, like a slow inundation. The body of sand thus rolling, is deep enough to bury houses and palaces in its bosom: travellers who are crossing those extensive deserts perceive its approach at a distance; and, in general, have time to avoid it, or turn out of its way, as it generally extends but to a moderate breadth. However, when it is extremely rapid, or very extensive, as sometimes is the case, no swiftness, no art, can avail nothing then remains, but to meet death with fortitude, and submit to be buried alive with resignation.

It is happy for us of Britain, that we have no such calamity to fear, for from this even some parts of Europe are not entirely free. We have an account given us in the History of the French Academy, of a miserable town in France, that is constantly in danger of being buried under a similar inundation; with which I will take leave to close this chapter. "In the neighbourhood of St. Paul de Leon, in Lower Brittany,* there lies a tract of country along the sea-side, which before the year 1666 was inhabited, but now lies deserted, by reason of the sands which cover it, to the height of twenty feet; and which every year advance more and more inland, and gain ground continually. From the time mentioned above, the sand has buried more than six leagues of the country inward; and it is now but half a league from the town of St. Paul; so that, in all appearance, the inhabitants must be obliged to abandon it entirely. In the country that has been overwhelmed, there are still to be seen the tops of some steeples peeping through the sand, and many chimnies that still remain above this sandy ocean. The inhabitants, however, had sufficient time to escape; but being deprived of their little all, they had no other resource but begging for their subsistence. This calamity chiefly owes its advancement to a north or an east wind, raising the sand, which is extremely fine, in such great quantities, and with such velocity, that M. Deslands, who gave the account, says, that while he was walking near the place, during a moderate breeze of wind, he was obliged, from time to time, to shake the sand from his clothes and his hat, on which it was lodged in great quantities, and made them too heavy to be easily borne. Still further, when the wind was violent, it drove the sand across a little arm of the sea, into the town of Roscoff, and covered the streets of that place two feet deep; so that they have been obliged to carry it off in carts. It may also be observed, that there are several particles of iron mixed with the sand, which are readily affected by the loadstone. The part of the coast that furnishes these sands, is a tract of about four leagues in length; and is upon a level with the sea at high-water. The shore lies in such a manner as to leave its sands subject only to the north and east winds, that bear them farther up the shore. It is easy to conceive how the same sand that has at one time been borne a short way inland, may, by some succeeding and stronger blast, be carried up much higher; and thus the whole may continue advancing forward, deluging the plain, and totally destroying its fertility. At the same time the sea, from whence this deluge of sand proceeds, may furnish it in inexhaustible quantities. This unhappy country, thus overwhelmed in so singular a manner, may well justify what the ancients and the moderns have reported concerning those tempests of sand in Africa, that are said to destroy villages and even armies, in their bosom."

* Histoire de l'Academie des Sciences, an. 1722.

CHAPTER XXI.

OF METEORS, AND SUCH APPEARANCES AS RESULT FROM A COMBINATION OF THE ELEMENTS.

In proportion as the substances of nature are more compounded and combined, their appearances become more inexplicable and amazing. The properties of water have been very nearly ascertained Many of the qualities of air, earth, and fire, have been discovered and estimated; but when these come to be united by nature, they often produce a result which no artificial combinations can imitate; and we stand surprised, that although we are possessed of all those substances which nature makes use of, she shows herself a much more various operator than the most skilful chymist ever appeared to be. Every cloud that moves, and every shower that falls, serves to mortify the philosopher's pride, and to show him hidden qualities in air and water, that he finds it difficult to explain. Dews, hail, snow, and thunder, are not less difficult for being more common. Indeed, when we reflect on the manner in which nature performs any one of these operations, our wonder increases. To see water, which is heavier than air, rising in air, and then falling in a form so very different from that in which it rose; to see the same fluid at one time descending in the form of hail, at another in that of snow; to see two clouds by dashing against each other, producing an electrical fire, which no watery composition that we know of can effect; these, I say, serve sufficiently to excite our wonder; and still the more, in proportion as the objects are ever pressing on our curiosity. Much, however, has been written concerning the manner in which nature operates in these productions; as nothing is so ungrateful to mankind as hopeless ignorance.

And first, with regard to the manner in which water evaporates, and rises to form clouds, much has been advanced, and many theories devised. All water,* say some, has a quantity of air mixed with it; and the heat of the sun darting down, disengages the particles of this air from the grosser fluid: the sun's rays being reflected back from the water, carry back with them those bubbles of air and water, which, being lighter than the condensed air, will ascend till they meet with a more rarefied air; and they will then stand suspended. Experience, however, proves nothing of all this. Particles of air or fire, are not thus known to ascend with a thin coat of water; and in fact we know the little particles of steam are solid drops of water. But besides this, water is known to evaporate more powerfully in the severest frost, than when the air is moderately warm.† Dr. Hamilton, therefore, of the university of Dublin, rejecting this theory, has endeavoured to establish another. According to him, as aquafortis is a menstruum that dissolves iron, and keeps it mixed in the fluid; as aquaregia is a menstruum that dissolves gold; or as water dissolves salts to a certain quantity; so air is a menstruum that corrodes and dissolves a certain quantity of water, and keeps it suspended above.

---

* Spectacle de la Nature, vol. iii. † Memoires de l'Academie des Sciences, an. 1705

But however ingenious this may be, it can hardly be admitted; as we know by Mariotte's experiment,* that if water and air be inclosed together, instead of the air's acting as a menstruum upon the water, the water will act as a menstruum upon the air, and take it all up. We know also, that of two bodies, that which is most fluid and penetrating, is most likely to be the menstruum of the other; but water is more fluid and penetrating than air, and therefore the most likely of the two to be the menstruum. We know that all bodies are more speedily acted upon, the more their parts are brought into contact with the menstruum that dissolves them; but water, inclosed with compressed air, is not the more diminished thereby.† In short, we know, that cold, which diminishes the force of other menstruums, is often found to promote evaporation. In this variety of opinion and uncertainty of conjecture, I cannot avoid thinking that a theory of evaporation may be formed upon very simple and obvious principles, and embarrassed, as far as I can conceive, with very few objections.

We know that a repelling power prevails in nature, not less than an attractive one. This repulsion prevails strongly between the body of fire and that of water. If I plunge the end of a red-hot bar of iron into a vessel of water, the fluid rises, and large drops of it fly up in all manner of directions, every part bubbling and steaming until the iron be cold. Why may we not for a moment compare the rays of the sun, darted directly upon the surface of the water, to so many bars of red-hot iron, each bar indeed infinitely small, but not the less powerful? In this case, wherever a ray of fire darts, the water, from its repulsive quality, will be driven on all sides; and, of consequence, as in the case of the bar of iron, a part of it will rise. The parts thus rising, however, will be extremely small; as the ray that darts is extremely so. The assemblage of the rays darting upon the water in this manner, will cause it to rise in a light thin steam above the surface; and as the parts of this steam are extremely minute, they will be lighter than air, and consequently float upon it. There is no need for supposing them bubbles of water filled with fire; for any substance, even gold itself, will float on air, if its parts be made small enough; or, in other words, if its surface be sufficiently increased. This water, thus disengaged from the general mass, will be still farther attenuated and broken by the reflected rays, and, consequently, more adapted for ascending.

From this plain account, every appearance in evaporation may be easily deduced. The quantity of heat increases evaporation, because it raises a greater quantity of steam. The quantity of wind increases evaporation; for, by waving the surface of the water, it thus exposes a greater surface to the evaporating rays. A dry frost, in some measure, assists the quantity of evaporation; as the quantity of rays are found to be no way diminished thereby. Moist weather alone prevents evaporation; for the rays being absorbed, refracted, and broken, by the intervening moisture, before they arrive at the surface, cannot produce the effect; and the vapour will rise in a small proportion.

Thus far we have accounted for the ascent of vapours; but to ac-

* Mariotte, de la Nature de l'Air, p. 97, 106. † See Boyle's Works, vol. ii. p 619

count for their falling again, is attended with rather more difficulty. We have already observed, that the particles of vapour, disengaged from the surface of the water, will be broken and attenuated in their ascent, by the reflected, and even the direct rays, that happen to strike upon their minute surfaces. They will, therefore, continue to ascend, till they rise above the operation of the reflected rays, which reaches but to a certain height above the surface of the earth. Being arrived at this region, which is cold for want of reflected heat, they will be condensed, and suspended in the form of clouds. Some vapours that ascend to great heights, will be frozen into snow; others, that are condensed lower down, will put on the appearance of a mist, which we find the clouds to be, when we ascend among them, as they hang along the sides of a mountain. These clouds of snow and rain, being blown about by winds, are either entirely scattered and dispersed above, or they are still more condensed by motion, like a snow-ball, that grows more large and solid as it continues to roll. At last, therefore, they will become too weighty for the air which first raised them to sustain; and they will descend with their excess of weight, either in snow or rain. But as they will fall precipitately, when they begin to descend, the air, in some measure, will resist the falling; for as the descending fluid gathers velocity in its precipitation, the air will increase its resistance to it, and the water will, therefore, be thus broken into rain; as we see, that water which falls from the tops of houses, though it begins in a spout, separates into drops before it has got to the bottom. Were it not for this happy interposition of the air, between us and the water falling from a considerable height above us, a drop of rain might fall with dangerous force, and a hailstone might strike us with fatal rapidity.

In this manner, evaporation is produced by day; but when the sun goes down, a part of that vapour which his rays had excited, being no longer broken, and attenuated by the reflecting rays, it will become heavier than the air, even before it has reached the clouds; and it will, therefore, fall back in dews, which differ only from rain in descending before they have had time to condense it into a visible form.

Hail, the Cartesians say, is a frozen cloud, half melted, and frozen again in its descent. A hoar-frost is but a frozen dew. Lightning we know to be an electrical flash, produced by the opposition of two clouds; and thunder to be the sound proceeding from the same, continued by an echo reverberated among them. It would be to very little purpose, to attempt explaining exactly how these wonders are effected: we have as yet but little insight into the manner in which these meteors are found to operate upon each other; and, therefore, we must be contented with a detail rather of their effects than their causes.

In our own gentle climate, where nature wears the mildest and kindest aspect, every meteor seems to befriend us. With us, rains fall in refreshing showers, to enliven our fields, and to paint the landscape with a more vivid beauty. Snows cover the earth, to preserve its tender vegetables from the inclemency of the departing winter. The dews descend with such an imperceptible fall as no way injures the constitution. Even thunder itself is seldom injurious: and it is often wished for by the husbandman, to clear the air, and to kill numberless insects that are noxious to vegetation. Hail is the most injurious meteor that is known in our climate; but it seldom visits us with violence, and then its fury is but transient.

One of the most dreadful storms we hear of,[*] was that of Hertfordshire, in the year 1697. It began by thunder and lightning, which continued for some hours, when suddenly a black cloud came forward, against the wind, and marked its passage with devastation. The hailstones which it poured down, being measured, were found to be many of them fourteen inches round, and, consequently, as large as a bowling-green ball. Wherever it came, every plantation fell before it; it tore up the ground, split great oaks, and other trees, without number; the fields of rye were cut down, as if levelled with a scythe; wheat, oats, and barley, suffered the same damage. The inhabitants found but a precarious shelter, even in their houses, their tiles and windows being broke by the violence of the hailstones, which, by the force with which they came, seemed to have descended from a great height. The birds, in this universal wreck, vainly tried to escape by flight; pigeons, crows, rooks, and many more of the smaller and feebler kinds, were brought down. An unhappy young man, who had not time to take shelter, was killed; one of his eyes was struck out of his head, and his body was all over black with the bruises; another had just time to escape, but not without the most imminent danger, his body being bruised all over. But what is most extraordinary, all this fell within the compass of a mile

Mezeray, in his history of France, tells us of a shower of hail much more terrible, which happened in the year 1510, when the French monarch invaded Italy. There was, for a time, a horrid darkness, thicker than that of midnight, which continued till the terrors of mankind were changed to still more terrible objects, by thunder and lightning breaking the gloom, and bringing on such a shower of hail, as no history of human calamities could equal. These hailstones were of a bluish colour; and some of them weighed not less than a hundred pounds. A noisome vapour of sulphur attended the storm. All the birds and beasts of the country were entirely destroyed. Numbers of the human race suffered the same fate. But what is still more extraordinary, the fishes found no protection from their native element; but were equal sufferers in the general calamity.

These, however, are terrors that are seldom exerted in our mild climates. They only serve to mark the page of history with wonder: and stand as admonitions to mankind, of the various stores of punishment in the hands of the Deity, which his power can treasure up, and his mercy can suspend.

In the temperate zones, therefore, meteors are rarely found thus terrible; but between the tropics, and near the poles, they assume very dreadful and various appearances. In those inclement regions, where cold and heat exert their chief power, meteors seem peculiarly to have fixed their residence. They are seen there in a thousand terrifying forms, astonishing to Europeans, yet disregarded by the na-

---

[*] Phil. Trans. vol. ii. p 147.

tives, from their frequency. The wonders of air, fire, and water, are there combined, to produce the most tremendous effects; and to sport with the labours and apprehensions of mankind. Lightnings, that flash without noise; hurricanes, that tear up the earth; clouds, that all at once pour down their contents, and produce an instant deluge; mock suns; northern lights, that illuminate half the hemisphere; circular rainbows; halos; fleeting balls of fire; clouds, reflecting back the images of things on earth, like mirrors; and water-spouts, that burst from the sea, to join with the mists that hang immediately above hem. These are but a part of the phænomena that are common in those countries; and from many of which our own climate is, in a great measure, exempted.

The meteors of the torrid zone, are different from those that are found near the polar circles; and it may readily be supposed, that in those countries where the sun exerts the greatest force in raising vapours of all kinds, there should be the greatest quantity of meteors. Upon the approach of the winter months, as they are called, under the Line, which usually begin about May, the sky, from a fiery brightness, begins to be overcast, and the whole horizon seems wrapt in a muddy cloud. Mists and vapours still continue to rise; and the air, which so lately before was clear and elastic, now becomes humid, obscure, and stifling; the fogs become so thick, that the light of the sun seems, in a manner, excluded; nor would its presence be known, but for the intense and suffocating heat of its beams, which dart through the gloom, and, instead of dissipating, only serve to increase the mist. After this preparation, there follows an almost continual succession of thunder, rain, and tempests. During this dreadful season, the streets of cities flow like rivers; and the whole country wears the appearance of an ocean. The inhabitants often make use of this opportunity to lay in a stock of fresh water, for the rest of the year; as the same cause which pours down the deluge at one season, denies the kindly shower at another. The thunder which attends the fall of these rains, is much more terrible than that we are generally acquainted with. With us, the flash is seen at some distance, and the noise shortly after ensues; our thunder generally rolls in one quarter of the sky, and one stroke pursues another. But here it is otherwise; the whole sky seems illuminated with unremitted flashes of lightning; every part of the air seems productive of its own thunders; and every cloud produces its own shock. The strokes come so thick, that the inhabitants can scarce mark the intervals; but all is one unremitted roar of elementary confusion. It should seem, however, that the lightning of those countries is not so fatal, or so dangerous, as with us; since, in this case, the torrid zone would be uninhabitable.

When these terrors have ceased, with which, however, the natives are familiar, meteors of another kind begin to make their appearance. The intense beams of the sun, darting upon stagnant waters, that generally cover the surface of the country, raise vapours of various kinds. Floating bodies of fire, which assume different names, rather from their accidental forms, than from any real difference between them, are seen without surprise. The *draco volans*, or flying dragon, as it is called ; the *ignis fatuus*, or wandering fire ; the *fires of St. Helmo*, or the mariner's light, are every where frequent ; and of these we have numberless descriptions. " As I was riding in Jamaica," says Mr. Barbham, " one morning, from my habitation, situated about three miles north-west from Jago de la Vega, I saw a ball of fire, appearing to me of the bigness of a bomb, swiftly falling down with a great blaze. At first I thought it fell into the town ; but when I came nearer, I saw many people gathered together, a little to the southward, in the savanna, to whom I rode up, to inquire the cause of their meeting they were admiring, as I found, the ground's being strangely broke up and ploughed by a ball of fire ; which, as they said, fell down there. I observed there were many holes in the ground ; one in the middle of the bigness of a man's head, and five or six smaller round about it, of the bigness of one's fist, and so deep as not to be fathomed by such implements as were at hand. It was observed, also, that all the green herbage was burnt up, near the holes ; and there continued a strong smell of sulphur near the place, for some time after."

Ulloa gives an account of one of a similar kind, at Quito.[*] " About nine at night," says he, " a globe of fire appeared to rise from the side of the mountain Pichinca, and so large, that it spread a light over all the part of the city facing that mountain. The house where I lodged, looking that way, I was surprised with an extraordinary light, darting through the crevices of the window-shutters. On this appearance, and the bustle of the people in the street, I hastened to the window, and came time enough to see it in the middle of its career ; which continued from west to south, till I lost sight of it, being intercepted by a mountain, that lay between me and it. It was round ; and its apparent diameter about a foot. I observed it to rise from the sides of Pichinca ; although to judge from its course, it was behind that mountain where this congeries of inflammable matter was kindled. In the first half of its visible course, it emitted a prodigious effulgence, then it began gradually to grow dim ; so that, upon its disappearing behind the intervening mountain, its light was very faint."

Meteors, of this kind, are very frequently seen between the tropics ; but they sometimes, also, visit the more temperate regions of Europe. We have the description of a very extraordinary one, given us by Montanari, that serves to shew to what great heights, in our atmosphere, these vapours are found to ascend. In the year 1676, a great globe of fire was seen at Bononia, in Italy, about three quarters of an hour after sun-set It passed westward, with a most rapid course, and at the rate of not less than a hundred and sixty miles in a minute, which is much swifter than the force of a cannon-ball, and at last stood over the Adriatic sea. In its course it crossed over all Italy ; and, by computation, it could not have been less than thirty-eight miles above the surface of the earth. In the whole line of its course, wherever it approached, the inhabitants below could distinctly hear it, with a hissing noise, resembling that of a fire-work. Having passed away to sea, towards Corsica, it was heard, at last, to go off with a most violent explosion, much louder than that of a cannon ;

[*] Ulloa, vol. i. p. 41.

and, immediately after, another noise was heard, like the rattling of a great cart, upon a stony pavement; which was, probably, nothing more than the echo of the former sound. Its magnitude, when at Bononia, appeared twice as long as the moon, one way, and as broad the other; so that, considering its height, it could not have been less than a mile long, and half a mile broad. From the height at which this was seen, and there being no volcano on that quarter of the world from whence it came, it is more than probable that this terrible globe was kindled on some part of the contrary side of the globe, in those regions of vapours, which we have been just describing; and thus, rising above the air, and passing in a course opposite to that of the earth's motion, in this manner it acquired its amazing rapidity.

To these meteors, common enough southward, we will add one more of a very uncommon kind, which was seen by Ulloa, at Quito, in Peru; the beauty of which will, in some measure, serve to relieve us, after the description of those hideous ones preceding. "At day-break," says he, "the whole mountain of Pambamarca, where we then resided, was encompassed with very thick clouds; which the rising of the sun dispersed so far, as to leave only some vapours, too fine to be seen. On the side opposite to the rising sun, and about ten fathoms distant from the place where we were standing, we saw, as in a look-ing-glass, each his own image; the head being, as it were, the centre of three circular rainbows, one without the other, and just near enough o each other as that the colours of the internal verged upon those more external; while round all was a circle of white, but with a greater space between. In this manner these circles were erected, like a mirror, before us; and as we moved, they moved, in disposition and order. But, what is most remarkable, though we were six in number, every one saw the phænomenon with regard to himself, and not that relating to others. The diameter of the arches gradually altered, as the sun rose above the horizon; and the whole, after continuing a long time, insensibly faded away. In the beginning, the diameter of the inward iris, taken from its last colour, was about five degrees and a half; and that of the white arch, which surrounded the rest, was not less than sixty-seven degrees. At the beginning of the phæ-nomenon, the arches seemed of an oval, or elliptical figure, like the disk of the sun; and afterwards became perfectly circular. Each of these was of a red colour, bordered with an orange; and the last bor-dered by a bright yellow, which altered into a straw-colour, and this turned to a green; but, in all, the external colour remained red." Such is the description of one of the most beautiful illusions that has ever been seen in nature. This alone seems to have combined al 'he splendours of optics in one view. To understand the manner, therefore, how this phænomenon was produced, would require a per-fect knowledge of optics, which it is not our present province to enter upon. It will be sufficient, therefore, only to observe, that all these appearances arise from the density of the cloud, together with 'ts uncommon and peculiar situation, with respect to the spectator and the sun. It may be observed, that but one of these three rainbows was real, the rest being only reflections thereof. It may also be observed, that whenever the spectator stands between the sun and a cloud of falling rain, a rainbow is seen, which is nothing more than the reflection of the different coloured rays of light from the bosom of the cloud. If, for instance, we take a glass globe, filled with water, and hang it up before us opposite the sun, in many situations it will appear transparent; but if it is raised higher, or sideways, to an angle of forty-five degrees, it will at first appear red; altered a very little higher, yellow; then green, then blue, then violet colour; in short, it will assume successively all the colours of the rainbow, but, if raised higher still, it will become transparent again. A falling shower may be considered as an infinite number of these little transparent globes, assuming different colours, by being placed at the proper heights. The rest of the shower will appear transparent; and no part of it will seem coloured, but such as are at angles of forty-five degrees from the eye, forty-five degrees upward, forty-five degrees on each side, and forty-five degrees downward, did not the plane of the earth prevent us. We therefore see only an arch of the rainbow, the lower part being cut off from our sight by the earth's interposition. However, upon the tops of very high mountains, circular rainbows are seen, because we can see to an angle of forty-five degrees downward, as well as upward, or sideways, and therefore we take in the rainbow's complete circle.

In those forlorn regions round the poles, the meteors, though of another kind, are not less numerous and alarming. When the winter begins, and the cold prepares to set in, the same misty appearance which is produced in the southern climates by the heat, is there produced by the contrary extreme.* The sea smokes like an oven, and a fog arises, which mariners call the *frost smoke*. This cutting mist commonly raises blisters on several parts of the body; and, as soon as it is wafted to some colder part of the atmosphere, it freezes to little icy particles, which are driven by the wind, and create such an intense cold on land, that the limbs of the inhabitants are sometimes frozen, and drop off.

There, also, hallos, or luminous circles round the moon, are oftener seen than in any other part of the earth, being formed by the frost smoke; although the air otherwise seems to be clear. A lunar rainbow also, is often seen there, though somewhat different from that which is common with us; as it appears of a pale white, striped with gray. In these countries also, the aurora borealis streams with peculiar lustre, and variety of colours. In Greenland it generally arises in the east, and darts its sportive fires, with variegated beauty, over the whole horizon. Its appearance is almost constant in winter; and, at those seasons when the sun departs, to return no more for half a year, this meteor kindly rises to supply its beams, and affords sufficient light for all the purposes of existence. However, in the very midst of their tedious night, the inhabitants are not entirely forsaken. The tops of the mountains are often seen painted with the red rays of the sun; and the poor Greenlander from thence begins to date his chronology. It would appear whimsical to read a Greenland calendar, in which we might be told, that one of their chiefs, having lived

* Paul Edgar's History of Greenland.

.orty days, died, at last, of a good old age; and that his widow continued for half a day to deplore his loss, with great fidelity, before she admitted a second husband.

The meteors of the day, in these countries, are not less extraordinary than those of the night: mock suns are often reflected upon an opposite cloud; and the ignorant spectator fancies that there are often three or four real suns in the firmament at the same time. In this splendid appearance the real sun is always readily known by its superior brightness, every reflection being seen with diminished splendour. The solar rainbow there, is often seen different from ours. Instead of a pleasing variety of colours, it appears of a pale white, edged with a stripe of dusky yellow; the whole being reflected from the bosom of a frozen cloud.

But, of all the meteors which mock the imagination with an appearance of reality, those strange illusions that are seen there, in fine serene weather, are the most extraordinary and entertaining. "Nothing," says Krantz, "ever surprised me more, than on a fine warm summer's day, to perceive the islands that lie four leagues west of our shore, putting on a form quite different from what they are known to have. As I stood gazing upon them, they appeared at first infinitely greater than what they naturally are; and seemed as if I viewed them through a large magnifying glass. They were not thus only made larger, but brought nearer to me. I plainly descried every stone upon the land, and all the furrows filled with ice, as if I stood close by. When this illusion had lasted for a while, the prospect seemed to break up, and a new scene of wonder to present itself. The islands seemed to travel to the shore, and represented a wood, or a tall cut hedge. The scene then shifted, and showed the appearance of all sorts of curious figures; as ships with sails, streamers, and flags; antique elevated castles, with decayed turrets; and a thousand forms, for which fancy found a resemblance in nature.—When the eye had been satisfied with gazing, the whole group of riches seemed to rise in air, and at length vanish into nothing. At such times the weather is quite serene and clear; but compressed with subtle vapours, as it is in very hot weather; and these appearing between the eye and the object, give it all that variety of appearances which glasses of different refrangibilities would have done." Mr. Krantz observes, that commonly a couple of hours afterwards, a gentle west wind and a visible mist follows, which puts an end to this *lusus naturæ*.

It were easy to swell this catalogue of meteors with the names of many others, both in our own climate and in other parts of the world. Such as falling stars, which are thought to be no more than unctuous vapours, raised from the earth to small heights, and continuing to shine till that matter which first raised, and supported them, being burnt out, they fall back again to the earth, with extinguished flame. burning spears, which are a peculiar kind of aurora borealis; bloody rains, which are said to be the excrements of an insect, that at that time has been raised into the air. Showers of stones, fishes, and ivy-berries, at first, no doubt, raised into the air by tempests in one country, and falling at some considerable distance in the manner of rain, to as tonise another. But omitting these, of which we know little more than what is thus briefly mentioned, I will conclude this chapter with the description of a water-spout; a most surprising phænomenon, not less dreadful to mariners than astonishing to the observer of nature.

These spouts are seen very commonly in the tropical seas, and sometimes in our own. Those seen by Tournefort, in the Mediterranean, he has described as follows: "The first of these," says this great botanist, "that we saw, was about a musket-shot from our ship. There we perceived the water began to boil, and to rise about a foot above its level. The water was agitated and whitish; and above its surface there seemed to stand a smoke, such as might be imagined to come from wet straw before it begins to blaze. It made a sort of a murmuring sound, like that of a torrent heard at a distance, mixed, at the same time, with a hissing noise like that of a serpent shortly after, we perceived a column of this smoke rise up to the clouds, at the same time whirling about with great rapidity. It appeared to be as thick as one's finger; and the former sound still continued. When this disappeared, after lasting for about eight minutes, upon turning to the opposite quarter of the sky, we perceived another, which began in the manner of the former; presently after a third appeared in the west; and instantly beside it, still another arose. The most distant of these three could not be above a musket-shot from the ship. They all continued like so many heaps of wet straw set on fire, that continued to smoke, and to make the same noise as before. We soon after perceived each, with its respective canal, mounting up in the clouds, and spreading where it touched; the cloud, like the mouth of a trumpet, making a figure, to express it intelligibly, as if the tail of an animal were pulled at one end by a weight. These canals were of a whitish colour, and so tinged, as I suppose, by the water which was contained in them; for previous to this, they were apparently empty, and of the colour of transparent glass. These canals were not straight, but bent in some parts, and far from being perpendicular, but rising in their clouds with a very inclined ascent. But what is very particular, the cloud to which one of them was pointed, happening to be driven by the wind, the spout still continued to follow its motion, without being broken; and passing behind one of the others, the spouts crossed each other, in the form of a St. Andrew's Cross. In the beginning they were all about as thick as one's finger, except at the top, where they were broader, and two of them disappeared; but shortly after, the last of the three increased considerably; and its canal, which was at first so small, soon became as thick as a man's arm, then as his leg, and at last thicker than his whole body. We saw distinctly, through this transparent body, the water, which rose up with a kind of spiral motion; and it sometimes diminished a little of its thickness, and again resumed the same; sometimes widening at top, and sometimes at bottom; exactly resembling a gut filled with water, pressed with the fingers to make the fluid rise, or fall; and I am well convinced, that this alteration in the spout was caused by the wind, which pressed the cloud, and impelled it to give up its contents. After some time, its bulk was so diminished as to be no thicker than a man's arm again: and thus swelling, and diminishing, it at last became very small. In the end, I observed the sea which was raised about it, to resume its level by degrees, and the end of the canal that touched it to become as small as if it had been tied round with a cord; and this continued till the light, striking through the cloud, took away the view. I still, however, continued to look, expecting that its parts would join again, as I had before seen in one of the others, in which the spout was more than once broken, and yet again came together; but I was disappointed, for the spout appeared no more."

Many have been the solutions offered for this surprising appearance Mr. Buffon supposes the spout here described, to proceed from the operation of fire, beneath the bed of the sea; as the waters at the surface are thus seen agitated. However, the solution of Dr. Stuart is not divested of probability; who thinks it may be accounted for by suction, as in the application of a cupping-glass to the skin.

Wherever spouts of this kind are seen, they are extremely dreaded by mariners; for if they happen to fall upon a ship, they most commonly dash it to the bottom. But, if the ship be large enough to sustain the deluge, they are at least sure to destroy its sails and rigging, and render it unfit for sailing. It is said that vessels of any force usually fire their guns at them, loaden with a bar of iron; and if so happy as to strike them, the water is instantly seen to fall from them, with a dreadful noise, though without any farther mischief.

I am at a loss whether we ought to reckon these spouts called *typhons*, which are sometimes seen at land, of the same kind with those so often described by mariners at sea, as they seem to differ in several respects. That, for instance, observed at Hatfield, in Yorkshire, in 1687, as it is described by the person who saw it, seems rather to have been a whirlwind than a water-spout. The season in which it appeared was very dry, the weather extremely hot, and the air very cloudy. After the wind had blown for some time, with considerable force, and condensed the black clouds one upon another, a great whirling of the air ensued; upon which the centre of the clouds, every now and then darted down, in the shape of a thick long black pipe; in which the relator could distinctly view a motion, like that of a screw, continually screwing up to itself, as it were, whatever it happened to touch. In its progress it moved slowly over a grove of young trees, which it violently bent, in a circular motion. Going forward to a barn, it in a minute stript it of all the thatch, and filled the whole air with the same. As it came near the relator, he perceived that its blackness proceeded from a gyration of the clouds, by contrary winds, meeting in a point, or a centre; and where the greatest force was exerted, there darting down, like an Archimedes' screw, to suck up all that came in its way. Another which he saw, some time after, was attended with still more terrible effects; levelling or tearing up great oak-trees, catching up the birds in its vortex, and dashing them against the ground. In this manner it proceeded, with an audible whirling noise, like that of a mill; and at length dissolved, after having done much mischief.

But we must still continue to suspend our assent as to the nature even of these land spouts, since they have been sometimes found to drop, in a great column of water, at once upon the earth, and produce an instant inundation,* which could not readily have happened had they been caused by the gyration of a whirlwind only. Indeed, every conjecture, regarding these meteors, seems to me entirely unsatisfactory. They sometimes appear in the calmest weather at sea, of which I have been an eyewitness; and, therefore, these are not caused by a whirlwind. They are always capped by a cloud; and, therefore, are not likely to proceed from fires at the bottom. They change place; and therefore suction seems impracticable. In short, we still want facts, upon which to build a rational theory; and, instead of knowledge, we must be contented with admiration. To be well acquainted with the appearances of nature, even though we are ignorant of their causes, often constitutes the most useful wisdom.

But among all the wonders that have lately engaged the attention of the philosopher and the chymist, is the circumstance, that after the explosion of certain luminous meteors, heavy stones, varying in bulk and number, have almost constantly fallen from them to the earth. Credibility in a fact, for which not even a conjectural cause in the remotest degree probable could be assigned, was for some time suspended; but the proofs are now so numerous, and of such respectable authority, that it can no longer be doubted.

In July, 1794, about twelve stones fell near Sienna in Tuscany, as related by the Earl of Bristol. December 13, 1795, a large stone of fifty-six pounds weight, fell at Wold cottage in Yorkshire, and is described by Captain Topham. February 19, 1796, a stone of ten pounds weight fell in Portugal, an account of which is given by Mr. Southey. December 19, 1798, showers of stones fell at Benares in the East Indies, upon the testimony of J. Lloyd Williams, Esq. April 26, 1803, according to M. Fourcroy, several stones, from ten to fourteen pounds weight, fell near L'Aigle in Normandy.

Various conjectures have been made, to account for their appearance; but such is the obscurity of the subject, that no opinion in the slightest degree probable has yet been advanced. It was at first supposed, that they had been thrown out of volcanoes, but the immense distance from all volcanoes renders this opinion of little value. Chaldni endeavoured to prove, that the meteors from which they fell, were bodies floating in space, unconnected with any planetary system, attracted by the earth in their progress, and kindled by their motion in the atmosphere. Laplace suggests the probability of their having been thrown off by the volcanoes of the moon; but the meteors which almost always accompany them, and the swiftness of their horizontal motion, persuade us to reject this opinion. Sir William Hamilton, and Mr. King, with greater probability, consider them as concretions actually formed in the atmosphere. After all, we must be content to leave this phænomenon (as also the showers of sulphur and the vast masses of iron said to have fallen in South America and Siberia, and supposed to have their origin from the same causes) to the accumulated wisdom of future ages.

* Phil. Trans. vol. iv. p. 2, 108.

CHAPTER XXII.

THE CONCLUSION.

Having thus gone through a particular description of the earth, let us now pause for a moment, to contemplate the great picture before us. The universe may be considered as the palace in which the Deity resides; and this earth as one of its apartments. In this, all the meaner races of animated nature mechanically obey him; and stand ready to execute his commands without hesitation. Man alone is found refractory; he is the only being endued with a power of contradicting these mandates. The Deity was pleased to exert superior power in creating him a superior being; a being endued with a choice of good and evil; and capable, in some measure, of co-operating with his own intentions. Man, therefore, may be considered as a limited creature, endued with powers imitative of those residing in the Deity. He is thrown into a world that stands in need of his help; and has been granted a power of producing harmony from partial confusion.

If, therefore, we consider the earth as allotted for our habitation, we shall find, that much has been given us to enjoy, and much to amend; that we have ample reasons for our gratitude, and still more for our industry. In those great outlines of nature, to which art cannot reach, and where our greatest efforts must have been ineffectual, God himself has finished these with amazing grandeur and beauty. Our beneficent Father has considered these parts of nature as peculiarly his own; as parts which no creature could have skill or strength to amend; and therefore made them incapable of alteration, or of more perfect regularity. The heavens and the firmament show the wisdom and the glory of the Workman. Astronomers, who are best skilled in the symmetry of systems, can find nothing there that they can alter for the better. God made these perfect, because no subordinate being could correct their defects.

When, therefore, we survey nature on this side, nothing can be more splendid, more correct, or amazing. We there behold a Deity residing in the midst of a universe, infinitely extended every way, animating all, and cheering the vacuity with his presence! We behold an immense and shapeless mass of matter, formed into worlds by his power, and dispersed at intervals, to which even the imagination cannot travel! In this great theatre of his glory, a thousand suns, like our own, animate their respective systems, appearing and vanishing at Divine command. We behold our own bright luminary fixed in the centre of its system, wheeling its planets in times proportioned to their distances, and at once dispensing light, heat, and action. The earth also is seen with its twofold motion; producing, by the one, the change of seasons; and by the other, the grateful vicissitudes of day and night. With what silent magnificence is all this performed! with what seeming ease! The works of art are exerted with interrupted force; and their noisy progress discovers the obstructions they receive: but the earth, with a silent steady rotation, successively presents every part of its bosom to the sun; at once imbibing nourishment and light from that parent of vegetation and fertility.

But not only provisions of heat and light are thus supplied, but its whole surface is covered with a transparent atmosphere, that turns with its motion, and guards it from external injury. The rays of the sun are thus broken into a genial warmth; and, while the surface is assisted, a gentle heat is produced in the bowels of the earth, which contributes to cover it with verdure. Waters also are supplied in healthful abundance, to support life, and assist vegetation. Mountains arise, to diversify the prospect, and give a current to the stream. Seas extend from one continent to the other, replenished with animals, that may be turned to human support; and also serving to enrich the earth with a sufficiency of vapour. Breezes fly along the surface of the fields, to promote health and vegetation. The coolness of the evening invites to rest, and the freshness of the morning renews for labour.

Such are the delights of the habitation that has been assigned to man! Without any one of these, he must have been wretched; and none of these could his own industry have supplied. But while many of his wants are thus kindly furnished on the one hand, there are numberless inconveniences to excite his industry on the other. This habitation, though provided with all the conveniences of air, pasturage, and water, is but a desert place, without human cultivation. The lowest animal finds more conveniences in the wilds of nature, than he who boasts himself their lord. The whirlwind, the inundation, and all the asperities of the air, are peculiarly terrible to man, who knows their consequences, and, at a distance, dreads their approach. The earth itself, where human art has not pervaded, puts on a frightful gloomy appearance. The forests are dark and tangled; the meadows overgrown with rank weeds; and the brooks stray without a determined channel. Nature, that has been kind to every lower order of beings, has been quite neglectful with regard to him; to the savage uncontriving man the earth is an abode of desolation, where his shelter is insufficient, and his food precarious.

A world thus furnished with advantages on one side, and inconveniences on the other, is the proper abode of reason; is the fittest to exercise the industry of a free and a thinking creature. These evils, which art can remedy, and prescience guard against, are a proper call for the exertion of his faculties; and they tend still more to assimilate him to his Creator. God beholds, with pleasure, that being which he has made, converting the wretchedness of his natural situation into a theatre of triumph; bringing all the headlong tribes of nature into subjection to his will; and producing that order and uniformity upon earth, of which his own heavenly fabric is so bright an example.

PART II.

OF ANIMALS.

CHAPTER I.

A COMPARISON OF ANIMALS WITH THE INFERIOR RANKS OF CREATION.

Having given an account of the earth in general, and the advantages and inconveniences with which it abounds, we now come to consider it more minutely. Having described the habitation, we are naturally led to inquire after the inhabitants. Amidst the infinitely different productions which the earth offers, and with which it is every where covered, animals hold the first rank; as well because of the finer formation of their parts, as of their superior power. The vegetable, which is fixed to one spot, and obliged to wait for its accidental supplies of nourishment, may be considered as the prisoner of nature. Unable to correct the disadvantages of its situation, or to shield itself from the dangers that surround it, every object that has motion, may be its destroyer.

But animals are endowed with powers of motion and defence. The greatest part are capable, by changing place, of commanding nature; and of thus obliging her to furnish that nourishment which is most agreeable to their state. Those few that are fixed on one spot, even in this seemingly helpless situation, are, nevertheless, protected from external injury by a hard shelly covering, which they often can close at pleasure, and thus defend themselves from every assault. And here, I think, we may draw the line between the animal and vegetable kingdoms. Every animal, by some means or other, finds protection from injury; either from its force, or courage, its swiftness, or cunning. Some are protected by hiding in convenient places, and others by taking refuge in a hard resisting shell. But vegetables are totally unprotected; they are exposed to every assailant, and patiently submissive in every attack. In a word, an animal is an organized being that is in some measure provided for its own security; a vegetable is destitute of every protection.

But though it is very easy, without the help of definitions, to distinguish a plant from an animal, yet both possess many properties so much alike, that the two kingdoms, as they are called, seem mixed with each other. Hence, it frequently puzzles the naturalist to tell exactly where animal life begins, and vegetative terminates; nor indeed, is it easy to resolve, whether some objects offered to view, be of the low est of the animal, or the highest of the vegetable race. The sensitive plant, that moves at the touch, seems to have as much perception as the fresh-water polypus, that is possessed of a still slower share of motion. Besides, the sensitive plant will not reproduce, upon cutting in pieces, which the polypus is known to do; so that the vegetable production seems to have the superiority. But, notwithstanding this, the polypus hunts for its food, as most other animals do. It changes its situation, and therefore possesses a power of choosing its food, or retreating from danger. Still, therefore, the animal kingdom is far removed above the vegetable; and its lowest denizen is possessed of very great privileges, when compared with the plants with which it is often surrounded.

However, both classes have many resemblances, by which they are raised above the unorganized and inert masses of nature. Minerals are mere inactive, insensible bodies, entirely motionless of themselves, and waiting some external force to alter their forms, or their properties. But it is otherwise with animals and vegetables; these are endued with life and vigour; they have their state of improvement and decay; they are capable of reproducing their kinds; they grow from seeds in some, and from cuttings in others; they seem all possessed of sensation, in a greater or less degree; they both have their enmities and affections; and as some animals are, by nature, impelled to violence, so some plants are found to exterminate all others, and make a wilderness of the places round them. As the lion makes a desert of the forest where it resides, thus no other plant will grow under the shade of the manchinel-tree. Thus, also, that plant in the West-Indies, called *caraguata*, clings round whatever tree it happens to approach: there it quickly gains the ascendant; and loading the tree with a verdure not its own, keeps away that nourishment designed to feed the trunk; and at last entirely destroys its supporter. As all animals are ultimately supported upon vegetables, so vegetables are greatly propagated, by being made a part of animal food. Birds distribute the seeds wherever they fly, and quadrupeds prune them into greater luxuriance. By these means the quantity of food, in a state of nature, is kept equal to the number of the consumers; and, lest some of the weaker ranks of animals should find nothing for their support, but all the provisions be devoured by the strong, different vegetables are appropriated to different appetites. If, transgressing this rule, the stronger ranks should invade the rights of the weak, and, breaking through all regard to appetite, should make an indiscriminate use of every vegetable, nature then punishes the transgression, and poison marks the crime as capital.

If, again, we compare vegetables and animals, with respect to the places where they are found, we shall find them bearing a still stronger similitude. The vegetables that grow in a dry and sunny soil, are strong and vigorous, though not luxuriant; so also are the animals of such a climate. Those, on the contrary, that are the joint product of heat and moisture are luxuriant and tender; and the animals assimilating to the vegetable food, on which they ultimately subsist, are much larger in such places than in others. Thus, in the internal parts of South America, and Africa, where the sun usually scorches all above, while inundations cover all below, the insects, reptiles, and other animals, grow to a prodigious size: the earth-worm of America is often a yard in length, and as thick as a walking cane; the boiguacu, which is the largest of the serpent kind, is sometimes forty feet in length; the bats, in those countries, are as big as a rabbit; the toads are bigger than a duck; and their spiders are as large as a sparrow. On the contrary, in the cold frozen regions of the north, where vegetable nature is stinted of its growth, the few animals in those climates partake of the diminution; all the wild animals, except the bear, are much smaller than in milder countries; and such of the domestic kinds as are carried thither, quickly degenerate, and grow less. Their very insects are of the minute kinds, their bees and spiders being not half so large as those in the temperate zone.

The similitude between vegetables and animals is no where more obvious than in those that belong to the ocean, where the nature of one is admirably adapted to the necessities of the other. This element, it is well known, has its vegetables, and its insects that feed upon them, in great abundance. Over many tracts of the sea, a weed is seen floating, which covers the surface, and gives the resemblance of a green and extensive meadow. On the under side of these unstable plants, millions of little animals are found, adapted to their situation. For as their ground, if I may so express it, lies over their heads, their feet are placed upon their backs; and as land animals have their legs below their bodies, these have them above. At land also, most animals are furnished with eyes to see their food; but at sea, almost all the reptile kinds are without eyes, which might only give them prospects of danger at a time when unprovided with the means of escaping it.*

Thus, in all places, we perceive an obvious similitude between the animals and the vegetables of every region. In general, however, the most perfect races have the least similitude to the vegetable productions on which they are ultimately fed; while, on the contrary, the meaner the animal, the more local it is found to be, and the more it is influenced by the varieties of the soil where it resides. Many of the more humble reptile kinds are not only confined to one country, but also to a plant; nay, even to a leaf. Upon that they subsist; increase with its vegetation, and seem to decay as it declines. They are merely the circumscribed inhabitants of a single vegetable; take them from that, and they instantly die; being entirely assimilated to the plant they feed on, assuming its colour, and even its medicinal properties. For this reason there are infinite numbers of the meaner animals that we have never an opportunity of seeing in this part of the world; they are incapable of living separate from their kindred vegetables, which grow only in a certain climate.

Such animals as are formed more perfect, lead a life of less dependence; and some kinds are found to subsist in many parts of the world at the same time. But of all the races of animated nature, man is the least affected by the soil where he resides, and least influenced by the variations of vegetable sustenance: equally unaffected by the

---

* Linnæi Amenitates, vol. v. p. 68.

luxuriance of the warm climates, or the sterility of the poles, he has spread his habitations over the whole earth; and finds subsistence as well amidst the ice of the north as the burning deserts under the line. All creatures of an inferior nature, as has been said, have peculiar propensities to peculiar climates; they are circumscribed to zones, and confined to territories where their proper food is found in the greatest abundance; but man may be called the animal of every climate, and suffers but very gradual alterations from the nature of any situation.

As to animals of a meaner rank, whom man compels to attend him in his migrations, these being obliged to live in a kind of constraint, and upon vegetable food often different from that of their native soil, they very soon alter their natures with the nature of their nourishment, assimilate to the vegetables upon which they feed, and thus assume very different habits as well as appearances. Thus man, unaffected himself, alters and directs the nature of other animals at his pleasure; increases their strength for his delight, or their patience for his necessities.

This power of altering the appearances of things, seems to have been given him for very wise purposes. The Deity, when he made the earth, was willing to give his favoured creature many opponents, that might at once exercise his virtues, and call forth his latent abilities. Hence we find, in those wide uncultivated wildernesses, where man, in his savage state, owns inferior strength, and the beasts claim divided dominion, that the whole forest swarms with noxious animals and vegetables; animals as yet undescribed, and vegetables which want a name. In those recesses, nature seems rather lavish than magnificent in bestowing life. The trees are usually of the largest kinds, covered round with parasite plants, and interwoven at the tops with each other. The boughs, both above and below, are peopled with various generations; some of which have never been upon the ground, and others have never stirred from the branches on which they were produced. In this manner millions of minute and loathsome creatures pursue a round of uninterrupted existence, and enjoy a life scarcely superior to vegetation. At the same time, the vegetables in those places are of the larger kinds, while the animal race is of the smaller: but man has altered this disposition of nature; having, in a great measure, levelled the extensive forests, cultivated the softer and finer vegetables, destroyed the numberless tribes of minute and noxious animals, and taken every method to increase a numerous breed of the larger kinds. He thus has exercised a severe control; unpeopled nature, to embellish it; and diminished the size of the vegetable, in order to improve that of the animal kingdom.

To subdue the earth to his own use, was, and ought to be, the aim of man; which was only to be done by increasing the number of plants, and diminishing that of animals: to multiply existence, *alone*, was that of the Deity. For this reason, we find, in a state of nature, that animal life is increased to the greatest quantity possible; and we can scarcely form a system that could add to its numbers. First, plants, or trees, are provided by nature of the largest kinds; and, consequently, the nourishing surface is thus extended. In the second place, there are animals peculiar to every part of the vegetable, so that no part of it is lost. But the greatest possible increase of life would still be deficient, were there not other animals that lived upon animals; and these are themselves, in turn, food for some other greater and stronger set of creatures. Were all animals to live upon vegetables alone, thousands would be extinct that now have existence, as the quantity of their provision would shortly fail. But, as things are wisely constructed, one animal now supports another; and thus, all take up less room than they would by living on the same food; as, to make use of a similar instance, a greater number of people may be crowded into the same space, if each is made to bear his fellow upon his shoulders.

To diminish the number of animals, and increase that of vegetables, has been the general scope of human industry; and, if we compare the utility of the kinds, with respect to man, we shall find, that of the vast variety in the animal kingdom, but very few are serviceable to him; and, in the vegetable, but very few are entirely noxious. How small a part of the insect tribes, for instance, are beneficial to mankind, and what numbers are injurious! In some countries they almost darken the air: a candle cannot be lighted without their instantly flying upon it, and putting out the flame.* The closest recesses are no safeguard from their annoyance; and the most beautiful landscapes of nature only serve to invite their rapacity. As these are injurious from their multitudes, so most of the larger kinds are equally dreadful to him from their courage and ferocity. In the most uncultivated parts of the forest these maintain an undisputed empire; and man invades their retreats with terror. These are dreadful; and there are still more which are utterly useless to him, that serve to take up the room which more beneficial creatures might possess; and incommode him rather with their numbers than their enmities. Thus, in a catalogue of land animals, that amounts to more than twenty thousand, we can scarcely reckon up a hundred that are any way useful to him; the rest being either all his open or his secret enemies, immediately attacking him in person, or intruding upon that food he has appropriated to himself. Vegetables, on the contrary, though existing in greater variety, are but few of them noxious. The most deadly poisons are often of great use in medicine; and even those plants that only seem to cumber the ground serve for food to that race of animals which he has taken into friendship, or protection. The smaller tribes of vegetables, in particular, are cultivated, as contributing either to his necessities or amusements; so that vegetable life is as much promoted by human industry, as animal life is controlled and diminished.

Hence, it was not without a long struggle, and various combinations of experience and art, that man acquired his present dominion. Almost every good that he possesses was the result of the contest; for, every day, as he was contending, he was growing more wise: and patience and fortitude were the fruits of his industry.

Hence, also, we see the necessity of some animals living upon each other, to fill up the plan of Providence; and we may, consequently

* Ulloa's Description of Guayaquil.

infer the expediency of man's living upon all. Both animals and vegetables seem equally fitted to his appetites; and, were any religious or moral motives to restrain him from taking away life, upon any account, he would only thus give existence to a variety of beings made to prey upon each other; and, instead of preventing, multiply mutual destruction.

CHAPTER II.

ON THE GENERATION OF ANIMALS

BEFORE we survey animals in their state of maturity, and performing the functions adapted to their respective natures, method requires that we should consider them in the more early periods of their existence. There has been a time when the proudest and the noblest animal was a partaker of the same imbecility with the meanest reptile; and, while yet a candidate for existence, equally helpless and contemptible. In their incipient state, all are upon a footing; the insect and the philosopher being equally insensible, clogged with matter, and unconscious of existence. Where, then, are we to begin with the history of those beings, that make such a distinguished figure in the creation? Or, where lie those peculiar characters in the parts that go to make up animated nature—that mark one animal as destined to creep in the dust, and another to glitter on the throne.

This has been a subject that has employed the curiosity of all ages, and the philosophers of every age have attempted the solution. In tracing nature to her most hidden recesses, she becomes too minute, or obscure, for our inspection; so that we find it impossible to mark her first differences, to discover the point where animal life begins, or the cause that conduces to set it in motion. We know little more than that the greatest number of animals require the concurrence of a male and female to reproduce their kind; and that these distinctly and invariable are found to beget creatures of their own species. Curiosity has, therefore, been active, in trying to discover the immediate result of this union; how far either sex contributes to the bestowing animal life, and whether it be to the male or female that we are most indebted for the privilege of our existence.

Hippocrates has supposed that fecundity proceeded from the mixture of the seminal liquor of both sexes, each of which equally contributes to the formation of the incipient animal. Aristotle, on the other hand, would have the seminal liquor in the male alone to contribute to this purpose, while the female supplied the proper nourishment for its support. Such were the opinions of these fathers of philosophy; and these continued to be adopted by the naturalists and schoolmen of succeeding ages with blind veneration. At length, Steno and Harvey, taking anatomy for their guide, gave mankind a nearer view of nature just advancing into animation. These perceived in all such animals as produced their young alive, two glandular bodies, near the womb, resembling that ovary, or cluster of small eggs, which is found in fowls; and, from the analogy between both, they gave these also the name of ovaria. These, as they resembled eggs, they naturally concluded had the same offices; and, therefore, they were induced to think that all animals, of what kind soever, were produced from eggs. At first, however, there was some altercation raised against this system: for, as these ovaria were separate from the womb, it was objected that they could not be any way instrumental in replenishing that organ, with which they did not communicate. But, upon more minute inspection, Fallopius, the anatomist, perceived two tubular vessels depending from the womb, which, like the horns of a snail, had a power of erecting themselves, of embracing the ovaria, and of receiving the eggs, in order to be fecundated by the seminal liquor. This discovery seemed, for a long time after, to fix the opinions of philosophers. The doctrine of Hippocrates was re-established, and the chief business of generation was ascribed to the female. This was for a long time the established opinion of the schools; but Leuwenhoeck, once more, shook the whole system, and produced a new schism among the lovers of speculation. Upon examining the seminal liquor of a great variety of male animals with microscopes, which helped his sight more than that of any of his predecessors; he perceived therein infinite numbers of little living creatures, like tadpoles, very brisk, and floating in the fluid with a seeming voluntary motion. Each of these, therefore, was thought to be the rudiments of an animal, similar to that from which it was produced; and this only required a reception from the female, together with proper nourishment, to complete its growth. The business of generation was now, therefore, given back to the male a second time, by many; while others suspended their assent, and chose rather to confess ignorance than to embrace error.*

In this manner has the dispute continued for several ages, some accidental discovery serving, at intervals, to renew the debate, and revive curiosity. It was a subject where speculation could find much room to display itself; and Mr. Buffon, who loved to speculate, would not omit such an opportunity of giving scope to his propensity. According to this most pleasing of all naturalists, the microscope discovers that the seminal liquor, not only of males, but of females also, abounds in these moving little animals, which have been mentioned above, and that they appear equally brisk in either fluid. These he takes not to be real animals, but organical particles, which, being simple, cannot be said to organize themselves, but go to the composition of all organized bodies whatsoever; in the same manner as a tooth in the wheel of a watch, cannot be called either a wheel, or the watch, and yet contributes to the sum of the machine. These organical particles are, according to him, diffused throughout all nature, and to be found not only in the seminal liquor, but in most other fluids in the parts of vegetables, and all parts of animated nature. As they happen, therefore, to be differently applied, they serve to constitute a part of the animal, or the vegetable, whose growth they serve to increase, while the superfluity is thrown off in the seminal liquor of both sexes, for the reproduction of other animals or vegetables of the same species. These particles assume different figures, according to the receptacle into which they enter; falling into the womb, they

* Bonet. Considerations sur les Corps Origines

unite into a fœtus; beneath the bark of a tree they pullulate into branches; and, in short, the same particles that first formed the animal in the womb, contribute to increase its growth when brought forth.*

To this system it has been objected, that it is impossible to conceive organical substances without being organized; and that, if devested of organization themselves, they could never make an organized body, as an infinity of circles could never make a triangle. It has been objected, that it is more difficult to conceive the transformation of these organical particles, than even that of the animal, whose growth we are inquiring after; and this system, therefore, attempts to explain one obscure thing by another still more obscure.

But an objection, still stronger than these, has been advanced by an ingenious countryman of our own; who asserts, that these little animals, which thus appear swimming and sporting in almost every fluid we examine with a microscope, are not real living particles, but some of the more opaque parts of the fluid, that are thus increased in size, and seem to have a much greater motion than they have in reality. For the motion being magnified with the object, the smallest degree of it will seem very considerable; and a being almost at rest may, by these means, be apparently put into violent action. Thus, for instance, if we look upon the sails of a windmill moving at a distance, they appear to go very slow; but, if we approach them, and thus magnify their bulk to our eye, they go round with great rapidity. A microscope, in the same manner, serves to bring our eye close to the object, and thus to enlarge it; and not only increase the magnitude of its parts, but of its motion. Hence therefore, it would follow, that these organical particles that are said to constitute the bulk of living nature, are but mere optical illusions; and the system founded on them must, like them, be illusive.

These, and many other objections, have been made to this system; which, instead of enlightening the mind, serve only to show, that too close a pursuit of nature very often leads to uncertainty. Happily, however, for mankind, the most intricate inquiries are generally the most useless. Instead, therefore, of balancing accounts between the sexes, and attempting to ascertain to which the business of generation most properly belongs, it will be more instructive, as well as amusing, to begin with animal nature, from its earliest retirements, and evanescent outlines, and pursue the incipient creature through all its changes in the womb, till it arrives into open day.

The usual distinction of animals, with respect to their manner of generation, has been into the oviparous and viviparous kinds; or, in other words, into those that bring forth an egg, which is afterwards hatched into life, and those that bring forth their young alive and perfect. In one of these two ways all animals were supposed to have been produced, and all other kinds of generation were supposed imaginary or erroneous. But later discoveries have taught us to be more cautious in making general conclusions, and have even induced many to doubt whether animal life may not be produced merely from putrefaction.†

* Mr. Buffon. † Bonet. Consid. p. 100.

Indeed, the infinite number of creatures that putrid substances seem to give birth to, and the variety of little insects seen floating in liquors, by the microscope, appear to favour this opinion. But however this may be, the former method of classing animals can now by no means be admitted, as we find many animals that are produced neither from the womb, nor from the shell, but merely from cuttings; so that to multiply life in some creatures, it is sufficient only to multiply the dissection. This being the simplest method of generation, and that in which life seems to require the smallest preparation for its existence, I will begin with it, and so proceed to the two other kinds, from the meanest to the most elaborate.

The earth-worm, the millipedes, the sea-worm, and many marine insects, may be multiplied by being cut in pieces; but the polypus is noted for its amazing fertility; and hence it will be proper to take the description. The structure of the polypus may be compared to the finger of a glove, open at one end, and closed at the other. The closed end represents the tail of the polypus, with which it serves to fix itself to any substance it happens to be upon; the open end may be compared to the mouth; and, if we conceive six or eight small strings issuing from this end, we shall have a proper idea of its arms, which it can erect, lengthen, and contract at pleasure, like the horns of a snail. This creature is very voracious, and makes use of its arms as a fisherman does of his net, to catch and entangle such little animals as happen to come within its reach. It lengthens these arms several inches, keeps them separate from each other, and thus occupies a large space in the water in which it resides. These arms, when extended, are as fine as threads of silk, and have a most exquisite degree of feeling. If a small worm happens to get within the sphere of their activity, it is quickly entangled by one of these arms, and, soon after, the other arms come to its aid: these altogether shortening, the worm is drawn into the animal's mouth, and quickly devoured, colouring the body as it is swallowed. Thus much is necessary to be observed of this animal's method of living, to show that it is not of the vegetable tribe, but a real animal, performing the functions which other animals are found to perform, and endued with powers that many of them are destitute of. But what is most extraordinary remains yet to be told; for, if examined with a microscope, there are seen several little specks, like buds, that seem to pullulate from different parts of its body; and these soon after appear to be young polypi, and like the large polypus, begin to cast their little arms about for prey, in the same manner. Whatever they happen to ensnare is devoured, and gives a colour not only to their own bodies, but to that of the parent; so that the same food is digested, and serves for the nourishment of both. The food of the little one passes into the large polypus, and colours its body; and this, in its turn, digests and swallows its food to pass into theirs. In this manner every polypus has a new colony sprouting from its body; and these new ones, even while attached to the parent animal, become parents themselves, having a smaller colony also budding from them; all, at the same time, busily employed in seeking for their prey, and the food of any one of them serving for the nourishment, and circulating through the bodies of all the rest. This society, however, is every hour dissolving; those newly produced are seen at intervals to leave the body of the large polypus, and become, shortly after, the head of a beginning colony themselves.

In this manner the polypus multiplies naturally; but one may take a much readier and shorter way to increase them, and this only by cutting them in pieces. Though cut into thousands of parts, each part still retains its vivacious quality, and each shortly becomes a distinct and a complete polypus; whether cut lengthways or crossways, it is all the same; this extraordinary creature seems a gainer by our endeavours, and multiplies by apparent destruction. The experiment has been tried, times without number, and still attended with the same success. Here, therefore, naturalists who have been blamed for the cruelty of their experiments upon living animals, may now boast of their increasing animal life, instead of destroying it. The production of the polypus is a kind of philosophical generation. The famous Sir Thomas Brown hoped one day to be able to produce children by the same method as trees are produced; the polypus is multiplied in this manner; and every philosopher may thus, if he please, boast of a very numerous, though, I should suppose, a very useless progeny.

This method of generation, from cuttings, may be considered as the most simple kind, and is a strong instance of the little pains Nature takes in the formation of her lower and humbler productions. As the removal of these from inanimate into animal existence is but small, there are but few preparations made for their journey. No organs of generation seem provided, no womb to receive, no shell to protect them in their state of transition. The little reptile is quickly fitted for all the offices of its humble sphere, and, in a very short time, arrives at the height of its contemptible perfection.

The next generation is of those animals that we see produced from the egg. In this manner all birds, most fishes, and many of the insect tribes, are brought forth. An egg may be considered as a womb, detached from the body of the parent animal, in which the embryo is but just beginning to be formed. It may be regarded as a kind of incomplete delivery, in which the animal is disburdened of its young before its perfect formation. Fishes and insects, indeed, most usually commit the care of their eggs to hazard; but birds, which are more perfectly formed, are found to hatch them into maturity by the warmth of their bodies. However, any other heat, of the same temperature, would answer the end as well; for either the warmth of the sun, or of a stove, is equally efficacious in bringing the animal in the egg to perfection. In this respect, therefore, we may consider generation from the egg as inferior to that in which the animal is brought forth alive. Nature has taken care of the viviparous animal in every stage of its existence. That force which separates it from the parent, separates it from life; and the embryo is shielded with unceasing protection till it arrives at exclusion. But it is different with the little animal in the egg; often totally neglected by the parent, and always separable from it, every accident may retard its growth, or even destroy its existence. Besides, art or accident, also, may bring this animal to a state of perfection; so that it can never be considered as a complete work of nature, in which so much is left for accident to finish or destroy.

But however inferior this kind of generation may be, the observation of it will afford great insight into that of nobler animals, as we can here watch the progress of the growing embryo, in every period of its existence, and catch it in those very moments when it first seems stealing into motion. Malpighi and Haller have been particularly industrious on this subject; and, with a patience almost equalling that of the sitting hen, have attended incubation in all its stages. From them, therefore, we have an amazing history of the chicken in the egg, and of its advances into complete formation.

It would be methodically tedious to describe those parts of the egg which are well known and obvious; such as its shell, its white, and its yolk; but the disposition of these is not so apparent. Immediately under the shell lies that common membrane, or skin, which lines it on the inside, adhering closely to it every where, except at the broad end, where a little cavity is left, that is filled with air, which increases as the animal within grows larger. Under this membrane are contained two whites, though seeming to us to be only one, each wrapped up in a membrane of its own, one white within the other. In the midst of all is the yolk, wrapt round likewise in its own membrane. At each end of this are two ligaments, called *chalazæ*, which are, as it were, the poles of this microcosm, being white dense substances, made from the membranes, and serving to keep the white and the yolk in their places. It was the opinion of Mr. Derham that they served also for another purpose; for a line being drawn from one ligament to the other, would not pass directly through the middle of the yolk, but rather towards one side, and would divide the yolk into two unequal parts, by which means these ligaments served to keep the smallest side of the yolk always uppermost; and in this part he supposed the cicatricula, or first speck of life, to reside; which, by being uppermost, and consequently next the hen, would be thus in the warmest situation. But this is rather fanciful than true, the incipient animal being found in all situations, and not particularly influenced by any.* The cicatricula, which is the part where the animal first begins to show signs of life, is not unlike a vetch, or a lentil, lying on one side of the yolk, and within its membrane. All these contribute to the little animal's convenience, or support; the outer membranes, and ligaments, preserve the fluids in their proper places; the white serves as nourishment; and the yolk, with its membranes, after a time, becomes a part of the animal's body.† This is the description of a hen's egg, and answers to that of all others, how large or how small soever.

Previous to putting the eggs to the hen, our philosophers first examined the cicatricula, or little spot, already mentioned; and which may be considered as the most important part of the egg. This was found, in those that were impregnated by the cock, to be large; but, in those laid without the cock, very small. It was found, by the microscope to be a kind of bag, containing a transparent liquor, in the

* Haller. † Ibid

midst of which the embryo was seen to reside. The embryo resembled a composition of little threads, which the warmth of future incubation tended to enlarge, by varying and liquifying the other fluids contained within the shell, and thus pressing them either into the pores or tubes of their substance.

Upon placing the eggs in a proper warmth,[*] either under the sun, or in a stove, after six hours the vital speck begins to dilate, like the pupil of the eye. The head of the chicken is instantly seen, with the back bone, something resembling a tadpole, floating in its ambient fluid, but as yet seeming to assume none of the functions of animal life. In about six hours more the little animal is seen more distinctly; the head becomes more plainly visible, and the vertebræ of the back more easily perceivable. All these signs of preparation for life are increased in six hours more: and at the end of twenty-four, the ribs begin to take their places, the neck begins to lengthen, and the head to turn to one side.

At this time,[†] also, the fluids in the egg seem to have changed place; the yolk, which was before in the centre of the shell, approaches nearer to the broad end. The watery part of the white is, in some measure, evaporated through the shell, and the grosser part sinks to the small end. The little animal appears to turn towards the part of the broad end, in which a cavity has been described, and with its yolk seems to adhere to the membrane there. At the end of forty hours the great work of life seems fairly begun, and the animal plainly appears to move; the back bone, which is of a whitish colour, thickens; the head is turned still more on one side; the first rudiments of the eyes begin to appear, the heart beats, and the blood begins already to circulate. The parts, however, as yet are fluid; but, by degrees, become more and more tenacious, and hardened into a kind of jelly. At the end of two days, the liquor in which the chicken swims, seems to increase; the head appears with two little bladders in the place of eyes; the heart beats in the manner of every embryo where the blood does not circulate through the lungs. In about fourteen hours after this, the chicken is grown more strong; its head, however, is still bent downwards; the veins and arteries begin to branch, in order to form the brain; and the spinal marrow is seen stretching along the back bone. In three days the whole body of the chicken appears bent; the head, with its two eye-balls, with their different humours, now distinctly appear; and five other vessels are seen, which soon unite to form the rudiments of the brain. The outlines also of the thighs and wings begin to be seen, and the body begins to gather flesh. At the end of the fourth day, the vesicles, that go to form the brain, approach each other; the wings and thighs appear more solid; the whole body is covered with a jelly-like flesh; the heart, that was hitherto exposed, is now covered up within the body, by a very thin transparent membrane; and, at the same time, the umbilical vessels, that unite the animal to the yolk, now appear to come forth from the abdomen. After the fifth and sixth days, the vessels of the brain begin to be covered over; the wings and thighs

---

[*] Malpighi.   [†] Harvey.

lengthen; the belly is closed up, and tumid; the liver is seen within it very distinctly, not yet grown red, but of a very dusky white; both the ventricles of the heart are discerned, as if they were two separate hearts, beating distinctly; the whole body of the animal is covered over; and the traces of the incipient feathers are already to be seen. The seventh day, the head appears very large; the brain is covered entirely over; the bill begins to appear betwixt the eyes; and the wings, the thighs, and the legs, have acquired their perfect figure.* Hitherto, however, the animal appears as if it had two bodies; the yolk is joined to it by the umbilical vessels that come from the belly, and is furnished with its vessels, through which the blood circulates, as through the rest of the body of the chicken, making a bulk greater than that of the animal itself. But towards the end of incubation, the umbilical vessels shorten the yolk, and with it the intestines are thrust up into the body of the chicken, by the action of the muscles of the belly; and the two bodies are thus formed into one. During this state, all the organs are found to perform their secretions; the bile is found to be separated, as in grown animals; but it is fluid, transparent, and without bitterness: and the chicken then also appears to have lungs. On the tenth, the muscles of the wings appear, and the feathers begin to push out. On the eleventh, the heart, which hitherto had appeared divided, begins to unite; the arteries which belong to it join into it, like the fingers into the palm of the hand. All these appearances only come more into view, because the fluids the vessels had hitherto secreted were more transparent; but as the colour of the fluids deepen, their operations and circulations are more distinctly seen. As the animal thus, by the eleventh day completely formed, begins to gather strength, it becomes more uneasy in its situation, and exerts its animal powers with increasing force. For some time before it is able to break the shell in which it is imprisoned, it is heard to chirrup, receiving a sufficient quantity of air for this purpose, from that cavity which lies between the membrane and the shell, and which must contain air to resist the external pressure. At length, upon the twentieth day, in some birds sooner, and later in others, the inclosed animal breaks the shell, within which it has been confined, with its beak; and by repeated efforts, at last procures its enlargement.

From this little history we perceive, that those parts which are most conducive to life, are the first that are begun: the head and the backbone, which, no doubt, inclose the brain, and the spinal marrow, though both are too limpid to be discerned, are the first that are seen to exist; the beating of the heart is perceived soon after; the less noble parts seem to spring from these: the wings, the thighs, the feet, and lastly the bill. Whatever, therefore, the animal has double, or whatever it can live without the use of, these are latest in production. Nature first sedulously applying to the formation of the nobler organs, without which life would be of short continuance, and would be begun in vain.

The resemblance between the beginning animal in the egg, and the embryo in the womb, is very striking; and this similitude has induced

* Haller.

many to assert, that all animals are produced from eggs, in the same manner. They consider an egg excluded from the body by some, and separated into the womb by others, to be actions merely of one kind; with this only difference, that the nourishment of the one is kept within the body of the parent, and increases as the embryo happens to want the supply; the nourishment of the other is prepared all at once, and sent out with the beginning animal, as entirely sufficient for its future support. But leaving this to the discussion of anatomists, let us proceed rather with facts than dissertations; and as we have seen the progress of an oviparous animal, or one produced from the shell, let us likewise trace that of a viviparous animal, which is brought forth alive. In this investigation, Graaf has, with a degree of patience characteristic of his nation, attended the progress and increase of various animals in the womb, and minutely marked the changes they undergo. Having dissected a rabbit, half an hour after impregnation, he perceived the horns of the womb, that go to embrace and communicate with the ovary, to be more red than before; but no other change in the rest of the parts. Having dissected another, six hours after, he perceived the follicules, or the membrane covering the eggs contained in the ovary, to become reddish. In a rabbit dissected after twenty-four hours, he perceived, in one of the ovaries, three follicules, and, in the other, five, that were changed; being become, from transparent, dark and reddish. In one dissected after three days, he perceived the horns of the womb very strictly to embrace the ovaries; and he observed three of the follicules in one of them, much longer and harder than before; pursuing his inquisition, he also found two of the eggs actually separated into the horns of the womb, and each about the size of a grain of mustard-seed; these little eggs were each of them inclosed in a double membrane, the inner parts being filled with a very limpid liquor. After four days, he found, in one of the ovaries, four, and, in the other, five follicules, emptied of their eggs; and in the horns correspondent to these, he found an equal number of eggs thus separated; these eggs were now grown larger than before, and somewhat of the size of sparrow-shot. In five days, the eggs were grown to the size of duck-shot, and could be blown from the part of the womb where they were, by the breath. In seven days, these eggs were found of the size of a pistol-bullet, each covered with its double membrane, and these much more distinct than before. In nine days, having examined the liquor contained in one of these eggs, he found it, from a limpid colour, less fluid, to have got a light cloud floating upon it. In ten days, this cloud began to thicken, and to form an oblong body, of the figure of a little worm: and, in twelve days, the figure of the embryo was distinctly to be perceived, and even its parts came into view. In the region of the breast he perceived two bloody specks; and two more that appeared whitish. Fourteen days after impregnation, the head of the embryo was become large and transparent, the eyes prominent, the mouth open, and the rudiments of the ears beginning to appear; the back-bone, of a whitish colour, was bent towards the breast; the two bloody specks being now considerably increased, appeared to be nothing less than the outlines of the two ventricles of the heart; and the two whitish specks on each side, now appeared to be the rudiments of the lungs; towards the region of the belly, the liver began to be seen, of a reddish colour, and a little intricate mass, like ravel'ed thread, discerned, which soon appeared to be the stomach and the intestines; the legs soon after began to be seen, and to assume their natural positions: and from that time forth, all the parts being formed, every day only served to develope them still more, until the thirty-first day, when the rabbit brought forth her young, completely fitted for the purposes of their humble happiness.

Having thus seen the stages of generation in the meaner animals, let us take a view of its progress in man; and trace the feeble beginnings of our own existence. An account of the lowliness of our own origin, if it cannot amuse, will at least serve to humble us; and it may take from our pride, though it fails to gratify our curiosity. We cannot here trace the variations of the beginning animal, as in the former instances; for the opportunities of inspection are but few and accidental: for this reason we must be content often to fill up the blanks of our history with conjecture. And, first, we are entirely ignorant of the state of the infant in the womb, immediately after conception; but we have good reason to believe that it proceeds, as in most other animals, from the egg.* Anatomists inform us, that four days after conception, there is found in the womb, an oval substance, about the size of a small pea, but longer one way than the other; this little body is formed by an extremely fine membrane, inclosing a liquor a good deal resembling the white of an egg: in this may, even then, be perceived, several small fibres, united together, which form the first rudiments of the embryo. Besides these, are seen another set of fibres, which soon after become the placenta, or that body by which the animal is supplied with nourishment.

Seven days after conception, we can readily distinguish, by the eye, the first lineaments of the child in the womb. However, they are as yet without form; showing at the end of seven days pretty much such an appearance as that of the chicken after four and twenty hours, being a small jelly-like mass, yet exhibiting the rudiments of the head; the trunk is barely visible: there likewise is to be discerned a small assemblage of fibres issuing from the body of the infant, which afterwards become the blood-vessels that convey nourishment from the placenta to the child, while inclosed in the womb.

Fifteen days after conception, the head becomes distinctly visible, and even the most prominent features of the visage begin to appear. The nose is a little elevated; there are two black specks in the place of the eyes; and two little holes, where the ears are afterwards seen. The body of the embryo also is grown larger; and, both above and below, are seen two little protuberances, which mark the places from whence the arms and thighs are to proceed. The length of the whole body, at this time, is less than half an inch.

At the end of three weeks, the body has received very little increase; but the legs and feet, with the hands and arms, are become

* The history of the child in the womb is translated from Mr. Buffon, with some alterations.

apparent. The growth of the arms is more speedy than that of the legs; and the fingers are sooner separated than the toes. About this time the internal parts are found, upon dissection, to become distinguishable. The places of the bones are marked by small threadlike substances, that are yet more fluid even than a jelly. Among them, the ribs are distinguishable, like threads also, disposed on each side of the spine; and even the fingers and toes scarce exceed hairs in thickness.

In a month the embryo is an inch long; the body is bent forward, a situation which it almost always assumes in the womb, either because a posture of this kind is the most easy, or because it takes up the least room. The human figure is now no longer doubtful: every part of the face is distinguishable; the body is sketched out; the bowels are to be distinguished as threads; the bones are still quite soft, but in some places beginning to assume a greater rigidity; the blood-vessels that go to the placenta, which, as was said, contributes to the child's nourishment, are plainly seen issuing from the navel, (being therefore called the *umbilical vessels*,) and going to spread themselves upon the placenta. According to Hippocrates, the male embryo developes sooner than the female: he adds, that at the end of thirty days, the parts of the body of the male are distinguishable; while those of the female are not equally so till ten days after.

In six weeks the embryo is grown two inches long; the human figure begins to grow every day more perfect; the head being still much larger, in proportion to the rest of the body; and the motion of the heart is perceived almost by the eye. It has been seen to beat in an embryo of fifty days old, a long time after it had been taken out of the womb.

In two months the embryo is more than two inches in length. The ossification is perceivable in the arms and thighs, and in the point of the chin, the under jaw being greatly advanced before the upper. These parts, however, may as yet be considered as bony points, rather than as bones. The umbilical vessels, which before went side by side, are now begun to be twisted, like a rope, one over the other, and go to join with the placenta, which, as yet, is but small.

In three months the embryo is above three inches long, and weighs about three ounces. Hippocrates observes, that not till then the mother perceives the child's motion; and he adds, that in female children, the motion is not observable till the end of four months. However, this is no general rule, as there are women who assert that they perceived themselves to be quick with child, as their expression is, at the end of two months; so that this quickness seems rather to arise from the proportion between the child's strength, and the mother's sensibility, than from any determinate period of time. At all times, however, the child is equally alive; and, consequently, those juries of matrons that are to determine upon the pregnancy of criminals, should not inquire whether the woman be quick, but whether she be with child; if the latter be perceivable, the former follows of course.

Four months and a half after conception, the embryo is from six to seven inches long. All the parts are so augmented, that even their proportions are now distinguishable. The very nails begin to appear upon the fingers and toes; and the stomach and intestines already begin to perform their functions of receiving and digesting. In the stomach is found a liquor similar to that in which the embryo floats; in one part of the intestines, a milky substance; and in the other, an excrementitious. There is found also a small quantity of bile in the gall bladder; and some urine in its own proper receptacle. By this time also, the posture of the embryo seems to be determined. The head is bent forward, so that the chin seems to rest upon its breast; the knees are raised up towards the head, and the legs bent backward, somewhat resembling the posture of those who sit on their haunches. Sometimes the knees are raised so high as to touch the cheeks, and the feet are crossed over each other; the arms are laid upon the breast, while one of the hands, and often both, touch the visage; sometimes the hands are shut, and sometimes also the arms are found hanging down by the body. These are the most usual postures which the embryo assumes; but these it is frequently known to change; and it is owing to these alterations that the mother so frequently feels those twitches which are usually attended with pain.

The embryo, thus situated, is furnished by nature with all things proper for its support; and, as it increases in size, its nourishment also is found to increase with it. As soon as it first begins to grow in the womb, that receptacle, from being very small, grows larger; and, what is more surprising, thicker every day. The sides of a bladder, as we know, the more they are distended, the more they become thin. But here the larger the womb grows, the more it appears to thicken. Within this the embryo is still farther involved, in two membranes, called the *chorion* and *amnios*; and floats in a thin transparent fluid, upon which it seems, in some measure, to subsist. However, the great storehouse, from whence its chief nourishment is supplied, is called the *placenta*; a red substance somewhat resembling a sponge, that adheres to the inside of the womb, and communicates, by the umbilical vessels, with the embryo. These umbilical vessels, which consist of a vein and two arteries, issue from the navel of the child, and are branched out upon the placenta; where they, in fact, seem to form its substance; and, if I may so express it, to suck up their nourishment from the womb, and the fluids contained therein. The blood thus received from the womb, by the placenta, and communicated by the umbilical vein to the body of the embryo, is conveyed to the heart; where, without ever passing into the lungs, as in the born infant, it takes a shorter course; for entering the right auricle of the heart, instead of passing up into the pulmonary artery, it seems to break this partition, and goes directly through the body of the heart, by an opening called the *foramen ovale*, and from thence to the aorta, or great artery; by which it is driven into all parts of the body. Thus we see the placenta, in some measure, supplying the place of lungs; for as the little animal can receive no air by inspiration, the lungs are therefore useless. But we see the placenta converting the fluid of the womb into blood, and sending it, by the umbilical vein, to the heart; from whence it is dispatched by a quicker and shorter circulation through the whole frame.

In this manner the embryo reposes in the womb; supplied with that nourishment which is fitted to its necessities, and furnished with those organs that are adapted to its situation. As its sensations are but few, its wants are in the same proportion; and it is probable that a sleep, with scarce any intervals, marks the earliest period of human life. As the little creature, however, gathers strength and size, it seems to become more wakeful and uneasy; even in the womb it begins to feel the want of something it does not possess; a sensation that seems coeval with man's nature, and never leaves him till he dies. The embryo even then begins to struggle for a state more marked by pleasure and pain, and, from about the sixth month, begins to give the mother warning of the greater pain she is yet to endure. The continuation of pregnancy in woman is usually nine months, but there have been many instances when the child has lived that was born at seven; and some are found to continue pregnant a month above the usual time. When the appointed time approaches, the infant, that has for some months been giving painful proofs of its existence, now begins to increase its efforts for liberty. The head is applied downwards, to the aperture of the womb, and by reiterated efforts it endeavours to extend the same: these endeavours produce the pain, which all women in labour feel in some degree; those of strong constitutions the least, those most weakly the most severely; since we learn, that the women of Africa always deliver themselves, and are well a few hours after; while those of Europe require assistance, and recover more slowly. Thus the infant, still continuing to push with its head forward, by the repetition of its endeavours, at last succeeds, and issues into life. The blood, which had hitherto passed through the heart, now takes a wider circuit; and the foramen ovale closes; the lungs, that had till this time been inactive, now first begin their functions; the air rushes in to distend them; and this produces the first sensation of pain, which the infant expresses by a shriek: so that the beginning of our lives, as well as the end, is marked with anguish.*

From comparing these accounts, we perceive that the most laboured generation is the most perfect; and that the animal, which, in proportion to its bulk, takes the longest time for production, is always the most complete when finished. Of all others, man seems the slowest in coming into life, as he is the slowest in coming to perfection; other animals of the same bulk, seldom remain in the womb above six months, while he continues nine; and even after his birth, appears more than any other to have his state of imbecility prolonged.

We may observe also, that that generation is the most complete, in which the fewest animals are produced: Nature, by attending to the production of one at a time, seems to exert all her efforts in bringing it to perfection; but, where this attention is divided, the animals so produced come into the world with partial advantages. In this manner twins are never, at least while infants, so large, or so strong, as those that come singly into the world; each having, in some measure, robbed the other of its right; as that support, which Nature meant for one, has been prodigally divided.

---

* Bonet. Contemplat. de la Nature, vol. i. p 212

In this manner, as those animals are the best that are produced singly, so we find that the noblest animals are ever the least fruitful. These are seen usually to bring forth but one at a time, and to place all their attention upon that alone. On the other hand, all the oviparous kinds produce an amazing plenty; and even the lower tribes of viviparous animals increase in a seeming proportion to their minuteness and imperfection. Nature seems lavish of life in the lower orders of the creation; and, as if she meant them entirely for the use of the nobler races, she appears to have bestowed greater pains in multiplying the number than in completing the kind. In this manner, while the elephant and the horse bring forth but one at a time, the spider and the beetle are seen to produce a thousand; and even among the smaller quadrupeds, all the inferior kinds are extremely fertile; any one of these being found, in a very few months, to become the parent of a numerous progeny.

In this manner, therefore, the smallest animals multiply in the greatest proportion; and we have reason to thank Providence that the most formidable animals are the least fruitful. Had the lion and the tiger the same degree of fecundity with the rabbit or the rat, all the arts of man would be unable to oppose these fierce invaders; and we should soon perceive them become the tyrants of those who claim the lordship of the creation. But Heaven, in this respect, has wisely consulted the advantage of all. It has opposed to man only such enemies as he has art and strength to conquer; and as large animals require proportional supplies, nature was unwilling to give new life, where it, in some measure, denied the necessary means of subsistence.

In consequence of this pre-established order, the animals that are endowed with the most perfect methods of generation, and bring forth but one at a time, seldom begin to procreate till they have almost acquired their full growth. On the other hand, those which bring forth many, engender before they have arrived at half their natural size. The horse and the bull come almost to perfection before they begin to generate; the hog and the rabbit scarcely leave the teat before they become parents themselves. In whatever light, therefore, we consider this subject, we shall find that all creatures approach most to perfection, whose generation most nearly resembles that of man. The reptile produced from cutting is but one degree above the vegetable. The animal produced from the egg is a step higher in the scale of existence; that class of animals which are brought forth alive, are still more exalted. Of these, such as bring forth one at a time are the most complete: and the foremost of these stands Man, *the great master of all*, who seems to have united the perfections of all the rest in his formation.

CHAPTER III.

THE INFANCY OF MAN.

WHEN we take a survey of the various classes of animals, and examine their strength, their beauty, or their structure, we shall find man to possess most of those advantages united, which the rest enjoy partially. Infinitely superior to all others in the power of the understanding, he is also superior to them in the fitness and proportions of his form. He would, indeed, have been one of the most miserable beings upon earth, if with a sentient mind he was so formed as to be incapable of obeying its impulse; but Nature has otherwise provided; as with the most extensive intellects to command, she has furnished him with a body the best fitted for obedience.

In infancy,* however, that mind and this body form the most helpless union in all animated nature; and, if any thing can give us a picture of complete imbecility, it is a man when just come into the world. The infant just born stands in need of all things, without the power of procuring any. The lower races of animals, upon being produced, are active, vigorous, and capable of self-support; but the infant is obliged to wait in helpless expectation; and its cries are its only aid to procure subsistence.

An infant just born may be said to come from one element into another: for, from the watery fluid in which it was surrounded, it now immerges into air; and its first cries seem to imply how greatly it regrets the change. How much longer it could have continued in a state of almost total insensibility in the womb, is impossible to tell: but it is very probable that it could remain there some hours more. In order to throw some light upon this subject, Mr. Buffon so placed a pregnant bitch, as that her puppies were brought forth in warm water, in which he kept them above half an hour at a time. However, he saw no change in the animals thus newly brought forth; they continued the whole time vigorous; and, during the whole time, it is very probable that the blood circulated through the same channels through which it passed while they continued in the womb.

Almost all animals have their eyes closed,† for some days after being brought into the world. The infant opens them the instant of its birth. However, it seems to keep them fixed and idle; they want that lustre which they acquire by degrees; and if they happen to move, it is rather an accidental gaze than an exertion of the act of seeing. The light alone seems to make the greatest impression upon them. The eyes of infants are sometimes found turned to the place where it is strongest; and the pupil is seen to dilate and diminish, as in grown persons, in proportion to the quantity it receives But still the infant is incapable of distinguishing objects; the sense of seeing, like the rest of the senses, requires a habit before it becomes any way serviceable. All the senses must be compared with each other, and must be made to correct the defects of one another, before they can give just information. It is probable, therefore, that if the infant could express its own sensations, it would give a very extraordinary description of the illusions which it suffers from them The sight might, perhaps, be represented as inverting objects, or multiplying them; the hearing, instead of conveying one uniform tone, might be said to bring up an interrupted succession of noises; and the touch apparently would divide one body into as many as there are fingers that graspod it. But all these errors are lost in one

* Buffon, vol. iv. p. 173.   † Buffon, vol. iv p. 173.

confused idea of existence; and it is happy for the infant that it can then make but very little use of its senses, when they could serve only to bring it false information.

If there be any distinct sensations, those of pain seem to be much more frequent and stronger than those of pleasure. The infant's cries are sufficient indications of the uneasiness it must, at every interval, endure ; while, in the beginning, it has got no external marks to testify its satisfactions. It is not till after forty days that it is seen to smile; and not till that time also, that tears begin to appear, its former expressions of uneasiness being always without them. As to any other marks of the passions, the infant being as yet almost without them, it can express none of them in its visage; which, except in the act of crying and laughing, is fixed in a settled serenity. All the other parts of the body seem equally relaxed and feeble: its motions are uncertain, and its postures without choice; it is unable to stand upright ; its hams are yet bent, from the habit which it received from its position in the womb; it has not strength enough in its arms to stretch them forward, much less to grasp any thing with its hands ; it rests just in the posture it is laid ; and, if abandoned, must continue in the same position.

Nevertheless, though this be the description of infancy among mankind in general, there are countries and races among whom infancy does not seem marked with such utter imbecility, but where the children, not long after they are born, appear possessed of a greater share of self-support. The children of negroes have a surprising degree of this premature industry: they are able to walk at two months ; or, at least, to move from one place to another : they also hang to the mother's back without any assistance, and seize the breast over her shoulder; continuing in this posture till she thinks proper to lay them down. This is very different in the children of our countries, that seldom are able to walk under a twelvemonth.

The skin of children newly brought forth is always red, proceeding from its transparency, by which the blood beneath appears more conspicuous. Some say that this redness is greatest in those children than are afterwards about to have the finest complexions; and it appears reasonable that it should be so, since the thinnest skins are always the fairest. The size of a new-born infant is generally about twenty inches, and its weight about twelve pounds. The head is large, and all the members delicate, soft, and puffy. These appearances alter with its age ; as it grows older, the head becomes less in proportion to the rest of the body ; the flesh hardens ; the bones, that before birth grew very thick in proportion, now lengthen by degrees, and the human figure more and more acquires its due dimensions. In such children, however, as are but feeble or sickly, the head always continues too big for the body ; the heads of dwarfs being extremely large in proportion.

Infants, when newly born, pass most of their time in sleeping, and awake with crying, excited either by sensations of pain or of hunger. Man, when come to maturity, but rarely feels the want of food as eating twice or thrice in the four and twenty hours is known to suffice the most voracious: but the infant may be considered as a little glut ton, whose only pleasure consists in its appetite; and this, except when it sleeps, it is never easy without satisfying. Thus nature has adapted different desires to the different periods of life; each as it seems most necessary for human support or succession. While the animal is yet forming, hunger excites it to that supply which is necessary for its growth; when it is completely formed, a different appetite takes place, that incites it to communicate existence.—These two desires take up the whole attention at different periods, but are very seldom found to prevail strongly together in the same age; one pleasure ever serving to repress the other: and, if we find a person of full age, placing a principal part of his happiness in the nature and quantity of his food, we have strong reasons to suspect, that with respect to his other appetites, he still retains a part of the imbecility of his childhood.

It is extraordinary, however, that infants, who are thus more voracious than grown persons, are nevertheless more capable of sustaining hunger. We have several instances, in accidental cases of famine, in which the child has been known to survive the parent, and seen clinging to the breast of its dead mother. Their little bodies also, are more patient of cold; and we have similar instances of the mother's perishing in the snow, while the infant has been found alive beside her. However, if we examine the internal structure of infants, we shall find an obvious reason for both these advantages. Their blood-vessels are known to be much larger than in adults; and their nerves much thicker and softer: thus being furnished with a more copious quantity of juices, both of the nervous and sanguinary kinds, the infant finds a temporary sustenance in this superfluity, and does not expire till both are exhausted. The circulation also being larger and quicker, supplies it with proportionable warmth, so that it is more capable of resisting the accidental rigours of the weather.

The first nourishment of infants is well known to be the mother's milk; and, what is remarkable, the infant has milk in its own breasts, which may be squeezed out by compression: this nourishment becomes less grateful as the child gathers strength; and perhaps, also, more unwholesome. However, in cold countries, which are unfavourable to propagation, and where the female has seldom above three or four children at the most, during her life, she continues to suckle the child for four or five years together. In this manner the mothers of Canada and Greenland are often seen suckling two or three children, of different ages, at a time.

The life of infants is very precarious, till the age of three or four, from which time it becomes more secure; and when a child arrives at its seventh year, it is then considered as a more certain life, as Mr. Buffon asserts, than at any other age whatever. It appears, from Simpson's Tables, that of a certain number of children born at the same time, a fourth part are found dead at the end of the first year: more than one-third at the end of the second; and at least half at the end of the third; so that those who live to be above three years old, are indulged a longer term than half the rest of their fellow creatures. Nevertheless, life, at that period, may be considered as mere animal existence; and rather a preparation for, than an enjoyment of, those satisfactions, both of mind and body, that make life of real value: and hence it is more natural for mankind to deplore a fellow-creature, cut off in the bloom of life, than one dying in early infancy. The one, by living up to youth, and thus wading through the disadvantageous parts of existence, seems to have earned a short continuance of its enjoyments: the infant, on the contrary, has served but a short apprenticeship to pain; and, when taken away, may be considered as rescued from a long continuance of misery.

There is something very remarkable in the growth of the human body.* The embryo in the womb continues to increase still more and more till it is born. On the other hand, the child's growth is less every year, till the time of puberty, when it seems to start up of a sudden. Thus, for instance, the embryo, which is an inch long in the first month, grows but one inch and a quarter in the second; it then grows one and a half in the third; two and a half in the fourth; and in this manner it keep increasing, till in the last month of its continuance it is actually found to grow four inches; and, in the whole, about eighteen inches long. But it is otherwise with the child when born; if we suppose it eighteen inches at that time, it grows in the first year six or seven inches; in the second year, it grows but four inches; in the third year, about three; and so on, at the rate of about an inch and a half, or two inches, each year, till the time of puberty, when nature seems to make one great last effort, to complete her work, and unfold the whole animal machine.

The growth of the mind in children seems to correspond with that of the body. The comparative progress of the understanding is greater in infants than in children of three or four years old. If we only reflect a moment on the amazing acquisitions that an infant makes in the first and second years of life, we shall have much cause for wonder. Being sent into a world where every thing is new and unknown, the first months of life are spent in a kind of torpid amazement; an attention distracted by the multiplicity of objects that press to be known. The first labour, therefore, of the little learner is, to correct the illusions of the senses, to distinguish one object from another, and to exert the memory, so as to know them again. In this manner a child of a year old has already made a thousand experiments; all which it has properly ranged, and distinctly remembers. Light, heat, fire, sweets, and bitters, sounds soft or terrible, are all distinguished at the end of a very few months. Besides this, every person the child knows, every individual object it becomes fond of, its rattles, or its bells, may be all considered as so many new lessons to the young mind, with which it has not become acquainted, without repeated exertions of the understanding. At this period of life, the knowledge of every individual object cannot be acquired without the same effort which, when grown up, is employed upon the most abstract idea: every thing the child hears or sees, all the marks and characters of nature, are as much unknown, and require the same attention to attain, as if the reader were set to understand the characters of an Ethiopic manuscript; and yet we see

* Buffon, vol. iv. p. 173.

in how short a time the little student begins to understand them all, and to give evident marks of early industry.

It is very amusing to pursue the young mind, while employed in its first attainments. At about a year old the same necessities that first engaged its faculties, increase as its acquaintance with nature enlarges. Its studies, therefore, if I may use the expression, are no way relaxed; for having experienced what gave pleasure at one time, it desires a repetition of it from the same object; and, in order to obtain this, that object must be pointed out; here, therefore, a new necessity arises, which, very often, neither its little arts nor importunities can remove; so that the child is at last obliged to set about naming the objects it desires to possess or avoid. In beginning to speak, which is usually about a year old, children find a thousand difficulties. It is not without repeated trials that they come to pronounce any one of the letters; nor without an effort of the memory, that they can retain them. For this reason, we frequently see them attempting a sound which they had learned, but forgot; and when they have failed, I have often seen their attempts attended with apparent confusion. The letters soonest learned, are those which are most easily formed; thus A and B require an obvious disposition of the organs, and their pronunciation is consequently soon attained. Z and R, which require a more complicated position, are learned with greater difficulty. And this may, perhaps, be the reason why the children in some countries speak sooner than in others; for the letters mostly occurring in the language of one country, being such as are of easy pronunciation, that language is of course more easily attained. In this manner the children of the Italians are said to speak sooner than those of the Germans, the language of the one being smooth and open; that of the other, crowded with consonants, and extremely guttural.

But be this as it will, in all countries children are found able to express the greatest part of their wants by the time they arrive at two years old; and from the moment the necessity of learning new words ceases, they relax their industry. It is then that the mind, like the body, seems every year to make slow advances; and, in order to spur up attention, many systems of education have been contrived.

Almost every philosopher who has written on the education of children, has been willing to point out a method of his own, chiefly professing to advance the health, and improve the intellects at the same time. These are usually found to begin with finding nothing right in the common practice, and by urging a total reformation. In consequence of this, nothing can be more wild or imaginary than their various systems of improvement. Some will have the children every day plunged in cold water, in order to strengthen their bodies; they will have them converse with the servants in nothing but the Latin language, in order to strengthen their minds; every hour of the day must be appointed for its own studies, and the child must learn to make these very studies an amusement; till about the age of ten or eleven it becomes a prodigy of premature improvement. Quite opposite to this, we have others, whom the courtesy of mankind also calls *philosophers*; and they will have the child learn nothing till the age of ten or eleven, at which the former has attained so much perfection; with them the mind is to be kept empty, until it has a proper distinction of some metaphysical ideas about truth, and the promising pupil is debarred the use of even his own faculties, lest they should conduct him into prejudice and error. In this manner some men whom fashion has celebrated for profound and fine thinkers, have given their hazarded and untried conjectures, upon one of the most important subjects in the world, and the most interesting to humanity. When men speculate at liberty upon innate ideas, or the abstracted distinctions between will and power, they may be permitted to enjoy their systems at pleasure, as they are harmless, although they may be wrong; but when they allege that children are to be every day plunged in cold water, and, whatever be their constitutions, indiscriminately inured to cold and moisture; that they are to be kept wet in the feet, to prevent their catching cold; and never to be corrected when young, for fear of breaking their spirits when old; these are such noxious errors, that all reasonable men should endeavour to oppose them. Many have been the children whom these opinions, begun in speculation, have injured or destroyed in practice; and I have seen many a little philosophical martyr, whom I wished, but was unable to relieve.

If any system be therefore necessary, it is one that would serve to show a very plain point; that very little system is necessary. The natural and common course of education is in every respect the best: I mean that in which the child is permitted to play among its little equals, from whose similar instructions it often gains the most useful stores of knowledge. A child is not idle because it is playing about the fields, or pursuing a butterfly; it is all this time storing its mind with objects upon the nature, the properties, and the relations of which future curiosity may speculate.

I have ever found it a vain task to try to make a child's learning its amusement; nor do I see what good end it would answer were it actually attained. The child, as was said, ought to have its share of play, and it will be benefited thereby; and for every reason also it ought to have its share of labour. The mind, by early labour, will be thus accustomed to fatigues and subordination; and whatever be the person's future employment in life, he will be better fitted to endure it: he will be thus enabled to support the drudgeries of office with content; or to fill up the vacancies of life with variety. The child, therefore, should by times be put to its duty; and be taught to know, that the task is to be done, or the punishment to be endured. I do not object against alluring it to duty by reward; but we well know that the mind will become more strongly stimulated by pain; and both may, upon some occasions, take their turn to operate. In this manner, a child, by playing with its equals abroad, and labouring with them at school, will acquire more health and knowledge, than by being bred up under the wing of any speculative system-maker; and will be thus qualified for a life of activity and obedience. It is true indeed, that when educated in this manner, the boy may not be so seemingly sensible and forward as one bred up under solitary instruction; and, perhaps, this early forwardness is more engaging than useful. It is well known that many of those children who have been such prodigies of literature before ten, have not made an adequate progress to twenty. It should seem that they only began learning manly things before their time; and, while others were busied in picking up that knowledge adapted to their age and curiosity, these were forced upon subjects unsuited to their years: and, upon that account alone, appearing extraordinary. The stock of knowledge in both may be equal, but with this difference, that each is yet to learn what the other knows.

But whatever may have been the acquisitions of children at ten or twelve, their greatest and most rapid progress is made when they arrive near the age of puberty. It is then that all the powers of nature seem at work in strengthening the mind and completing the body; the youth acquires courage, and the virgin modesty; the mind, with new sensations, assumes new powers; it conceives with greater force, and remembers with greater tenacity. About this time, therefore, which is various in different countries, more is learned in one year than in any two of the preceding; and on this age in particular, the greatest weight of instruction ought to be thrown.

CHAPTER IV.

OF PUBERTY.

It has been often said, that the season of youth is the season of pleasures: but this can only be true in savage countries, where but little preparation is made for the perfection of human nature, and where the mind has but a very small part in the enjoyment. It is otherwise in those places where nature is carried to the highest pitch of refinement, in which this season of the greatest sensual delight is wisely made subservient to the succeeding, and more rational one of manhood. Youth, with us, is but a scene of preparation; a drama, upon the right conduct of which all future happiness is to depend. The youth who follows his appetites, too soon seizes the cup, before it has received its best ingredients; and, by anticipating his pleasures, robs the remaining parts of life of their share; so that his eagerness only produces a manhood of imbecility, and an age of pain.

The time of puberty is different in various countries, and always more late in men than in women. In the warm countries of India, the women are marriageable at nine or ten, and the men at twelve or thirteen. It is also different in cities, where the inhabitants lead a more soft, luxurious life, from the country, where they work harder, and fare less delicately. Its symptoms are seldom alike in different persons; but it is usually known by a swelling of the breasts in one sex, and a roughness of the voice in the other. At this season, also, the women seem to acquire new beauty, while the men lose all that delicate effeminacy of countenance which they had when boys.

All countries, in proportion as they are civilized or barbarous, improve or degrade the nuptial satisfaction. In those miserable regions, where strength makes the only law, the stronger sex exerts its power, and becomes the tyrant over the weaker: while the inhabitant of Negroland is indolently taking his pleasure in the fields, his wife is obliged to till the grounds that serve for their mutual support. It is thus in all barbarous countries, where the men throw all the laborious duties of life upon the women; and, regardless of beauty, put the softer sex to those employments that must effectually destroy it.

But in countries that are half barbarous, particularly wherever Mahometanism prevails, the men run into the very opposite extreme. Equally brutal with the former, they exert their tyranny over the weaker sex, and consider that half of the human creation as merely made to be subservient to the depraved desires of the other. The chief, and indeed the only aim of an Asiatic, is to be possessed of many women; and to be able to furnish a seraglio is the only tendency of his ambition. As the savage was totally regardless of beauty, he, on the contrary, prizes it too highly; he excludes the person who is possessed of such personal attractions from any share in the duties or employments of life; and, as if willing to engross all beauty to himself, increases the number of his captives in proportion to the progress of his fortune. In this manner he vainly expects to augment his satisfactions, by seeking from many that happiness which he ought to look for in the society of one alone. He lives a gloomy tyrant, amidst wretches of his own making; he feels none of those endearments which spring from affection, none of those delicacies which arise from knowledge. His mistresses, being shut out from the world, and totally ignorant of all that passes there, have no arts to entertain his mind, or calm his anxieties; the day passes with them in sullen silence, or languid repose; appetite can furnish but few opportunities of varying the scene: and all that falls beyond it must be irksome expectation.

From this avarice of women, if I may be allowed to express it so, has proceeded that jealousy and suspicion which ever attends the miser: hence those low and barbarous methods of keeping the women of those countries guarded, and of making and procuring eunuchs to attend them. These unhappy creatures are of two kinds, the white and the black. The white are generally made in the country where they reside, being but partly deprived of the marks of virility; the black are generally brought from the interior parts of Africa, and are made entirely bare. These are chiefly chosen for their deformity; the thicker the lips, the flatter the nose, and the more black the teeth, the more valuable the eunuch; so that the vile jealousy of mankind here inverts the order of nature, and the poor wretch finds himself valued in proportion to his deficiencies. In Italy, where this barbarous custom is still retained, and eunuchs are made in order to improve the voice, the laws are severely aimed against such practice; so that being entirely prohibited, none but the poorest and most abandoned of the people still secretly practise it upon their children. Of those served in this manner, not one in ten is found to become a singer; but such is the luxurious folly of the times, that the success of one amply compensates for the failure of the rest. It is very difficult to account for the alterations which castration makes in the voice, and the other parts of the body. The eunuch is shaped differently from others. His legs are of an equal thickness above and below; his knees weak; his shoulders narrow; and his beard thin and downy. In this manner his person is rendered more deformed; but his desires, as I am told, still continue the same; and actually, in Asia, some of them are found to have their seraglios, as well as their masters. Even in our country, we have an instance of a very fine woman's being married to one of them, whose appearance was the most unpromising; and what is more extraordinary still, I am told, that this couple continue perfectly happy in each other's society.

The mere necessities of life seem the only aim of the savage; the sensual pleasures are the only study of the semi-barbarian; but the refinement of sensuality, by reason, is the boast of real politeness. Among the merely barbarous nations, such as the natives of Madagascar, or the inhabitants of Congo, nothing is desired so ardently as to prostitute their wives or daughters to strangers, for the most trifling advantages; they will account it a dishonour not to be among the foremost who are thus received into favour: on the other hand, the Mahometan keeps his wife faithful by confining her person, and would instantly put her to death if he but suspected her chastity. With the politer inhabitants of Europe, both these barbarous extremes are avoided; the woman's person is left free, and no constraint is imposed but upon her affections. The passion of love, which may be considered as the nice conduct of ruder desire, is only known and practised in this part of the world; so that what other nations guard as their right, the more delicate European is contented to ask as a favour. In this manner the concurrence of mutual appetite contributes to increase mutual satisfaction; and the power on one side of refusing, makes every blessing more grateful when obtained by the other. In barbarous countries, woman is considered merely as a useful slave; in such as are somewhat more refined, she is regarded as a desirable toy; in countries entirely polished, she enjoys juster privileges; the wife being considered as a useful friend, and an agreeable mistress. Her mind is still more prized than her person; and without the improvement of both, she can never expect to become truly agreeable; for her good sense alone can preserve what she has gained by her beauty.

Female beauty, as was said, is always seen to improve about the age of puberty; but if we should attempt to define in what this beauty consists, or what constitutes its perfection, we should find nothing more difficult to determine. Every country has its peculiar way of thinking, in this respect; and even the same country thinks differently at different times. The ancients had a very different taste from what prevails at present. The eye-brows joining in the middle was considered as a very peculiar grace by Tibullus, in the enumeration of the charms of his mistress. Narrow foreheads were approved of, and scarce any of the Roman ladies that are celebrated for their other perfections, but are also praised for the redness of their hair. The nose also of the Grecian Venus was such as would appear at present an actual deformity, as it fell in a straight line from the forehead, without the smallest sinking between the eyes, without which we never see a face at present

Among the moderns, every country seems to have peculiar ideas of beauty.* The Persians admire large eye-brows, joining in the middle; the edges and corners of the eyes are tinctured with black, and the size of the head is increased by a great variety of bandages formed into a turban. In some parts of India, black teeth and white hair are desired with ardour; and one of the principal employments of the women of Thibet, is to redden the teeth with herbs, and to make their hair white by a certain preparation. The passion for coloured teeth obtains also in China and Japan; where to complete their idea of beauty, the object of desire must have little eyes, nearly closed, feet extremely small, and a waist far from being shapely. There are some nations of the American Indians that flatten the heads of their children, by keeping them, while young, squeezed between two boards, so as to make the visage much larger than it would naturally be. Others flatten the head at top; and others make it as round as they possibly can. The inhabitants along the western coasts of Africa have a very extraordinary taste for beauty. A flat nose, thick lips, and jet-black complexion, are there the most indulgent gifts of nature. Such, indeed, they are all, in some degree, found to possess. However, they take care, by art, to increase their natural deformities, as they should seem to us; and they have many additional methods of rendering their persons still more frightfully pleasing. The whole body and visage is often scarred with a variety of monstrous figures; which is not done without great pain, and repeated incision: and even sometimes parts of the body are cut away. But it would be endless to remark the various arts which caprice or custom has employed to distort and disfigure the body, in order to render it more pleasing; in fact, every nation, how barbarous soever, seems unsatisfied with the human figure, as nature has left it, and has its peculiar arts of heightening beauty. Painting, powdering, cutting, boring the nose and the ears, lengthening the one, and depressing the other, are arts practised in many countries; and, in some degree, admired in all. These arts might have been at first introduced to hide epidemic deformities; custom, by degrees, reconciles them to the view; till, from looking upon them with indifference, the eye at length begins to gaze with pleasure.

CHAPTER V

OF THE AGE OF MANHOOD †

The human body attains to its full height during the age of puberty; or, at least, a short time after. Some young people are found to cease growing at fourteen or fifteen; others continue their growth till two or three and twenty. During this period they are all of a slender

* Buffon.
† This chapter is translated from Mr. Buffon, whose description is very excellent. Whatever I have added, is marked by inverted commas. "thus." And in whatever trifling points I have differed, the notes will serve to show

make; their thighs and legs small, and the muscular parts are yet unfilled. But, by degrees, the fleshy fibres augment; the muscles swell, and assume their figure; the limbs become proportioned and rounder; and before the age of thirty, the body in men, has acquired the most perfect symmetry. In women, the body arrives at perfection much sooner, as they arrive at the age of maturity more early; the muscles, and all the other parts being weaker, less compact and solid, than those of man, they require less time in coming to perfection; and, as they are less in size, that size is sooner completed. Hence the persons of women are found to be as complete at twenty, as those of men are found to be at thirty.

The body of a well-shaped man ought to be square; the muscles should be expressed with boldness, and the lines of the face strongly marked. In the woman, all the muscles should be rounder, the lines softer, and the features more delicate. Strength and majesty belong to the man; grace and softness are the peculiar embellishments of the other sex. In both, every part of their form declares their sovereignty over other creatures. Man supports his body erect; his attitude is that of command: and his face, which is turned towards the heavens, displays the dignity of his station. The image of his soul is painted in his visage; and the excellence of his nature penetrates through the material form in which it is inclosed. His majestic port, his sedate and resolute step, announce the nobleness of his rank. He touches the earth only with his extremity; and beholds it as if at a disdainful distance. His arms are not given him, as to other creatures, for pillars of support; nor does he lose, by rendering them callous against the ground, that delicacy of touch which furnishes him with so many of his enjoyments. His hands are made for very different purposes · to second every intention of his will, and to perfect the gifts of Nature.

When the soul is at rest, all the features of the visage seem settled in a state of profound tranquillity. Their proportion, their union, and their harmony, seem to mark the sweet serenity of the mind, and give a true information of what passes within. But when the soul is excited, the human visage becomes a living picture;-where the passions are expressed with as much delicacy as energy, where every motion is designed by some correspondent feature, where every impression anticipates the will, and betrays those hidden agitations, that he would often wish to conceal.

It is particularly in the eyes that the passions are painted; and in which we may most readily discover their beginning. The eye seems to belong to the soul more than any other organ; it seems to participate of all its emotions; as well the most soft and tender, as the most tumultuous and forceful. It not only receives, but transmits them by sympathy; the observing eye of one catches the secret fire from another: and the passion thus often becomes general.

Such persons as are short-sighted labour under a particular disadvantage in this respect. They are, in a manner, entirely cut off from the language of the eyes; and this gives an air of stupidity to the face, which often produces very unfavourable prepossessions. However intelligent we find such persons to be, we can scarcely be brought back from our first prejudice, and often continue in the first erroneous opinion. In this manner we are too much induced to judge of men by their physiognomy; and having, perhaps, at first, caught up our judgments prematurely, they mechanically influence us all our lives after. This extends even to the very colour or the cut of people's clothes; and we should for this reason be careful, even in such trifling particulars, since they go to make up a part of the total judgment which those we converse with may form to our advantage.

The vivacity, or the languid motion of the eyes, gives the strongest marks to physiognomy; and their colour contributes still more to enforce the expression. The different colours of the eye are the dark hazle, the light hazle, the green, the blue and grey, the whitish grey, "and also the red." These different colours arise from the different colours of the little muscles that serve to contract the pupil; "and they are very often found to change colour with disorder and with age."

The most ordinary colours are the hazle and the blue, and very often both these colours are found in the eyes of the same person. Those eyes which are called black, are only of the dark hazle, which may be easily seen upon closer inspection; however, those eyes are reckoned the most beautiful where the shade is the deepest: and either in these, or the blue eyes, the fire, which gives its finest expression to the eye, is more distinguishable in proportion to the darkness of the tint. For this reason, the black eyes, as they are called, have the greatest vivacity; but, probably, the blue have the most powerful effect in beauty, as they reflect a greater variety of lights, being composed of more various colours.

This variety, which is found in the colour of the eyes, is peculiar to man, and one or two other kinds of animals; but, in general, the colour in any one individual is the same in all the rest. The eyes of oxen are brown; those of sheep of a water colour; those of goats are gray; "and it may also be, in general, remarked, that the eyes of most white animals are red; thus the rabbit, the ferret, and, even in the human race, the white Moor, all have their eyes of a red colour."

Although the eye, when put into motion, seems to be drawn on one side, yet it only moves round the centre; by which its coloured part moves nearer, or farther from the angle of the eye-lids, or is elevated or depressed. The distance between the eyes is less in man than in any other animal; and in some of them it is so great, that it is impossible that they should ever view the same object with both eyes at once, unless it be very far off. "This, however, in them is rather an advantage than an inconvenience, as they are thus able to watch round them, and guard against the dangers of their precarious situation."

Next to the eyes, the features which most give a character to the face, are the eye-brows, which being, in some measure, more apparent than the other features, are most readily distinguished at a distance. "Le Brun, in giving a painter directions with regard to the passions, places the principal expression of the face in the eye-brows." From their elevation and depression, most of the furious passions are characterized; and such as have this feature extremely moveable, are usually known to have an expressive face. By means of these we can imitate all the other passions, as they are raised and depressed, at command; the rest of the features are generally fixed; or, when put into motion, they do not obey the will; the mouth and eyes, in an actor, for instance, may, by being violently distorted, give a very different expression from what he would intend; but the eye-brows can scarcely be exerted improperly; their being raised, denotes all those passions which pride or pleasure inspire; and their depression marks those which are the effects of contemplation and pain; and such who have this feature, therefore, most at command, are often found to excel as actors."

The eye-lashes have an effect, in giving expression to the eye, particularly when long and close; they soften its glances, and improve its sweetness. Man and apes are the only animals that have eye-lashes both upon the upper and lower lids; all other animals want them on the lid below.

The eye-lids serve to guard the ball of the eye, and to furnish it with a proper moisture. The upper lid rises and falls; the lower has scarce any motion; and although their being moved depends on the will, yet it often happens that the will is unable to keep them open, when sleep or fatigue oppresses the mind. In birds, and amphibious quadrupeds, the lower lid alone has motion; fishes and insects have no eye-lids whatever.

The forehead makes a large part of the face, and a part which chiefly contributes to its beauty. It ought to be justly proportioned; neither too round nor too flat; neither too narrow nor too low; and the hair should come thick upon its extremities. It is known to every body how much the hair tends to improve the face; and how much the being bald serves to take away from beauty. The highest part of the head is that which becomes bald the soonest, as well as that part which lies immediately above the temples. The hair under the temples, and at the back of the head, is very seldom known to fail, " and women are much less apt to become bald than men;" Mr. Buffon seems to think they never become bald at all; but we have too many instances of the contrary among us, not to contradict very easily the assertion. Of all parts or appendages of the body, the hair is that which is found most different in different climates; and often not only contributes to mark the country, but also the disposition of the man. It is in general thickest where the constitution is strongest; and more glossy and beautiful where the health is most permanent. The ancients held the hair to be a sort of excrement, produced like the nails; the part next the root pushing out that immediately contiguous. But the moderns have found that every hair may be truly said to live, to receive nutriment, to fill and distend itself like the other parts of the body. The roots, they observe, do not turn gray sooner than the extremities, but the whole hair changes colour at once; and we have many instances of persons who have grown gray in one night's time.[*]
Each hair, if viewed with a microscope, is found to consist of five or six lesser ones, all wrapped up in one common covering; it appears

---

[*] Mr Buffon says, that the hair begins to grow gray at the points, but the fact is otherwise.

knotted, like some sorts of grass, and sends forth branches at the joints. It is bulbous at the root, by which it imbibes its moisture from the body, and it is split at the points; so that a single hair, at its end resembles a brush. Whatever be the size, or the shape of the pore, through which the hair issues, it accommodates itself to the same; being either thick, as they are large; small, as they are less; round, triangular, and variously formed, as the pores happen to be various. The hair takes its colour from the juices flowing through it, and it is found that this colour differs in different tribes and races of people The Americans, and the Asiatics, have their hair black, thick, straight, and shining. The inhabitants of the torrid climates of Africa have it black, short, and woolly. The people of Scandinavia have it red, long, and curled; and those of our own, and the neighbouring countries, are found with hair of various colours. However, it is supposed by many, that every man resembles in his disposition the inhabitants of those countries whom he resembles in the colour and the nature of his hair; so that the black are said, like the Asiatics, to be grave and acute; the red, like the Gothic nations, to be choleric and bold. However this may be, the length and the strength of the hair is a general mark of a good constitution; and as that hair which is strongest is most commonly curled, so curled hair is generally regarded among us as a beauty. The Greeks, however, had a very different idea of beauty in this respect; and seem to have taken one of their peculiar national distinctions from the length and the straightness of the hair.

The nose is the most prominent feature in the face; but, as it has scarce any motion, and that only in the strongest passions, it rather adds to the beauty than to the expression of the countenance. "However, I am told by the skilful in this branch of knowledge, that wide nostrils add a great deal to the bold and resolute air of the countenance; and where they are narrow, though it may constitute beauty, it seldom improves expression." The form of the nose, and its advanced position, are peculiar to the human visage alone. Other animals, for the most part, have nostrils, with a partition between them: but none of them have an elevated nose. Apes themselves have scarce any thing else of this feature, but the nostrils; the rest of the feature lying flat upon the visage, and scarce higher than the cheek-bones. "Among all the tribes of savage men also, the nose is very flat; and I have seen a Tartar who had scarce any thing else but two holes through which to breathe."

The mouth and lips, next to the eyes, are found to have the greatest expression. The passions have great power over this part of the face; and the mouth marks its different degrees by its different forms. The organ of speech still more animates this part, and gives it more life than any other feature in the countenance. The ruby colour of the lips, and the white enamel of the teeth, give it such a superiority over every other feature, that it seems to make the principal object of our regard. In fact, the whole attention is fixed upon the lips of the speaker; however rapid his discourse, however various the subject, the mouth takes correspondent situations; and deaf men have been often found to see the force of those reasonings which they could not hear understanding every word as it was spoken.

"The under jaw in man possesses a great variety of motions; while the upper has been thought by many to be quite immoveable."* However, that it moves in man, a very easy experiment will suffice to convince us. If we keep the head fixed, with any thing between our teeth, the edge of a table, for instance, and then open our mouths, we shall find that both jaws recede from it at the same time; the upper jaw rises, and the lower falls, and the table remains untouched between them. The upper jaw has motion as well as the under; and, what is remarkable, it has its proper muscles behind the head for thus raising and depressing it. Whenever, therefore, we eat, both jaws move at the same time, though very unequally; for the whole head moving with the upper jaw, of which it makes a part, its motions are thus less observable." In the human embryo, the under jaw is very much advanced before the upper. "In the adult, it hangs a good deal more backward; and those whose upper and under row of teeth are equally prominent, and strike directly against each other, are what the painters call under-hung; and they consider this as a great defect in beauty.† The under jaw in a Chinese face falls greatly more backward than with us; and, I am told, the difference is half an inch, when the mouth is shut naturally." In instances of the most violent passion, the under jaw has often an involuntary quivering motion; and often also, a state of languor produces another, which is that of yawning. "Every one knows how very sympathetic this kind of languid motion is: and that for one person to yawn, is sufficient to set all the rest of the company a yawning. A ridiculous instance of this was commonly practised upon the famous M'Laurin, one of the professors at Edinburgh. He was very subject to have his jaws dislocated; so that when he opened his mouth wider than ordinary, or when he yawned, he could not shut it again. In the midst of his harangues, therefore, if any of his pupils began to be tired of his lecture, he had only to gape or yawn, and the professor instantly caught the sympathetic affection; so that he thus continued to stand speechless, with his mouth wide open, till his servant, from the next room, was called in to set his jaw again."‡

When the mind reflects with regret upon some good unattained or lost, it feels an internal emotion, which acting upon the diaphragm, and that upon the lungs, produces a sigh; this, when the mind is strongly affected, is repeated; sorrow succeeds these first emotions; and tears are often seen to follow: sobbing is the sigh still more invigorated; and lamentation, or crying, proceeds from the continuance of the plaintive tone of the voice, which seems to implore pity. "There is yet a silent agony, in which the mind appears to disdain all external help, and broods over its distresses with gloomy reserve.

---

* Mr. Buffon is of this opinion. He says that the upper jaw is immoveable in all animals. However, the parrot is an obvious exception; and so is man himself, as shown above.

† Mr. Buffon says, that both jaws, in a perfect face, should be on a level: but this is denied by the best painters.

‡ Since the publication of this work, the editor has been credibly informed that the professor had not the defect here mentioned.

"This is the most dangerous state of mind: accidents or friendship may lessen the louder kinds of grief; but all remedies for this must be had from within: and there, despair too often finds the most deadly enemy."

Laughter is a sound of the voice, interrupted and pursued for some continuance. The muscles of the belly and the diaphragm are employed in its slightest exertions; but those of the ribs are strongly agitated in the louder: and the head sometimes is thrown backward, in order to raise them with greater ease. The smile is often an indication of kindness and good will: it is also often used as a mark of contempt and ridicule.

Blushing proceeds from different passions, being produced by shame, anger, pride, and joy. Paleness is often also the effect of anger; and almost ever attendant on fright and fear. These alterations in the colour of the countenance are entirely involuntary; all the other expressions of the passions are, in some small degree, under control; but blushing and paleness betray our secret purposes; and we might as well attempt to stop them as the circulation of the blood, by which they are caused.

The whole head, as well as the features of the face, takes peculiar attitudes from its passions: it bends forward to express humility, shame, or sorrow; it is turned to one side, in languor or in pity; it is thrown with the chin forward, in arrogance and pride; erect in self conceit and obstinacy; it is thrown backwards in astonishment; and combines its motions to the one side, and the other, to express contempt, ridicule, anger, and resentment. " Painters, whose study leads to the contemplation of external forms, are much more adequate judges of these than any naturalist can be; and it is with these a general remark, that no one passion is regularly expressed on different countenances in the same manner: that grief often sits upon the face like joy; and pride assumes the air of passion. It would be vain, therefore, in words, to express their general effect, since they are often as various as the countenances they sit upon; and in making this distinction nicely, lies all the skill of the physiognomist. In being able to distinguish what part of the face is marked by nature, and what by the mind; what part has been originally formed, and what is made by habit, constitutes this science, upon which the ancients so much valued themselves, and which we at present so little regard. Some, however, of the most acute men among us, have paid great attention to this art; and, by long practice, have been able to give some character of every person whose face they examined. Montaigne is well known to have disliked those men who shut one eye in looking upon any object; and Fielding asserts, that he never knew a person with a steady glavering smile, but he found him a rogue. However, most of these observations, tending to a discovery of the mind by the face, are merely capricious; and nature has kindly hid our hearts from each other, to keep us in good humour with our fellow-creatures."

The parts of the head which give the least expression to the face, are the ears; and they are generally found hidden under the hair. These, which are immoveable, and make so small an appearance in man, are very distinguishing features in quadrupeds. They serve in them as the principal marks of the passions; the ears discover their joys or their terrors with tolerable precision; and denote all their internal agitations. The smallest ears, in men, are said to be the most beautiful; but the largest are found to be the best for hearing. There are some savage nations who bore their ears, and so draw that part down, that the tips of the ears are seen to rest upon their shoulders.

The strange variety in the different customs of men appears still more extravagant in their manner of wearing their beards. Some, and among others, the Turks, cut the hair off their heads, and let their beards grow. The Europeans, on the contrary, shave their beards, and wear their hair. The negroes shave their heads in figures at one time, in stars at another, in the manner of friars; and still more commonly in alternate stripes; and their little boys are shaved in the same manner. The Talapoins of Siam shave the heads and the eye-brows of such children as are committed to their care. Every nation seems to have entertained different prejudices at different times, in favour of one part or another of the beard. Some have admired the hair upon the cheeks on each side, as we see with some low-bred men among ourselves, who want to be fine. Some like the hair lower down; some choose it curled, and others like it straight. " Some have cut it into a peak, and others shave all but the whisker. This particular part of the beard was highly prized among the Spaniards: till of late, a man without whiskers was considered as unfit for company; and where nature had denied them, art took care to supply the deficiency. We are told of a Spanish general who, when he borrowed a large sum of money from the Venetians, pawned his whiskers, which he afterwards took proper care to release. Kingson assures us, that a considerable part of the religion of the Tartars consists in the management of their whiskers; and that they waged a long and bloody war with the Persians, declaring them infidels, merely because they would not give their whiskers the orthodox cut. The kings of Persia carried the care of their beards to a ridiculous excess, when they chose to wear them matted with gold thread: and even the kings of France, of the first races, had them knotted and buttoned with gold. But of al nations, the Americans take the greatest pains in cutting their hair, and plucking their beards. The under part of the beard, and all but the whisker, they take care to pluck up by the roots, so that many have supposed them to have no hair naturally growing on that part; and even Linnæus has fallen into that mistake. Their hair is also cut into bands; and no small care employed in adjusting the whisker. In fact, we have a very wrong idea of savage finery, and are apt to suppose that, like the beasts of the forest, they rise, and are dressed with a shake: but the reverse is true; for no birth-night beauty takes more time or pains in the adorning her person than they. I remember when the Cherokee kings were over here, that I have waited for three hours, during the time they were dressing. They never would venture to make their appearance till they had gone through the tedious ceremonies of the toilet; they had their boxes of oil and ochre, their fat and their perfumes, like the most effeminate beau, and generally took up four hours in dressing, before they considered themselves as fit to be seen We must not, therefore, consider a delicacy in poin of dress, as a mark of refinement, since savages are much more difficult in this particular than the most fashionable or tawdry European. The more barbarous the people, the fonder of finery. In Europe, the lustre of jewels, and the splendour of the most brilliant colours, are generally given up to women, or to the weakest part of the other sex, who are willing to be contemptibly fine: but in Asia, these trifling fineries are eagerly sought after by every condition of men; and, as the proverb has it, we find the richest jewels in an Ethiop's ear. The passion for glittering ornaments is still stronger among the absolute barbarians, who often exchange their whole stock of provisions, and whatever else they happen to be possessed of, with our seamen, for a glass bead, or a looking-glass.

Although fashions have arisen in different countries from fancy and caprice, these, when they become general, deserve examination. Mankind have always considered it as a matter of moment, and they will ever continue desirous of drawing the attention of each other, by such ornaments as mark the riches, the power, or the courage of the wearer. The value of those shining stones which have at all times been considered as precious ornaments, is entirely founded upon their scarceness or their brilliancy. It is the same likewise with respect to those shining metals, the weight of which is so little regarded, when spread over our clothes. These ornaments are rather designed to draw the attention of others, than to add to any enjoyments of our own; and few there are that these ornaments will not serve to dazzle, and who can coolly distinguish between the metal and the man.

All things rare and brilliant will, therefore, ever continue to be fashionable, while men derive greater advantage from opulence than virtue; while the means of appearing considerable, are more easily acquired than the title to be considered. The first impression we generally make, arises from our dress; and this varies, in conformity to our inclinations, and the manner in which we desire to be considered. The modest man, or he who would wish to be thought so, desires to show the simplicity of his mind by the plainness of his dress; the vain man, on the contrary, takes a pleasure in displaying his superiority, " and is willing to incur the spectator's dislike, so he does but excite his attention."

Another point of view which men have in dressing, is to increase the size of their figure, and to take up more room in the world than nature seems to have allotted them. We desire to swell out our clothes by the stiffness of art, and raise our heels, while we add to the largeness of our heads. How bulky soever our dress may be, our vanities are still more bulky. The largeness of the doctor's wig arises from the same pride with the smallness of the beau's queue. Both want to have the size of their understanding measured by the size of their heads.

There are some modes that seem to have a more reasonable origin, which is to hide or to lessen the defects of nature. To take men altogether, there are many more deformed and plain than beautiful and shapely. The former, as being the most numerous, give law to fashion, and their laws are generally such as are made in their own favour. The women begin to colour their cheeks with red, when the natural roses are faded; and the younger are obliged to submit, though not compelled by the same necessity. In all parts of the world, this custom prevails more or less; and powdering and frizzing the hair, though not so general, seems to have arisen from a similar control.

But leaving the draperies of the human picture, let us return to the figure, unadorned by art. Man's head, whether considered externally or internally, is differently formed from that of all other animals, the monkey-kind only excepted, in which there is a striking similitude. There are some differences, however, which we shall take notice of in another place. The bodies of all quadruped animals are covered with hair; but the head of man seems the part most adorned, and that more abundantly than in any other animal.

There is a very great variety in the teeth of all animals; some have them above and below; others have them in the under jaw only: in some they stand separate from each other; while in some they are continued and united. The palate of some fishes is nothing else but a bony plate studded with points, which perform the office of teeth. All these substances, in every animal, derive their origin from the nerves; the substance of the nerves hardens by being exposed to the air; and the nerves that terminate in the mouth, being thus exposed, acquire a bony solidity. In this manner, the teeth and nails are formed in man; and in this manner also, the beak, the hoofs, the horns, and the talons of other animals are found to be produced.

The neck supports the head, and unites it to the body. This part is much more considerable in the generality of quadrupeds, than in man. But fishes and other animals that want lungs similar to ours, have no neck whatsoever. Birds, in general, have the neck longer than any other kind of animals; those of them, which have short claws, have also short necks; those, on the contrary, that have them long, are found to have the neck in proportion. "In men, there is a lump upon the wind-pipe, formed by the thyroid cartilage, which is not to be seen in women; an Arabian fable says, that this is a part of the original apple, that has stuck in the man's throat by the way, but that the woman swallowed her part of it down."

The human breast is outwardly formed in a very different manner from that of other animals. It is larger in proportion to the size of the body; and none but man, and such animals as make use of their fore-feet as hands, such as monkeys, bats, and squirrels, and such quadrupeds as climb trees, are found to have those bones called the *clavicles*, or, as we usually term them, the *collar bones*.* The breasts in women are larger than in men; however, they seem formed in the same manner; and sometimes milk is found in the breasts of men as well as in those of women. Among animals, there is a great variety in this part of the body. The teats of some, as in the ape and the elephant, are like those of men, being but two, and placed on each side of the breast. The teats of the bear amount to four. The sheep has but two, placed between the hinder legs. Other animals, such as the bitch, and the sow, have them all along the belly; and as they produce many young, they have a great many teats for their support. The form also of the teats varies in different animals, and in the same

* Mr. Buffon says that none but monkeys have them; but this is an over-sight.

animal at different ages. The bosom, in females, seems to unite all our ideas of beauty, where the outline is continually changing, and the gradations are soft and regular.

"The graceful fall of the shoulders, both in man and woman, constitute no small part of beauty. In apes, though otherwise made like us, the shoulders are high, and drawn up on each side towards the ears. In man they fall by a gentle declivity; and the more so, in proportion to the beauty of his form. In fact, being high-shouldered, is not without reason considered as a deformity, for we find very sickly persons are always so; and people when dying are ever seen with their shoulders drawn up in a surprising manner. The muscles that serve to raise the ribs, mostly rise near the shoulders; and the higher we raise the shoulders, we the more easily raise the ribs likewise. It happens, therefore, in the sickly and the dying, who do not breathe without labour, that to raise the ribs, they are obliged to call in the assistance of the shoulders; and thus their bodies assume from habit, that form which they are so frequently obliged to assume. Women with child also are usually seen to be high-shouldered; for the weight of the inferior parts drawing down the ribs, they are obliged to use every effort to elevate them, and thus they raise the shoulders of course. During pregnancy, also, the shape, not only of the shoulders, but also of the breast, and even the features of the face, are greatly altered; for the whole upper fore-part of the body is covered with a broad thin skin, called the myoides, which, being at that time drawn down, it also draws down with it the skin, and consequently the features of the face. By these means the visage takes a particular form; the lower eye-lids, and the corners of the mouth, are drawn downwards, so that the eyes are enlarged, and the mouth lengthened; and women, in these circumstances, are said by the midwives, to be *all mouth and eyes.*"

The arms of men but very little resemble the fore-feet of quadrupeds, and much less the wings of birds. The ape is the only animal that is possessed of hands and arms; but these are much more rudely fashioned, and with less exact proportion, than in men; "the thumb not being so well opposed to the rest of the fingers in their hands, as in ours."

The form of the back is not much different in man from that of other quadruped animals, only that the reins are more muscular in him and stronger. The buttock, however, in man, is different from that of all other animals whatsoever. What goes by that name in other creatures, is only the upper part of the thigh; man being the only animal that supports himself perfectly erect, the largeness of this part is owing to the peculiarity of his position.

Man's feet, also, are different from those of all other animals, those even of apes not excepted. The foot of the ape is rather a kind of awkward hand; its toes, or rather fingers, are long, and that of the middle, longest of all. This foot also wants the heel, as in man; the sole is narrower, and less adapted to maintain the equilibrium of the body in walking, dancing, or running.

The nails are less in man than in any other animal. If they were much longer than the extremities of the fingers, they would rather be prejudicial than serviceable, and obstruct the management of the hand. Such savages as let them grow long, make use of them in flaying animals, in tearing their flesh, and such like purposes; however, though their nails are considerably larger than ours, they are by no means to be compared to the hoofs or the claws of other animals. "They may sometimes be seen longer, indeed, than the claws of any animal whatsoever: as we learn that the nails of some of the learned men in China are longer than their fingers. But these want that solidity which might give force to their exertions, and could never, in a state of nature, have served them for annoyance or defence."

There is little known exactly with regard to the proportion of the human figure; and the beauty of the best statues is better conceived, by observing than by measuring them. The statues of antiquity, which were at first copied after the human form, are now become the models of it; nor is there one man found whose person approaches to those inimitable performances that have thus, in one figure, united the perfections of many. It is sufficient to say, that from being at first models, they are now become originals, and are used to correct the deviations in that form from whence they were taken." I will not, however, pretend to give the proportions of the human body as taken from these, there being nothing more arbitrary, and which good painters themselves so much contemn. Some, for instance, who have studied after these, divide the body into ten times the length of the face, and others into eight. Some pretend to tell us, that there is a similitude of proportion in different parts of the body. Thus, that the hand is the length of the face, the thumb the length of the nose, the space between the eyes is the breadth of an eye; that the breadth of the thigh, at thickest, is double that of the thickest part of the leg, and treble the smallest; that the arms extended, are as long as the figure is high; that the legs and thighs are half the length of the figure. All this, however, is extremely arbitrary; and the excellence of a shape, or the beauty of a statue, results from the attitude and position of the whole, rather than any established measurements, begun without experience, and adopted by caprice. In general, it may be remarked that the proportions alter in every age, and are obviously different in the two sexes. In women, the shoulders are narrower, and the neck proportionably longer than in men. The hips, also, are considerably larger, and the thighs much shorter than in men. These proportions, however, vary greatly at different ages. In infancy, the upper parts of the body are much larger than the lower; the legs and thighs do not constitute any thing like half the height of the whole figure; in proportion as the child increases in age, the inferior parts are found to lengthen; so that the body is not equally divided until it has acquired its full growth.

The size of men varies considerably. Men are said to be tall who are from five feet eight inches to six feet high. The middle stature is from five feet five to five feet eight: and those are said to be of small stature who fall under these measures. "However, it ought to be remarked, that the same person is always taller when he rises in the morning, than upon going to bed at night; and sometimes there is an inch difference; and I have seen more. Few persons are sensible of this remarkable variation; and, I am told, it was first perceived in England by a recruiting officer. He often found that those men whom he had enlisted for soldiers, and answered to the appointed standard at one time, fell short of it when they came to be measured before the colonel at the head-quarters. This diminution in their size proceeded from the different times of the day, and the different states of the body, when they happened to be measured. If, as was said, they were measured in the morning, after the night's refreshment, they were found to be commonly half an inch, and very often a whole inch, taller than if measured after the fatigues of the day; if they were measured when fresh in the country, and before a long fatiguing march to the regiment, they were found to be an inch taller than when they arrived at their journey's end. All this is now well known among those who recruit for the army; and the reason of this difference of stature is obvious. Between all the joints of the back-bone, which is composed of several pieces, there is a glutinous liquor deposited, which serves, like oil in a machine, to give the parts an easy play upon each other. This lubricating liquor, or synovia, as the anatomists call it, is poured in during the season of repose, and is consumed by exercise and employment; so that in a body, after hard labour, there is scarce any of it remaining; but all the joints grow stiff, and their motion becomes hard and painful. It is from hence, therefore, that the body diminishes in stature. For this moisture being drained away from between the numerous joints of the back-bone, they lie closer upon each other; and their whole length is thus very sensibly diminished; but sleep, by restoring the fluid again, swells the spaces between the joints, and the whole is extended to its former dimensions.

" As the human body is thus often found to differ from itself in size, so it is found to differ in its weight also; and the same person, without any apparent cause, is found to be heavier at one time than another. If, after having eaten a hearty dinner, or having drank hard, the person should find himself thus heavier, it would appear no way extraordinary; but the fact is, the body is very often found heavier some hours after eating a hearty meal than immediately succeeding it. If, for instance, a person fatigued by a day's hard labour, should eat a plentiful supper, and then get himself weighed upon going to bed, after sleeping soundly, if he is again weighed, he will find himself considerably heavier than before; and this difference is often found to amount to a pound, or sometimes to a pound and a half. From whence this adventitious weight is derived is not easy to conceive; the body, during the whole night, appears rather plentifully perspiring than imbibing any fluid; rather losing than gaining moisture: however, we have no reason to doubt but that either by the lungs, or perhaps, by a peculiar set of pores, it is all this time inhaling a quantity of fluid, which thus increases the weight of the whole body upon being weighed the next morning."*

---

* From this experiment, also, the learned may gather upon what a weak foundation the whole doctrine of the Sanctorian perspiration is built; but this disquisition more properly belongs to medicine than natural history.

Although the human body is externally more delicate than any of the quadruped kind, it is, notwithstanding, extremely muscular; and perhaps, for its size, stronger than that of any other animal. If we should offer to compare the strength of the lion with that of man, we should consider that the claws of this animal give us a false idea of its power; we ascribe to its force what is only the effects of its arms. Those which man has received from Nature are not offensive; happy had Art never furnished him with any more terrible than those which arm the paws of the lion.

But there is another manner* of comparing the strength of man with that of other animals; namely, by the weights which either can carry. We are assured that the porters of Constantinople carry burdens of nine hundred pounds weight. Mr. Desaguliers tells us of a man who, by distributing weights in such a manner as that every part of his body bore its share, he was thus able to raise a weight of two thousand pounds. A horse, which is about seven times our bulk, would be thus able to raise a weight of fourteen thousand pounds, if its strength were in the same proportion.† "But the truth is, a horse will not carry upon his back above a weight of two or three hundred pounds; while a man of confessedly inferior strength, is thus able to support two thousand. Whence comes this seeming superiority? The answer is obvious. Because the load upon the man's shoulders is placed to the greatest advantage; while, upon the horse's back, it is placed at the greatest disadvantage. Let us suppose, for a moment, the man standing as upright as possible, under the great load above mentioned. It is obvious that all the bones of his body may be compared to a pillar supporting a building, and that his muscles have scarce any share in this dangerous duty. However, they are not entirely inactive; as man, let him stand never so upright, will have some bending in the different parts of his body. The muscles, therefore, give the bones some assistance, and that with the greatest possible advantage. In this manner a man has been found to support two thousand weight, but may be capable of supporting a still greater. The manner in which this is done, is by strapping the load round the shoulders of the person who is to bear it, by a machine, something like that by which milk-vessels or water-buckets are carried. The load being thus placed on a scaffold, on each side, contrived for that purpose, and the man standing erect in the midst, all parts of the scaffold, except that where the man stands, are made to sink; and thus the man maintaining his position, the load, whatever it is, becomes suspended, and the column of his bones may be fairly said to support it. If, however, he should but ever so little give way, he must inevitably drop; and no power of his can raise the weights again. But the case is very different with regard to a load laid upon a horse. The column of the bones there lies a different way; and a weight of five hundred pounds, as I am told, would break the back of the strongest horse that could be found. The great force of a horse, and other quadrupeds, is ex-

* Mr. Buffon calls it a better manner; but this is not the case.

† Mr. Buffon carries this subject no farther; and thus far, without explanation, it i erroneous

erted when the load is in such a position as that the column of the bones can be properly applied, which is lengthwise. When, therefore, we are to estimate the comparative strength of a horse, we are not to try what he can carry, but what he can draw; and, in this case, his amazing superiority over man is easily discerned; for one horse can draw a load that ten men cannot move. And in some cases it happens that a draught-horse draws the better for being somewhat loaded: for, as the peasants say, the load upon the back keeps him the better to the ground."

There is still another way of estimating human strength, by the perseverance and agility of our motions. Men who are exercised in running, outstrip horses, or at least hold their speed for a longer continuance. In a journey, also, a man will walk down a horse; and, after they have both continued to proceed for several days, the horse will be quite tired, and the man will be fresher than in the beginning. The king's messengers of Ispahan, who are runners by profession, go thirty-six leagues in fourteen hours. Travellers assure us, that the Hottentots outstrip lions in the chace; and that the savages who hunt the elk, pursue with such speed, that they at last tire down, and take it. We are told many very surprising things of the great swiftness of the savages, and of the long journeys they undertake on foot, through the most craggy mountains, where there are no paths to direct, nor houses to entertain them. They are said to perform a journey of twelve hundred leagues in less than six weeks. "But notwithstanding what travellers report of this matter, I have been assured, from many of our officers and soldiers, who compared their own swiftness with that of the native Americans, during the last war, that although the savages held out, and as the phrase is, had better bottoms, yet, for a spurt, the Englishmen were more nimble and speedy."

Nevertheless, in general, civilized man is ignorant of his own powers: he is ignorant how much he loses by effeminacy, and what might be acquired by habit and exercise. Here and there, indeed, men are found among us of extraordinary strength; but that strength, for want of opportunity, is seldom called into exertion. "Among the ancients, it was a quality of much greater use than at present; as in war the same man that had strength sufficient to carry the heaviest armour, had strength sufficient also to strike the most fatal blow. In this case, his strength was at once his protection and his power. We ought not to be surprised, therefore, when we hear of one man terrible to an army, and irresistible in his career, as we find some generals represented in ancient history. But we may be very certain that this prowess was exaggerated by flattery, and exalted by terror. An age of ignorance is ever an age of wonder. At such times, mankind, having no just ideas of the human powers, are willing rather to represent what they wish, than what they know; and exalt human strength, to fill up the whole sphere of their limited conceptions. Great strength is an accidental thing; two or three in a country may possess it, and these may have a claim to heroism. But what may lead us to doubt of the veracity of these accounts is, that the heroes of antiquity are represented as the sons of heroes; their amazing strength is delivered down from father to son; and this we know to be contrary to the course of nature. Strength is not hereditary, although titles are: and I am very much induced to believe that this great tribe of heroes, who were all represented as the descendants of heroes, are more obliged to their titles than to their strength for their characters. With regard to the shining characters in Homer, they are all represented as princes, and as the sons of princes, while we are told of scarce any share of prowess in the meaner men of the army, who are only brought into the field for these to protect or to slaughter. But nothing can be more unlikely, than that those men who were bred in the luxury of courts, should be strong, while the whole body of the people, who receive a plainer and simpler education, should be comparatively weak. Nothing can be more contrary to the general laws of nature, than that all the sons of heroes should thus inherit not only the kingdoms, but the strength of their forefathers; and we may conclude, that they owe the greatest share of their imputed strength rather to the dignity of their stations than the force of their arms; and, like all fortunate princes, their flatterers happened to be believed. In latter ages, indeed, we have some accounts of amazing strength, which we can have no reason to doubt of. But in these, nature is found to pursue her ordinary course, and we find their strength accidental. We find these strong men among the lowest of the people, and gradually rising into notice, as this superiority had more opportunity of being seen. Of this number was the Roman tribune, who went by the name of the second Achilles, who, with his own hand, is said to have killed, at different times, three hundred of the enemy; and when treacherously set upon, by twenty-five of his own countrymen, although then past his sixtieth year, killed fourteen of them before he was slain. Of this number was Milo, who, when he stood upright, could not be forced out of his place. Pliny also tells us of one Athanatus, who walked across the stage at Rome, loaded with a breastplate weighing five hundred pounds, and buskins of the same weight. But of all the prodigies of strength, of whom we have any accounts in Roman history, Maximin, the emperor, is to be reckoned the foremost. Whatever we are told relative to him is well attested; his character was too exalted not to be thoroughly known; and that very strength for which he was celebrated, at last procured him no less a reward than the empire of the world. Maximin was above nine feet in height, and the best proportioned man in the whole empire. He was by birth a Thracian; and, from being a simple herdsman, rose through the gradations of office, until he came to be emperor of Rome. The first opportunity he had of exerting his strength, was in the presence of all the citizens, in the theatre, where he overthrew twelve of the strongest men in wrestling, and outstript two of the fleetest horses in running, all in one day. He could draw a chariot loaden, that two strong horses could not move; he could break a horse's jaw with a blow of his fist; and its thigh with a kick. In war he was always foremost and invincible: happy had it been for him and his subjects, if, from being formidable to his enemies, he had not become still more so to his subjects; he reigned for some time with all the world his enemy; all mankind wishing him dead, yet none daring to strike the blow. As if fortune had resolved that through life he should continue unconquerable, he was killed at

...ast by his own soldiers, while he was sleeping. We have many other instances in later ages of very great strength, and not fewer of amazing swiftness; but these merely corporeal perfections are now considered as of small advantage, either in war or in peace. The invention of gunpowder has, in some measure, levelled all force to one standard, and has wrought a total change in martial education through all parts of the world. In peace also, the invention of new machines every day, and the application of the strength of the lower animals to the purposes of life, have rendered human strength less valuable. The boast of corporeal force is therefore consigned to savage nations, where those arts not being introduced, it may still be needful; but in more polite countries, few will be proud of that strength which other animals can be taught to exert to as useful purposes as they.

" If we compare the largeness and thickness of our muscles with those of any other animal, we shall find that, in this respect, we have the advantage; and if strength or swiftness depended upon the quantity of muscular flesh alone, I believe that, in this respect, we should be more active and powerful than any other. But this is not the case; a great deal more than the size of the muscles goes to constitute activity or force; and it is not he who has the thickest legs that can make the best use of them. Those, therefore, who have written elaborate treatises on muscular force, and have estimated the strength of animals by the thickness of their muscles, have been employed to very little purpose. It is in general observed, that thin and raw-boned men are always stronger and more powerful than such as are seemingly more muscular, as in the former all the parts have better room for their exertions."

Women want much of the strength of men; and, in some countries, the stronger sex have availed themselves of this superiority, in cruelly and tyrannically enslaving those who were made with equal pretensions to a share in all the advantages life can bestow. Savage nations oblige their women to a life of continual labour; upon them rest all the drudgeries of domestic duty, while the husband, indolently reclined in his hammock, is first served from the fruits of her industry. From this negligent situation he is seldom roused, except by the calls of appetite, when it is necessary, either by fishing or hunting, to make a variety in his entertainments. A savage has no idea of taking pleasure in exercise; he is surprised to see a European walk forward for his amusement, and then return back again. As for his part, he could be contented to remain for ever in the same situation, perfectly satisfied with sensual pleasures and undisturbed repose. The women of these countries are the greatest slaves upon earth; sensible of their weakness, and unable to resist, they are obliged to suffer those hardships which are naturally inflicted by such as have been taught that nothing but corporeal force ought to give pre-eminence. It is not, therefore, till after some degree of refinement, that women are treated with lenity; and not till the highest degree of politeness, that they are permitted to share in all the privileges of man. The first impulse of savage nature is to confirm their slavery; the next, of half barbarous nations, is to appropriate their beauty; and that of the perfectly polite, to engage their affections. In civilized countries, therefore, women have united the force of modesty to the power of their natural charms, and thus obtain that superiority over the mind which they are unable to extort by their strength.

CHAPTER VI.

OF SLEEP AND HUNGER.

As man, in all the privileges he enjoys, and the powers he is invested with, has a superiority over all other animals, so in his necessities he seems inferior to the meanest of them all. Nature has brought him into life with a greater variety of wants and infirmities than the rest of her creatures, unarmed in the midst of enemies. The lion has natural arms; the bear natural clothing; but man is destitute of all such advantages; and, from the superiority of his mind alone, he is to supply the deficiency. The number of his wants, however, were merely given in order to multiply the number of his enjoyments, since the possibility of being deprived of any good, teaches him the value of his possession. Were men born with those advantages which he learns to possess by industry, he would very probably enjoy them with a blunter relish; it is by being naked that he knows the value of a covering; it is by being exposed to the weather, that he learns the comforts of a habitation. Every want thus becomes a means of pleasure in the redressing; and the animal that has most desires, may be said to be capable of the greatest variety of happiness.

Besides the thousand imaginary wants peculiar to man, there are two which he has in common with all other animals, and which he feels in a more necessary manner than they. These are the wants of sleep and hunger. Every animal that we are acquainted with, seems to endure the want of these with much less injury to health than man; and some are most surprisingly patient in enduring both. The little domestic animals that we keep about us, may often set a lesson of calm resignation, in supporting want and watchfulness, to the boasted philosopher. They receive their pittance at uncertain intervals, and wait its coming with cheerful expectation. We have instances of the dog and the cat living in this manner, without food, for several days, and yet still preserving their attachment to the tyrant that oppresses them; still ready to exert their little services for his amusement or defence. But the patience of these is nothing to what the animals of the forest endure. As these mostly live upon accidental carnage, so they are often known to remain without food for several weeks together. Nature, kindly solicitous for their support, has also contracted their stomachs, to suit them for their precarious way of living; and kindly, while it abridges the banquet, lessens the necessity of providing for it. But the meaner tribes of animals are made still more capable of sustaining life without food, many of them remaining in a state of torpid indifference till their prey approaches, when they jump upon and seize it. In this manner, the snake or the spider continue for several months together, to subsist upon a single meal;

and some of the butterfly kind live upon little or nothing. But it is very different with man: his wants daily make their importunate demands; and it is known that he cannot continue to live many days without eating, drinking, and sleeping.

Hunger is a much more powerful enemy to man than watchfulness, and kills him much sooner. It may be considered as a disorder that food removes, and that would quickly be fatal, without its proper antidote. In fact, it is so terrible to man, that to avoid it he even encounters certain death; and rather than endure its tortures, exchanges them for immediate destruction. However, by what I have been told, it is much more dreadful in its approaches, than in its continuance; and the pains of a famishing wretch decrease, as his strength diminishes. In the beginning, the desire of food is dreadful indeed, as we know by experience, for there are few who have not, in some degree, felt its approaches. But, after the first or second day, its tortures become less terrible, and a total insensibility at length comes kindly in to the poor wretch's assistance. I have talked with the captain of a ship, who was one of six that endured it in its extremities, and who was the only person that had not lost his senses, when they received accidental relief. He assured me his pains at first were so great as to be often tempted to eat a part of one of the men who died, and which the rest of his crew actually for some time lived upon: he said, that during the continuance of this paroxysm, he found his pains insupportable, and was desirous at one time of anticipating that death which he thought inevitable; but his pains, he said, gradually decreased, after the sixth day, (for they had water in the ship, which kept them alive so long) and then he was in a state rather of languor than desire; nor did he much wish for food, except when he saw others eating; and that for a while revived his appetite, though with diminished importunity. The latter part of the time, when his health was almost destroyed, a thousand strange images rose upon his mind, and every one of his senses began to bring him wrong information. The most fragrant perfumes appeared to him to have a fœtid smell; and every thing he looked at took a greenish hue, and sometimes a yellow. When he was presented with food by the ship's company that took him and his men up, (four of whom died shortly after,) he could not help looking upon it with loathing, instead of desire; and it was not till after four days that his stomach was brought to its natural tone, when the violence of his appetite returned with a sort of canine eagerness.

Thus dreadful are the effects of hunger; and yet when we come to assign the cause that produces them, we find the subject involved in doubt and intricacy. This longing eagerness is, no doubt, given for a very obvious purpose; that of replenishing the body, wasted by fatigue and perspiration. Were not men stimulated by such a pressing monitor, they might be apt to pursue other amusements with a perseverance beyond their power, and forget the useful hours of refreshment, in those tempting ones of pleasure. But hunger makes a demand that will not be refused; and, indeed, the generality of mankind seldom await the call.

Hunger has been supposed by some to arise from the rubbing of the coats of the stomach against each other, without having any intervening substance to prevent their painful attrition. Others have imagined that its juices, wanting their necessary supply, turn acrid, or as some say, pungent, and thus fret its internal coats, so as to produce a train of the most uneasy sensations. Boerhaave, who established his reputation in physic by uniting the conjectures of all those that preceded him, ascribes hunger to the united effects of both these causes, and asserts, that the pungency of the gastric juices, and the attrition of its coats against each other, cause those pains, which nothing but food can remove. These juices continuing still to be separated in the stomach, and every moment becoming more acrid, mix with the blood, and infect the circulation; the circulation being thus contaminated, becomes weaker, and more contracted; and the whole nervous frame sympathising, a hectic fever, and sometimes madness, is produced, in which state the faint wretch expires. In this manner, the man who dies of hunger may be said to be poisoned by the juices of his own body, and is destroyed less by the want of nourishment than by the vitiated qualities of that which he had already taken.

However this may be, we have but few instances of men dying, except at sea, of absolute hunger. The decline of those unhappy creatures who are destitute of food, at land, being more slow and unperceived. These, from often being in need, and as often receiving an accidental supply, pass their lives between surfeiting and repining; and their constitution is impaired by insensible degrees. Man is unfit for a state of precarious expectation. That share of provident precaution which incites him to lay up stores for a distant day, becomes his torment, when totally unprovided against an immediate call. The lower race of animals, when satisfied, for the instant moment, are perfectly happy: but it is otherwise with man; his mind anticipates distress, and feels the pangs of want even before it arrests him. Thus the mind, being continually harassed by the situation, it at length influences the constitution, and unfits it for all its functions. Some cruel disorder, but no way like hunger, seizes the unhappy sufferer; so that almost all those men who have thus long lived by chance, and whose every day may be considered as a happy escape from famine, are known at last to die in reality, of a disorder caused by hunger, but which, in the common language, is often called a *broken heart*. Some of these I have known myself, when very little able to relieve them: and I have been told by a very active and worthy magistrate, that the number of such as die in London for want, is much greater than one would imagine—I think he talked of two thousand in a year!

But how numerous soever those who die of hunger may be, many times greater, on the other hand, are the number of those who die by repletion. It is not the province of the present page to speculate with the physician upon the danger of surfeits; or with the moralist, upon the nauseousness of gluttony; it will only be proper to observe, that as nothing is so prejudicial to health as hunger by constraint, so nothing is more beneficial to the constitution than voluntary abstinence. It was not without reason that religion enjoined this duty; since it answered the double purpose of restoring the health oppressed by luxury, and diminished the consumption of provisions, so that a part might come to the poor. It should be the business of the legislature, therefore, to enforce this divine precept; and thus, by restraining one part of mankind in the use of their superfluities, to consult for the benefit of those who want the necessaries of life. The injunctions for abstinence are strict over the whole continent; and were rigorously observed, even among ourselves, for a long time after the Reformation. Queen Elizabeth, by giving her commands upon this head the air of a political injunction, lessened, in a great measure, and, in my opinion, very unwisely, the religious force of the obligation. She enjoined that her subjects should fast from flesh on Fridays and Saturdays, but at the same time declared, that this was not commanded from motives of religion, as if there were any differences in meats, but merely to favour the consumption of fish, and thus to multiply the number of mariners, and also to spare the stock of sheep, which might be more beneficial in another way. In this manner the injunction defeated its own force; and this most salutary law became no longer binding, when it was supposed to come purely from man. How far it may be enjoined in the scriptures, I will not take upon me to say; but this may be asserted, that if the utmost benefit to the individual, and the most extensive advantage to society, serve to mark any institution as of Heaven, this of abstinence may be reckoned among the foremost.

Were we to give a history of the various benefits that have arisen from this command, and how conducive it has been to long life, the instances would fatigue with their multiplicity. It is surprising to what a great age the primitive Christians of the East, who retired from persecution in the deserts of Arabia, continued to live in all the bloom of health, and yet all the rigours of abstemious discipline. Their common allowance, as we are told, for four and twenty hours, was twelve ounces of bread, and nothing but water. On this simple beverage St. Anthony is said to have lived a hundred and five years; James the hermit, a hundred and four; Arsenius, tutor to the emperor Arcadius, a hundred and twenty; St. Epiphanius, a hundred and fifteen; Simeon, a hundred and twelve; and Rombald, a hundred and twenty. In this manner did these holy, temperate men live to an extreme old age, kept cheerful by strong hopes, and healthful by moderate labour.

Abstinence which is thus voluntary, may be much more easily supported than constrained hunger. Man is said to live without food for seven days, which is the usual limit assigned him; and perhaps in a state of constraint, this is the longest time he can survive the want of it. But in cases of voluntary abstinence, of sickness, or sleeping, he has been known to live much longer.

In the records of the Tower, there is an account of a Scotchman, imprisoned for felony, who, for the space of six weeks, took not the least sustenance, being exactly watched during the whole time; and for this he received the king's pardon.

When the American Indians undertake long journies, and when, consequently, a stock of provisions sufficient to support them the whole way, would be more than they could carry, in order to obviate this inconvenience, instead of carrying the necessary quantity, the contrive a method of palliating their hunger by swallowing pills made of calcined shells and tobacco. These pills take away all appetite, by producing a temporary disorder in the stomach; and, no doubt, the frequent repetition of this wretched expedient must at last be fatal. By this means, however, they continue several days without eating, cheerfully bearing such extremes of fatigue and watching, as would quickly destroy men bred up in a greater state of delicacy. For those arts by which we learn to obviate our necessities, do not fail to unfit us for their accidental encounter.

Upon the whole, therefore, man is less able to support hunger than any other animal; and he is not better qualified to support a state of watchfulness. Indeed, sleep seems much more necessary to him, than to any other creature: as, when awake, he may be said to exhaust a greater proportion of the nervous fluid, and consequently to stand in need of an adequate supply. Other animals, when most awake, are but little removed from a state of slumber; their feeble faculties, imprisoned in matter, and rather exerted by impulse than deliberation require sleep rather as a cessation from motion than from thinking. But it is otherwise with man; his ideas, fatigued with their various excursions, demand a cessation, not less than the body, from toil; and he is the only creature that seems to require sleep from double motives: not less for the refreshment of the mental, than of the bodily frame.

There are some lower animals, indeed, that seem to spend the greatest part of their lives in sleep; but, properly speaking, the sleep of such may be considered as a kind of death, and their waking a resurrection. Flies and insects are said to be asleep at a time when all the vital motions have ceased; without respiration, without any circulation of their juices, if cut in pieces, they do not awake, nor does any fluid ooze out at the wound. These may be considered rather as congealed than as sleeping animals; and their rest, during winter, rather as a cessation from life than a necessary refreshment; but in the higher races of animals, whose blood is not thus congealed and thawed by heat, these all bear the want of sleep much better than man; and some of them continue a long time without seeming to take any refreshment from it whatsoever.

But man is more feeble; he requires its due return; and if it fails to pay the accustomed visit, his whole frame is in a short time thrown into disorder; his appetite ceases; his spirits are dejected; his pulse becomes quicker and harder; and his mind, abridged of its slumbering visions, begins to adopt waking dreams. A thousand strange phantoms arise, which come and go without his will: these, which are transient in the beginning, at last take firm possession of the mind, which yields to their dominion, and after a long struggle, runs into confirmed madness. In that horrid state, the mind may be considered as a city without walls, open to every insult, and paying homage to every invader; every idea that then starts with any force, becomes a reality; and the reason, over-fatigued with its former importunities, makes no head against the tyrannical invasion, but submits to it from mere imbecility.

But it is happy for mankind, that this state of inquietude is seldom driven to an extreme; and that there are medicines which seldom fail to give relief. However, man finds it more difficult than any other animal to procure sleep; and some are obliged to court its approaches for several hours together before they incline to rest. It is in vain that all light is excluded; that all sounds are removed; that warmth and softness conspire to invite it: the restless and busy mind still retains its former activity; and reason, that wishes to lay down the reins, in spite of herself, is obliged to maintain them. In this disagreeable state, the mind passes from thought to thought, willing to lose the distinctness of perception by increasing the multitude of the images. At last, when the approaches of sleep are near, every object of the imagination begins to mix with that next it; their outlines become in a manner rounder; a part of their distinctions fade away; and sleep, that ensues, fashions out a dream from the remainder.

If then it should be asked from what cause this state of repose proceeds, or in what manner sleep thus binds us for several hours together, I must fairly confess my ignorance, although it is easy to tell what philosophers say upon the subject. Sleep, says one of them,* consists in a scarcity of spirits, by which the orifices or pores of the nerves in the brain, through which the spirits used to flow into the nerves, being no longer kept open by the frequency of the spirits, shut of themselves; thus the nerves, wanting a new supply of spirits, become lax, and unfit to convey any impression to the brain. All this, however, is explaining a very great obscurity by somewhat more obscure; leaving, therefore, those spirits to open and shut the entrances to the brain, let us be contented with simply enumerating the effects of sleep upon the human constitution.

In sleep, the whole nervous frame is relaxed, while the heart and the lungs seem more forcibly exerted. This fuller circulation produces also a swelling of the muscles, as they always find who sleep with ligatures on any part of their body. The increased circulation also, may be considered as a kind of exercise, which is continued through the frame; and, by this, the perspiration becomes more copious, although the appetite for food is entirely taken away. Too much sleep dulls the apprehension, weakens the memory, and unfits the body for labour. On the contrary, sleep too much abridged, emaciates the frame, produces melancholy, and consumes the constitution. It requires some care, therefore, to regulate the quantity of sleep, and just to take as much as will completely restore nature, without oppressing it. The poor, as Otway says, sleep little; forced, by their situation, to lengthen out their labour to their necessities, they have but a short interval for this pleasing refreshment; and I have ever been of opinion, that bodily labour demands a less quantity of sleep than mental. Labourers and artizans are generally satisfied with about seven hours; but I have known some scholars who usually slept nine, and perceived their faculties no way impaired by oversleeping.

The famous Philip Barretiere, who was considered as a prodigy of learning at the age of fourteen, was known to sleep regularly twelve

* Rohault.

hours in the twenty-four; the extreme activity of his mind, when awake, in some measure called for an adequate alternation of repose: and, I am apt to think, that when students stint themselves in this particular, they lessen the waking powers of the imagination, and weaken its most strenuous exertions. Animals that seldom think, as was said, can very easily dispense with sleep; and of men, such as think least, will, very probably, be satisfied with the smallest share. A life of study, it is well known, unfits the body for receiving this gentle refreshment; the approaches of sleep are driven off by thinking: when, therefore, it comes at last, we should not be too ready to interrupt its continuance.

Sleep is, indeed, to some, a very agreeable period of their existence and it has been a question in the schools, which was most happy, the man who was a beggar by night, and a king by day; or he who was a beggar by day, and a king by night? It is given in favour of the nightly monarch, by him who first started the question: "For the dream," says he, "gives the full enjoyment of the dignity, without its attendant inconveniences: while, on the other hand, the king, who supposes himself degraded, feels all the misery of his fallen fortune, without trying to find the comforts of his humble situation. Thus, by day, both states have their peculiar distresses: but, by night, the exalted beggar is perfectly blessed, and the king completely miserable." All this, however, is rather fanciful than just; the pleasure dreams can give us, seldom reaches to our waking pitch of happiness: the mind often, in the midst of its highest visionary satisfactions, demands of itself, whether it does not owe them to a dream; and frequently awakes with the reply.

But it is seldom, except in cases of the highest delight, or the most extreme uneasiness, that the mind has power thus to disengage itself from the dominion of fancy. In the ordinary course of its operations, it submits to those numberless phantastic images that succeed each other, and which, like many of our waking thoughts, are generally forgotten. Of these, however, if any, by their oddity, or their continuance, affect us strongly, they are then remembered; and there have been some who felt their impressions so strongly as to mistake them for realities, and to rank them among the past actions of their lives.

There are others upon whom dreams seem to have a very different effect, and who, without seeming to remember their impressions the next morning, have yet shown by their actions during sleep, that they were very powerfully impelled by their dominion. We have numberless instances of such persons, who, while asleep, have performed many of the ordinary duties to which they have been accustomed when waking; and, with a ridiculous industry, have completed by night what they failed doing by day. We are told, in the German Ephemerides, of a young student who, being enjoined a severe exercise by his tutor, went to bed despairing of accomplishing it. The next morning awaking, to his great surprise, he found the task fairly written out, and finished in his own hand-writing.

He was at first, as the account has it, induced to ascribe this strange production to the operation of an infernal agent; but his tutor, willing to examine the affair to the bottom, set him another exercise still more severe than the former, and took precautions to observe his conduct the whole night. The young gentleman, upon being so severely tasked, felt the same inquietude that he had done on the former occasion; went to bed gloomy and pensive, pondering on the next day's duty, and after some time, fell asleep. But shortly after, his tutor, who continued to observe him from a place that was concealed, was surprised to see him get up, and very deliberately go to the table, where he took out pen, ink, and paper, drew himself a chair, and sat very methodically to thinking: it seems, that his being asleep only served to strengthen the powers of his imagination; for he very quickly and easily went through the task assigned him, put his chair aside, and then returned to bed to take out the rest of his nap. What credit we are to give to this account, I will not pretend to determine; but this may be said, that the book from whence it is taken, has some good marks of veracity; for it is very learned, and very dull, and is written in a country noted, if not for truth, at least for want of invention.

The ridiculous history of Arlotto is well known, who has had a volume written, containing a narrative of the actions of his life, not one of which was performed while he was awake. He was an Italian Franciscan friar, extremely rigid in his manners, and remarkably devout and learned in his daily conversation. By night, however, and during his sleep, he played a very different character from what he did by day, and was often detected in very atrocious crimes. He was at one time detected in actually attempting a rape, and did not awake till the next morning, when he was surprised to find himself in the hands of justice. His brothers of the convent often watched him while he went very deliberately into the chapel, and there attempted to commit sacrilege. They sometimes permitted him to carry the chalice and the vestments away into his own chamber, and the next morning amused themselves at the poor man's consternation for what he had done. But of all his sleeping transgressions, that was the most ridiculous in which he was called to pray for the soul of a person departed. Arlotto, after having devoutly performed his duty, retired to a chamber which was shown him to rest; but there he had no sooner fallen asleep than he began to reflect that the dead body had got a ring upon one of the fingers, which might be useful to him: accordingly, with a pious resolution of stealing it, he went down, undressed as he was, into a room full of women, and, with great composure, endeavoured to seize the ring. The consequence was, that he was taken before the Inquisition for witchcraft; and the poor creature had like to have been condemned, till his peculiar character accidentally came to be known: however, he was ordered to remain for the rest of life in his own convent, and upon no account whatsoever to stir abroad.

What are we to say of such actions as these: or how account for this operation of the mind in dreaming? It should seem, that the imagination, by day, as well as by night, is always employed; and that, often against our wills, it intrudes where it is least commanded or desired. While awake, and in health, this busy principle cannot much delude us: it may build castles in the air, and raise a thousand phantoms before us; but we have every one of the senses alive to bear testimony to its falsehood. Our eyes show us that the prospect is not present; our hearing and our touch depose against its reality; and our taste and smelling are equally vigilant in detecting the impostor. Reason, therefore, at once gives judgment upon the cause, and the vagrant intruder, Imagination, is imprisoned, or banished from the mind. But in sleep it is otherwise; having, as much as possible, put our senses from their duty, having closed the eyes from seeing, and the ears, taste, and smelling, from their peculiar functions, and having diminished even the touch itself, by all the arts of softness, the imagination is then left to riot at large, and to lead the understanding without an opposer. Every incursive idea then becomes a reality and the mind, not having one power that can prove the illusion, takes them for truths. As in madness, the senses, from struggling with the imagination, are at length forced to submit; so in sleep, they seem for a while soothed into the like submission: the smallest violence exerted upon any one of them, however, rouses all the rest in their mutual defence; and the imagination, that had for a while told its thousand falsehoods, is totally driven away, or only permitted to pass under the custody of such as are every moment ready to detect its imposition.

CHAPTER VII.

OF SEEING.*

" HAVING mentioned the senses as correcting the errors of the imagination, and as forcing it, in some measure, to bring us just information, it will naturally follow, that we should examine the nature of those senses themselves: we shall thus be enabled to see how far they also impose on us, and how far they contribute to correct each other. Let it be observed, however, that in this we are neither giving a treatise of optics, or phonics, but an history of our own perceptions: and to those we chiefly confine ourselves."

The eyes very soon begin to be formed in the human embryo, and in the chicken also. Of all the parts which the animal has double, the eyes are produced the soonest, and appear the most prominent. It is true, indeed, that in viviparous animals, and particularly in man, they are not so large in proportion, at first, as in the oviparous kinds; nevertheless, they are more speedily developed, when they begin to appear, than any other parts of the body. It is the same with the organ of hearing; the little bones that compose the internal parts of the ear are entirely formed before the other bones, though much larger, have acquired any part of their growth or solidity. Hence it appears, that those parts of the body which are furnished with the

---

* This chapter is taken from M. Buffon. I believe the reader will readily excuse any apology; and, perhaps, may wish that I had taken this liberty much more frequently. What I add is marked, as in a former instance, with inverted commas.

greatest quantity of the nerves, are the first in forming. Thus the brain and the spinal marrow are the first seen begun in the embryo; and, in general, it may be said, that wherever the nerves go, or send their branches in great numbers, there the parts are soonest begun, and the most completely finished.

If we examine the eyes of a child some hours, or even some days after its birth, it will be easily discerned that it, as yet, makes no use of them. The humours of the organ not having acquired a sufficient consistence, the rays of light strike but confusedly upon the retina, or expansion of the nerves at the back of the eye. It is not till about a month after they are born, that children fix them upon objects; for, before that time, they turn them indiscriminately every where, without appearing to be affected by any. At six or seven weeks old, they plainly discover a choice in the objects of their attention; they fix their eyes upon the most brilliant colours, and seem peculiarly desirous of turning them towards the light. Hitherto, however, they only seem to fortify the organ for seeing distinctly; but they have still many illusions to correct.

The first great error in vision is, that the eye inverts every object: and it in reality appears to the child, until the touch has served to undeceive it, turned upside down. A second error in vision is, that every object appears double. The same object forms itself distinctly upon each eye, and is consequently seen twice. This error, also, can only be corrected by the touch; and although, in reality, every object we see appears inverted and double, yet the judgment and habit have so often corrected the sense, that we no longer submit to its imposition, but see every object in its just position, the very instant it appears. Were we, therefore, deprived of feeling, our eyes would not only misrepresent the situation, but also the number of all things around us.

To convince us that we see objects inverted, we have only to observe the manner in which images are represented, coming through a small hole, in a darkened room. If such a small hole be made in a dark room, so that no light can come in, but through it, all the objects without will be painted on the wall behind, but in an inverted position, their heads downwards. For as all the rays which pass from the different parts of the object without, cannot enter the hole in the same extent which they had in leaving the object; since, if so, they would require the aperture to be as large as the object; and, as each part, and every point of the object, sends forth the image of itself on every side, and the rays, which form these images, pass from all points of the object as from so many centres, so such only can pass through the small aperture as come in opposite directions. Thus the little aperture becomes a centre for the entire object; through which the rays from the upper parts, as well as from the lower parts of it, pass in converging directions; and, consequently, they must cross each other in the central point, and thus paint the objects behind, upon the wall, in an inverted position.

It is in like manner easy to conceive, that we see all objects double, whatever our present sensations may seem to tell us to the contrary. For, to convince us of this, we have only to compare the situation of any one object on shutting one eye, and then compare the same situation by shutting the other. If, for instance, we hold up a finger, and shut the right eye, we shall find it hide a certain part of the room; if again reshutting the other eye, we shall find that part of the room visible, and the finger seeming to cover a part of the room that had been visible before. If we open both eyes, however, the part covered will appear to lie between the two extremes. But, the truth is, we see the object our finger had covered, one image of it to the right, and the other to the left; but, from habit, suppose that we see but one image placed between both, our sense of feeling having corrected the errors of sight. And thus, also, if instead of two eyes we had two hundred, we should fancy the objects increased in proportion, until our sense had corrected the errors of another.

"The having two eyes might thus be said to be rather an inconvenience than a benefit, since one eye would answer the purposes of sight as well, and be less liable to illusion. But it is otherwise; two eyes greatly contribute, if not to distinct, at least extensive vision.* When an object is placed at a moderate distance, by the means of both eyes we see a larger share of it than we possibly could with one; the right eye seeing a greater portion of its right side, and the left eye of its correspondent side. Thus both eyes, in some measure, see round the object; and it is this that gives it, in nature, that bold relieve, or swelling, with which they appear, and which no painting, how exquisite soever, can attain to. The painter must be contented with shading on a flat surface; but the eyes, in observing nature, do not behold the shading only, but a part of the figure also, that lies behind these very shadings, which gives it that swelling which painters so ardently desire, but can never fully imitate.

"There is another defect, which either of the eyes, taken singly, would have, but which is corrected by having the organ double. In either eye there is a point, which has no vision whatsoever; so that if one of them only is employed in seeing, there is a part of the object to which it is always totally blind. This is that part of the optic nerve where its vein and artery run; which being insensible, that point of the object that is painted there must continue unseen. To be convinced of this, we have only to try a very easy experiment. If we take three black patches, and stick them upon a white wall, about a foot distant from each other, each about as high as the eye that is to observe them; then retiring six or seven feet back, and shutting one eye, by trying for some time, we shall find, that while we distinctly behold the black spots that are to the right and left, that which is in the middle remains totally unseen. Or, in other words, when we bring that part of the eye, where the optic artery runs, to fall upon the object, it will then become invisible. This defect, however, in either eye, is always corrected by both, since the part of the object that is unseen by one, will be very distinctly perceived by the other."

Beside the former defects, we can have no idea of distances from the sight, without the help of touch. Naturally, every object we see appears to be within our eyes; and a child, who has as yet made but

* Leonardo da Vinci.

little use of the sense of feeling, must suppose that every thing it sees makes a part of itself. Such objects are only seen more or less bulky as they approach or recede from its eyes; so that a fly that is near will appear larger than an ox at a distance. It is experience alone that can rectify this mistake; and a long acquaintance with the real size of every object, quickly assures us of the distance at which it is seen.—The last man in a file of soldiers appears in reality much less, perhaps ten times more diminutive, than the man next to us; however, we do not perceive this difference, but continue to think him of equal stature; for the numbers we have seen thus lessened by distance, and have found, by repeated experience, to be of the natural size when we come closer, instantly corrects the sense, and every object is perceived with nearly its natural proportion. But it is otherwise, if we observe objects in such situations as we have not had sufficient experience to correct the errors of the eye; if, for instance, we look at men from the top of a high steeple, they, in that case, appear very much diminished, as we have not had a habit of correcting the sense in that position.

Although a small degree of reflection will serve to convince us of the truth of these positions, it may not be amiss to strengthen them by an authority which cannot be disputed. Mr. Cheselden having couched a boy of thirteen of a cataract, who had hitherto been blind, and thus at once having restored him to sight, curiously marked the progress of his mind upon that occasion. This youth, though he had been till then incapable of seeing, yet was not totally blind, but could tell day from night, as persons in his situation always may. He could also, with a strong light, distinguish black from white, and either from the vivid colour of scarlet: however, he saw nothing of the form of bodies; and, without a bright light, not even colours themselves. He was, at first, couched only in one of his eyes; and when he saw for the first time, he was so far from judging of distances, that he supposed his eyes touched every object that he saw, in the same manner as his hands might be said to feel them. The objects that were most agreeable to him were such as were of plain surfaces and regular figures; though he could as yet make no judgment whatever of their different forms, nor give a reason why one pleased him more than another. Although he could form some idea of colours during his state of blindness, yet that was not sufficient to direct him at present; and he could scarcely be persuaded that the colours he now saw were the same with those he had formerly conceived such erroneous ideas of. He delighted most in green; but black objects, as if giving him an idea of his former blindness, he regarded with horror. He had, as was said, no idea of forms; and was unable to distinguish one object from another, though never so different. When those things were shown him, which he had been formerly familiarized to, by his feeling, he beheld them with earnestness, in order to remember them a second time; but, as he had too many to recollect at once, he forgot the greatest number; and for one he could tell, after seeing, there was a thousand he was totally unacquainted with. He was very much surprised to find, that those things and persons he loved best, were not the most beautiful to be seen; and even testified displeasure in not finding his parents so handsome as he conceived them to be. It was near two months before he could find that a picture resembled a solid body. Till then he only considered it as a flat surface, variously shadowed; but, when he began to perceive that these kind of shadings actually represented human beings, he then began to examine, by his touch, whether they had not the usual qualities of such bodies, and was greatly surprised to find, what he expected a very unequal surface, to be smooth and even. He was then shown a miniature picture of his father, which was contained in his mother's watch-case, and he readily perceived the resemblance; but asked, with great astonishment, how so large a face could be contained in so small a compass? It seemed as strange to him as if a bushel was contained in a pint vessel. At first, he could bear but a very small quantity of light and he saw every object much greater than the life; but, in proportion as he saw objects that were really large, he seemed to think the former were diminished; and although he knew the chamber where he was contained in the house, yet until he saw the latter, he could not be brought to conceive how a house could be larger than a chamber. Before the operation he had no great expectations from the pleasure he should receive from a new sense; he was only excited by the hopes of being able to read and write; he said, for instance, that he could have no greater pleasure in walking in the garden with his sight, than he had without it, for he walked there at his ease, and was acquainted with all the walks. He remarked also, with great justice, that his former blindness gave him one advantage over the rest of mankind, which was that of being able to walk in the night with confidence and security. But when he began to make use of his new sense, he seemed transported beyond measure. He said that every new object was a new source of delight, and that his pleasure was so great as to be past expression. About a year after, he was brought to Epsom, where there is a very fine prospect, with which he seemed greatly charmed; and he called the landscape before him a new method of seeing. He was couched in the other eye, a year after the former, and the operation succeeded equally well: when he saw with both eyes, he said that objects appeared to him twice as large as when he saw but with one; however, he did not see them doubled, or at least he showed no marks as if he saw them so. Mr. Cheselden mentions instances of many more that were restored to sight in this manner; they all seemed to concur in their perceptions with this youth; and they all seemed particularly embarassed in learning how to direct their eyes to the objects they wished to observe.

In this manner it is that our feeling corrects the sense of seeing, and that objects which appear of very different sizes, at different distances, are all reduced, by experience, to their natural standard. "But not the feeling only, but also the colour, and brightness of the object, contributes, in some measure, to assist us in forming an idea of the distance at which it appears.* Those which we see most strongly marked with light and shade, we readily know to be nearer than those

---

* Mr. Buffon gives a different theory, for which I must refer the reader to the original. That I have given, I take to be easy, and satisfactory enough.

on which the colours are more faintly spread, and that, in some measure, take a part of their hue from the air between us and them. Bright objects, also, are seen at a greater distance than such as are obscure; and, most probably, for this reason, that being less similar in colour to the air which interposes, their impressions are less effaced by it, and they continue more distinctly visible. Thus, a black and distant object is not seen so far off as a bright and glittering one. and a fire by night is seen much farther off than by day."

The power of seeing objects at a distance is very rarely equal in both eyes. When this inequality is in any great degree, the person so circumstanced then makes use only of one eye, shutting that which sees the least, and employing the other with all its power. And hence proceeds that awkward look which is known by the name of *strabism*

There are many reasons to induce us to think, that such as are near-sighted see objects larger than other persons; and yet the contrary is most certainly true, for they see them less. Mr. Buffon informs us that he himself is short-sighted, and that his left eye is stronger than his right. He has very frequently experienced, upon looking at any object, such as the letters of a book, that they appear less to the weakest eye; and that when he places the book, so as that the letters appear double, the images of the left eye, which is strongest, are greater than those of the right, which is the most feeble. He has examined several others, who were in similar circumstances, and has always found that the best eye saw every object the largest. This he ascribes to habit; for near-sighted people being accustomed to come close to the object, and view but a small part of it at a time, the habit ensues, when the whole of an object is seen, and it appears less to them than to others.

Infants having their eyes less than those of adults, must see objects also smaller in proportion. For the image formed on the back of the eye will be large, as the eye is capacious; and infants, having it not so great, cannot have so large a picture of the object. This may be a reason also why they are unable to see so distinctly, or at such distances as persons arrived at maturity.

Old men, on the contrary, see bodies close to them very indistinctly, but bodies at a great distance from them with more precision; and this may happen from an alteration in the coats, or, perhaps, humours of the eye; and not, as is supposed, from their diminution. The cornea, for instance, may become too rigid to adapt itself, and take a proper convexity for seeing minute objects; and its very flatness will be sufficient to fit it for distant vision.

When we cast our eyes upon an object extremely brilliant, or when we fix and detain them too long upon the same object, the organ is hurt and fatigued, its vision becomes indistinct, and the image of the body, which has thus too violently, or too perseveringly employed us, is painted upon every thing we look at, and mixes with every object that occurs. "And this is an obvious consequence of the eye taking in too much ight, either immediately, or by reflection. Every body exposed to the light, for a time, drinks in a quantity of its rays, which, being brought into darkness, it cannot instantly discharge. Thus the hand, if it be exposed to broad day-light for some time, and then remediately snatched into a dark room, will appear still luminous: and it will be some time before it is totally darkened. It is thus with the eye; which, either by an instant gaze at the sun, or a steady continuance upon some less brilliant object, has taken in too much light; its humours are, for a while, unfit for vision, until that be discharged, and room made for rays of a milder nature." How dangerous the looking upon bright and luminous objects is to the sight, may be easily seen, from such as live in countries covered for most part of the year with snow, who become generally blind before their time. Travellers who cross these countries, are obliged to wear a crape before their eyes, to save their eyes, which would otherwise be rendered totally unserviceable; and it is equally dangerous in the sandy plains of Africa. The reflection of the light is there so strong, that it is impossible to sustain the effect, without incurring the danger of losing one's sight entirely. Such persons, therefore, as read or write for any continuance, should choose a moderate light, in order to save their eyes; and, although it may seem insufficient at first, the eye will accustom itself to the shade, by degrees, and be less hurt by the want of light than the excess.

"It is, indeed, surprising how far the eye can accommodate itself to darkness, and make the best of a gloomy situation. When first taken from the light, and brought into a dark room, all things disappear; or, if any thing is seen, it is only the remaining radiations that still continue in the eye. But, after a very little time, when these are spent, the eye takes the advantage of the smallest ray that happens to enter; and this alone would, in time, serve for many of the purposes of life. There was a gentleman of great courage and understanding, who was a major under King Charles I.; this unfortunate man, sharing in his master's misfortunes, and being forced abroad, ventured at Madrid to do his king a signal service; but, unluckily, failed in the attempt. In consequence of this, he was instantly ordered to a dark and dismal dungeon, into which the light never entered, and into which there was no opening but by a hole at the top, down which the keeper put his provisions, and presently closed it again on the other side. In this manner the unfortunate loyalist continued for some weeks, distressed and disconsolate; but at last he began to think he saw some little glimmering of light. This internal dawn seemed to increase from time to time, so that he could not only discover the parts of his bed, and such other large objects, but, at length, he even began to perceive the mice that frequented his cell; and saw them as they ran about the floor, eating the crumbs of bread that happened to fall. After some months' confinement he was at last set free; but such was the effect of the darkness upon him, that he could not for some days venture to leave his dungeon, but was obliged to accustom himself by degrees to the light of the day.

CHAPTER VIII

ON HEARING.*

As the sense of hearing, as well as of sight, gives us notice of remote objects, so, like that, it is subject to similar errors, being capable of imposing on us upon all occasions, where we cannot rectify it by the sense of feeling. We can have from it no distinct intelligence of the distance from whence a sounding body is heard; a great noise far off, and a small one very near, produce the same sensation; and, unless we receive information from some other sense, we can never distinctly tell whether the sound be a great or a small one. It is not till we have learned, by experience, that the particular sound which is heard, is of a peculiar kind; then we can judge of the distance from whence we hear it. When we know the tone of the bell, we can then judge how far it is from us.

Every body that strikes against another produces a sound, which is simple, and but one in bodies which are not elastic, but which is often repeated in such as are. If we strike a bell, or a stretched string, for instance, which are both elastic, a single blow produces a sound, which is repeated by the undulations of the sonorous body, and which is multiplied as often as it happens to undulate or vibrate. These undulations each strike their own peculiar blow; but they succeed so fast, one behind the other, that the ear supposes them one continued sound: whereas, in reality, they make many. A person who should, for the first time, hear the toll of the bell, would, very probably, be able to distinguish these breaks of sound; and, in fact, we can readily ourselves perceive an intension and remission in the sound.

In this manner, sounding bodies are of two kinds; those unelastic ones, which being struck, return but a single sound; and those more elastic, returning a succession of sound; which uniting together form a tone. This tone may be considered as a great number of sounds, all produced one after the other, by the same body, as we find in a bell, or the string of a harpsichord, which continues to sound for some time after it is struck. A continuing tone may be also produced from a non-elastic body, by repeating the blow quick and often, as when we beat a drum, or when we draw a bow along the string of a fiddle.

Considering the subject in this light, if we should multiply the number of blows, or repeat them at quicker intervals upon the sounding body, as upon the drum, for instance, it is evident that this will have no effect in altering the tone; it will only make it either more even, or more distinct. But it is otherwise, if we increase the force of the blow: if we strike the body with double weight, this will produce a tone twice as loud as the former. If, for instance, I strike a table with a switch, this will be very different from the sound produced by striking 't with a cudgel. Hence, therefore, we may infer, that all bodies give a louder and graver tone, not in proportion to the number of times they are struck, but in proportion to the force that strikes

---

* This chapter is taken from Mr. Buffon, except where marked by inverted commas

.hem And, if this be so, those philosophers who make the tone of a sonorous body, of a bell, or the string of a harpsichord, for instance, to depend upon the number only of its vibrations, and not the force, have mistaken what is only an effect for a cause. A bell, or an elastic string, can only be considered as a drum beaten; and the frequency of the blows can make no alteration whatever in the tone. The largest bells, and the longest and thickest strings, have the most forceful vibrations; and, therefore, their tones are the most loud and the most grave.

To know the manner in which sounds thus produced become pleasing, it must be observed, no one continuing tone, how loud and swelling soever, can give us satisfaction; we must have a succession of them, and those in the most pleasing proportion. The nature of this proportion may be thus conceived. If we strike a body incapable of vibration with a double force, or, what amounts to the same thing, with a double mass of matter, it will produce a sound that will be doubly grave. Music has been said, by the ancients, to have been first invented from the blows of different hammers on an anvil. Suppose then we strike an anvil with a hammer of one pound weight, and again with a hammer of two pounds, it is plain that the two pound hammer will produce a sound twice as grave as the former. But if we strike with a two pound hammer, and then with a three pound, it is evident that the latter will produce a sound one third more grave than the former. If we strike the anvil with a three pound hammer, and then with a four pound, it will likewise follow that the latter will be a quarter part more grave than the former. Now, in the comparing between all those sounds, it is obvious that the difference between one and two is more easily perceived, than between two and three, three and four, or any numbers succeeding in the same proportion. The succession of sounds will be, therefore, pleasing in proportion to the ease with which they may be distinguished.—That sound which is double the former, or, in other words, the octave to the preceding tone, will, of all others, be the most pleasing harmony. The next to that, which is as two to three, or, in other words, the third, will be most agreeable. And thus, universally, those sounds whose difference may be most easily compared, are the most agreeable.

" Musicians, therefore, have contented themselves with seven different proportions of sound, which are called *notes*, and which sufficiently answer all the purposes of pleasure. Not but that they might adopt a greater diversity of proportions; and some have actually done so; but, in these, the differences of the proportion are so imperceptible, that the ear is rather fatigued than pleased in making the distinction. In order, however, to give variety, they have admitted half tones; but in all the countries where music is yet in its infancy, they have rejected such; and they can find music in none but the obvious ones. The Chinese, for instance, have neither flats nor sharps in their music; but the intervals between their other notes, are in the same proportion with ours.

" Many more barbarous nations have their peculiar instruments of music; and what is remarkable, the proportion between their notes is in all the same as in ours. This is not the place for entering into the nature of these sounds, their effects upon the air, or their consonances with each other. We are not now giving a history of sound, but of human perception.

"All countries are pleased with music; and if they have not skill enough to produce harmony, at least they seem willing to substitute noise. Without all question, noise alone is sufficient to operate powerfully on the spirits; and, if the mind be already predisposed to joy, I have seldom found noise fail of increasing it into rapture. The mind feels a kind of distracted pleasure in such powerful sounds, braces up every nerve, and riots in the excess. But, as in the eye, an immediate gaze upon the sun will disturb the organs, so, in the ear, a loud unexpected noise disorders the whole frame, and sometimes disturbs the sense ever after. The mind must have time to prepare for the expected shock, and to give its organs the proper tension for its arrival.

"Musical sounds, however, seem of a different kind. Those are generally most pleasing which are most unexpected. It is not from bracing up the nerves, but from the grateful succession of the sounds, that these become so charming. There are few, how indifferent soever, but have at times felt their pleasing impression; and, perhaps, even those who have stood out against the powerful persuasion of sounds, only wanted the proper tune, or the proper instrument, to allure them.

"The ancients give us a thousand strange instances of the effects of music upon men and animals. The story of Arion's harp that gathered the dolphins to the ship side, is well known; and what is remarkable, Schotteus assures us,* that he saw a similar instance of fishes being allured by music. They tell us of diseases that have been cured, unchastity corrected, seditions quelled, passions removed, and sometimes excited even to madness. Dr. Wallis has endeavoured to account for these surprising effects, by ascribing them to the novelty of the art. For my own part, I can scarce hesitate to impute them to the exaggeration of the writers. They are as hyperbolical in the effects of their oratory; and yet, we well know, there is nothing in the orations which they have left us, capable of exciting madness, or of raising the mind to that ungovernable degree of fury which they describe. As they have exaggerated, therefore, in one instance, we may naturally suppose that they have done the same in the other; and, indeed, from the few remains we have of their music, collected by Meibomius, one might be apt to suppose, there was nothing very powerful in what is lost. Nor does any one of the ancient instruments, such as we see them represented in statues, appear comparable to our fiddle.

"However this be, we have many odd accounts, not only among them, but the moderns, of the power of music; and it must not be denied, but that, on some particular occasions, musical sounds may have a very powerful effect. I have seen all the horses and cows in a field, where there were above a hundred, gather round a person that was blowing the French horn, and seeming to testify an awkward

---

* Quod oculis meis spectavi. Schotti Magic. universalis, pars ii. L 1 p. 26.

kind of satisfaction. Dogs are well known to be very sensible of different tones in music; and I have sometimes heard them sustain a very ridiculous part in a concert, where their assistance was neither expected nor desired.

"We are told of Henry IV. of Denmark,* that being one day desirous of trying in person whether a musician, who boasted that he could excite men to madness, was not an impostor, he submitted to the operation of his skill: but the consequence was much more terrible than he expected; for, becoming actually mad, he killed four of his attendants in the midst of his transports. A contrary effect of music we have† in the cure of a madman of Alais, in France, by music. This man, who was a dancing master, after a fever of five days, grew furious, and so ungovernable that his hands were obliged to be tied to his sides: what at first was rage, in a short time was converted into silent melancholy, which no arts could exhilirate, nor no medicine remove. In this sullen and dejected state, an old acquaintance accidentally came to inquire after his health; he found him sitting up in bed, tied, and totally regardless of every external object round him. Happening, however, to take up a fiddle that lay in the room, and touching a favourite air, the poor madman instantly seemed to brighten up at the sound; from a recumbent posture, he began to sit up; and, as the musician continued playing, the patient seemed desirous of dancing to the sound; but he was tied, and incapable of leaving his bed, so that he could only humour the tune with his head, and those parts of his arms which were at liberty. Thus the other continued playing, and the dancing-master practised his own art, as far as he was able, for about a quarter of an hour, when suddenly falling into a deep sleep, in which his disorder came to a crisis, he awaked perfectly recovered.

"A thousand other instances might be added, equally true: let it suffice to add one more, which is not true; I mean that of the tarantula. Every person who has been in Italy now well knows that the bite of this animal, and its being cured by music, is all a deception When strangers come into that part of the country, the country people are ready enough to take money for dancing to the tarantula. A friend of mine had a servant who suffered himself to be bit; the wound, which was little larger than the puncture of a pin, was uneasy for a few hours, and then became well without any farther assistance. Some of the country people, however, still make a tolerable livelihood of the credulity of strangers, as the musician finds his account in it no less than the dancer."

Sounds, like light, are not only extensively diffused, but are frequently reflected. The laws of this reflection, it is true, are not as well understood as those of light; all we know is, that sound is principally reflected by hard bodies; and their being hollow, also, sometimes increases the reverberation. "No art, however, can make an echo; and some who have bestowed great labour and expense upon such a project, have only erected shapeless buildings, whose silence was a mortifying lecture upon their presumption."

---

* Olai Magni, l. 15. hist. c. 28 † Hist. de l'Acad. 1708. p. 21.

The internal cavity of the ear seems to be fitted up for the purpose of echoing sound with the greatest precision. This part is fashioned out in the temporal bone, like a cavern cut into a rock. " In this the sound is repeated and articulated; and, as some anatomists tell us, (for we have as yet but very little knowledge on this subject,) is beaten against the tympanum, or drum of the ear, which moves four little bones joined thereto; and these move and agitate the internal air which lies on the other side; and lastly, this air strikes and affects the auditory nerves, which carry the sound to the brain."

One of the most common disorders in old age is deafness, which probably proceeds from the rigidity of the nerves in the labyrinth of the ear. This disorder, also, sometimes proceeds from a stoppage of the wax, which art may easily remedy. In order to know whether the defect be an internal or an external one, let the deaf person put a repeating-watch into his mouth, and if he hears it strike, he may be assured that his disorder proceeds from an external cause, and is, in some measure, curable : " for there is a passage from the ears into the mouth, by what anatomists call the *eustachian tube;* and, by this passage, people often hear sounds, when they are utterly without hearing through the larger channel : and this also is the reason that we often see persons who listen with great attention, hearken with their mouths open, in order to catch all the sound at every aperture."

It often happens that persons hear differently with one ear from the other; and it is generally found that these have what is called by musicians, *a bad ear.* Mr. Buffon, who has made many trials upon persons of this kind, always found that their defect in judging properly of sounds proceeded from the inequality of their ears; and receiving by both at the same time unequal sensations, they form an unjust idea. In this manner, as those people hear falsely, they also, without knowing it, sing false. Those persons also frequently deceive themselves with regard to the side from whence the sound comes, generally supposing the noise to come on the part of the best ear.

Such as are hard of hearing find the same advantage in the trumpet made for this purpose, that short-sighted persons do from glasses. These trumpets might be easily improved, so as to increase sounds, in the same manner that the telescope does objects; however, they could be used to advantage only in a place of solitude and stillness, as the neighbouring sounds would mix with the more distant, and the whole would produce in the ear nothing but tumult and confusion.

Hearing is a much more necessary sense to man than to animals. With these it is only a warning against danger, or an encouragement to mutual assistance. In man, it is the source of most of his pleasures, and without which the rest of his senses would be of little benefit. A man born deaf, must necessarily be dumb; and his whole sphere of knowledge must be bounded only by sensual objects. We have an instance of a young man, who being born deaf, was restored at the age of twenty-four to perfect hearing: the account is given in the Memoires of the Academy of Sciences, 1703, page 18.

A young man of the town of Chartres, between the age of twenty three and twenty-four, the son of a tradesman, and deaf and dumb from his birth, began to speak all of a sudden, to the great astonish ment of the whole town. He gave them to understand, that about three or four months before, he had heard the sound of the bells for the first time, and was greatly surprised at this new and unknown sensation. After some time, a kind of water issued from his left ear, and he then heard perfectly well with both. During these three months, he was sedulously employed in listening, without saying a word, and accustoming himself to speak softly (so as not to be heard) the words pronounced by others. He laboured hard also in perfecting himself in the pronunciation, and in the ideas attached to every sound. At length, having supposed himself qualified to break silence, he declared that he could now speak, although as yet but imperfectly. Soon after, some able divines questioned him concerning his ideas of his past state, and principally with respect to God, his soul, the mortality or turpitude of actions. The young man, however, had not driven his solitary speculations into that channel. He had gone to mass indeed with his parents; had learned to sign himself with the cross; to kneel down and assume all the grimaces of a man that was praying; but he did all this without any manner of knowledge of the intention or the cause; he saw others do the like, and that was enough for him; he knew nothing even of death, and it never entered into his head; he led a life of pure animal instinct; entirely taken up with sensible objects, and such as were present, he did not seem even to make as many reflections upon these as might reasonably be expected from his improving situation; and yet the young man was not in want of understanding; but the understanding of a man deprived of all commerce with others, is so very confined, that the mind is in some measure totally under the control of its immediate sensations.

Notwithstanding, it is very possible to communicate ideas to deaf men, which they previously wanted, and even give them very precise notions of some abstract subjects, by means of signs, and of letters. A person born deaf, may, by time, and sufficient pains, be taught to write and read, to speak, and, by the motions of the lips, to understand what is said to him; however, it is probable that, as most of the motions of speech are made within the mouth by the tongue, the knowledge from the motion of the lips is but very confined: " nevertheless, I have conversed with a gentleman thus taught, and in all the commonly occurring questions, and the usual salutations, he was ready enough, merely by attending to the motion of the lips alone. When I ventured to speak for a short continuance, he was totally at a loss, although he understood the subject when written extremely well." Persons taught in this manner, were at first considered as prodigies; but there have been so many instances of success of late, and so many are skilful in the art of instructing in this way, that, though still a matter of some curiosity, it ceases to be an object of wonder.

CHAPTER IX.

OF SMELLING, FEELING, AND TASTING.

An animal may be said to fill up that sphere which he can reach by his senses, and is actually large in proportion to the sphere to which its organ extends. By sight, man's enjoyments are diffused into a wide circle; that of hearing, though less widely diffused, nevertheless extends his powers; the sense of smelling is more contracted still; and the taste and touch are the most confined of all. Thus man enjoys very distant objects but with one sense only; more nearly he brings two senses at once to bear upon them; his sense of smelling assists the other two at its own distance, and of such objects as a man, he may be said to be in perfect possession.

Each sense, however, the more it acts at a distance, the more capable it is of making combinations, and is consequently the more improveable. Refined imaginations, and men of strong minds, take more pleasure, therefore, in improving the delights of the distant senses, than in enjoying such as are scarce capable of improvement.

By combining the objects of the extensive senses, all the arts of poetry, painting, and harmony, have been discovered; but the closer senses, if I may so call them, such as smelling, tasting, and touching, are, in some measure, as simple as they are limited, and admit of little variety. The man of imagination makes a great and an artificial happiness, by the pleasure of altering and combining; the sensualist just stops where he began, and cultivates only those pleasures which he cannot improve. The sensualist is contented with those enjoyments that are already made to his hand; but the man of pleasure is best pleased with growing happiness.

Of all the senses, perhaps, there is not one in which man is more inferior to other animals than in that of smelling. With man, it is a sense that acts in a narrow sphere, and disgusts almost as frequently as it gives him pleasure. With many other animals it is diffused to a very great extent: and never seems to offend them. Dogs not only trace the steps of other animals, but also discover them by the scent at a very great distance; and while they are thus exquisitely sensible of all smells, they seem no way disgusted by any.

But, although this sense is, in general, so very inferior in man, it is much stronger in those nations that abstain from animal food, than among Europeans. The Bramins of India have a power of smelling, as I am informed, equal to what it is in most other creatures. They can smell the water which they drink, that to us seems quite inodorous; and have a word, in their language, which denotes a country of fine water. We are told, also, that the negroes of the Antilles, by the smell alone, can distinguish between the footsteps of a Frenchman and a negro. It is possible, therefore, that we may dull this organ by our luxurious way of living; and sacrifice to the pleasures of taste, those which might be received from perfume.

However, it is a sense that we can, in some measure, dispense with; and I have known many that wanted it entirely, with but very little inconvenience from its loss. In a state of nature it is said to be useful in guiding us to proper nourishment, and deterring us from that which is unwholesome; but, in our present situation, such information is but little wanted; and, indeed, but little attended to. In fact, the sense of smelling gives us very often false intelligence. Many things that have a disagreeable odour, are, nevertheless, wholesome and pleasant to the taste; and such as make eating an art, seldom think a meal fit to please the appetite till it begins to offend the nose. On the other hand, there are many things that smell most gratefully, and yet are noxious, or fatal to the constitution. Some physicians think that perfumes in general are unwholesome; that they relax the nerves, produce head-aches, and even retard digestion. The manchineel apple, which is known to be deadly poison, is possessed of the most grateful odour. Some of those mineral vapours that are often found fatal in the stomach, smell like the sweetest flowers, and continue thus to flatter till they destroy. This sense, therefore, as it should seem, was never meant to direct us in the choice of food, but appears rather as an attendant than a necessary pleasure.

Indeed, if we examine the natives of different countries, or even different natives of the same, we shall find no pleasure in which they differ so widely as that of smelling. Some persons are pleased with the smell of a rose; while I have known others that could not abide to have it approach them. The savage nations are highly delighted with the smell of assafœtida, which is to us the most nauseous stink in nature. It would in a manner seem that our delight in perfumes was made by habit; and that a very little industry could bring us totally to invert the perception of odours.

Thus much is certain, that many bodies which at one distance are an agreeable perfume, when nearer are a most ungrateful odour. Musk and ambergrise, in small quantities, are considered by most persons as highly fragrant; and yet, when in larger masses, their scent is insufferable. From a mixture of two bodies, each whereof is, of itself, void of all smell, a very powerful smell may be drawn. Thus, by grinding quick-lime with sal-ammoniac, may be produced a very fœtid mixture. On the contrary, from a mixture of two bodies, that are separately disagreeable, a very pleasant aromatic odour may be gained. A mixture of aqua-fortis with spirit of wine produces this effect. But not only the alterations of bodies by each other, but the smallest change in us, makes a very great alteration in this sense, and frequently deprives us of it totally. A slight cold often hinders us from smelling; and as often changes the nature of odours. Some persons, from disorder, retain an incurable aversion to those smells which most pleased them before: and many have been known to have an antipathy to some animals, whose presence they instantly perceive by the smell. From all this, therefore, the sense of smelling appears to be an uncertain monitor, easily disordered, and not much missed when totally wanting.

The sense most nearly allied to smelling is that of tasting. This, some have been willing to consider merely as a nicer kind of touch, and have undertaken to account, in a very mechanical manner, for the difference of savours. "Such bodies," said they, "as are pointed, happening to be applied to the papillæ of the tongue, excite a very powerful sensation, and give us the idea of saltness. Such, on the contrary, as are of a rounder figure, slide smoothly along the papillæ, and are perceived to be sweet." In this manner they have, with minute labour, gone through the variety of imagined forms in bodies, and have given them as imaginary effects. All we can precisely determine upon the nature of tastes is, that the bodies to be tasted must be either somewhat moistened, or, in some measure, dissolved by the saliva, before they can produce a proper sensation: when both the tongue itself, and the body to be tasted, are extremely dry, no taste whatever ensues. The sensation is then changed; and the tongue, instead of tasting, can only be said, like any other part of the body, to feel the object.

It is for this reason that children have a stronger relish of tastes than those who are more advanced in life. This organ with them, from the greater moisture of their bodies, is kept in greater perfection; and is, consequently, better adapted to perform its functions. Every person remembers how great a pleasure he found in sweets, while a child; but his taste growing more obtuse with age, he is obliged to use artificial means to excite it. It is then that he is found to call in the assistance of poignant sauces, and strong relishes of salts and aromatics; all which the delicacy of his tender organ, in childhood, were unable to endure. His taste grows callous to the natural relishes, and is artificially formed to others more unnatural; so that the highest epicure may be said to have the most depraved taste; as it is owing to the bluntness of his organ, that he is obliged to have recourse to such a variety of expedients to gratify his appetite.

As smells are often rendered agreeable by habit, so also tastes may be. Tobacco and coffee, so pleasing to many, are yet, at first, very disagreeable to all. It is not without perseverance that we begin to have a relish for them; we force nature so long, that what was constraint in the beginning, at last becomes inclination.

The grossest, and yet the most useful of all the senses, is that of feeling. We are often seen to survive under the loss of the rest; but of this we can never be totally deprived, but with life. Although this sense is diffused over all parts of the body, yet it most frequently happens that those parts which are most exercised in touching, acquire the greatest degree of accuracy. Thus the fingers, by long habit, become greater masters in the art than any others, even where the sensation is more delicate and fine.[*] It is from this habit, therefore, and their peculiar formation, and not, as is supposed, from their being furnished with a greater quantity of nerves, that the fingers are thus perfectly qualified to judge of forms. Blind men, who are obliged to use them much oftener, have this sense much finer; so that the delicacy of the touch arises rather from the habit of constantly employing the fingers, than from any fancied nervousness in their conformation.

All animals that are furnished with hands[†] seem to have more understanding than others. Monkeys have so many actions like those of men, that they appear to have similar ideas of the form of bodies

---

[*] Buffon, vol. vi. p. 80.   [†] Ibid. vol. v. p. 22.

All other creatures, deprived of hands, can have no distinct ideas of the shape of the objects by which they are surrounded, as they want this organ, which serves to examine and measure their forms, their risings, and depressions. A quadruped, probably, conceives as erroneous an idea of any thing near him, as a child would of a rock or a mountain that it beheld at a distance. It may be for this reason, that we often see them frighted at things with which they ought to be better acquainted. Fishes, whose bodies are covered with scales, and who have no organs for feeling, must be the most stupid of all animals. Serpents, that are likewise destitute, are yet, by winding round several bodies, better capable of judging of their form. All these, however, can have but very imperfect ideas from feeling; and we have already seen, when deprived of this sense, how little the rest of the senses are to be relied on.

The feeling, therefore, is the guardian, the judge, and the examiner of all the rest of the senses. It establishes their information, and detects their errors. All the other senses are altered by time, and contradict their former evidence; but the touch still continues the same; and, though extremely confined in its operations, yet it is never found to deceive. The universe, to a man who had only used the rest of his senses, would be but a scene of illusion; every object misrepresented, and all its properties unknown. Mr. Buffon has imagined a man just newly brought into existence, describing the illusion of his first sensations, and pointing out the steps by which he arrived at reality. He considers him as just created, and awaking amidst the productions of nature; and, to animate the narrative still more strongly, has made his philosophical man a speaker. The reader will no doubt recollect Adam's speech in Milton as being similar. All that I can say to obviate the imputation of plagiarism is, that the one treats the subject more as a poet, the other more as a philosopher. The philosopher's man describes his first sensations in the following manner.*

I well remember that joyful anxious moment when I first became acquainted with my own existence. I was quite ignorant of what I was, how I was produced, or from whence I came. I opened my eyes: what an addition to my surprise! the light of the day, the azure vault of heaven, the verdure of the earth, the crystal of the waters, all employed me at once, and animated and filled me with inexpressible delight. I at first imagined that all those objects were within me, and made a part of myself.

Impressed with this idea, I turned my eyes to the sun; its splendour dazzled and overpowered me; I shut them once more; and, to my great concern, I supposed that during this short interval of darkness, I was again returning to nothing.

Afflicted, seized with astonishment, I pondered a moment on this great change, when I heard a variety of unexpected sounds. The whistling of the wind, and the melody of the groves, formed a concert, the soft cadence of which sunk upon my soul. I listened for some time, and was persuaded that all this music was within me.

* Buffon, vol. vi. p. 88.

Quite occupied with this new kind of existence, I had already forgotten the light which was my first inlet into life; when I once more opened my eyes, and found myself again in possession of my former happiness. The gratification of the two senses at once, was a pleasure too great for utterance.

I turned my eyes upon a thousand various objects; I soon found that I could lose them, and restore them at will; and amused myself more at leisure with a repetition of this new-made power.

I now began to gaze without emotion, and to hearken with tranquillity, when a light breeze, the freshness of which charmed me, wafted its perfumes to my sense of smelling, and gave me such satisfaction as even increased my self-love.

Agitated, roused by the various pleasures of my new existence, I instantly arose, and perceived myself moved along, as if by some unknown and secret power.

I had scarce proceeded forward, when the novelty of my situation once more rendered me immoveable. My surprise returned; I supposed that every object around me had been in motion: I gave to them that agitation which I produced by changing place; and the whole creation seemed once more in disorder.

I lifted my hand to my head; I touched my forehead; I felt my whole frame; I then supposed that my hand was the principal organ of my existence; all its informations were distinct and perfect; and so superior to the senses I had yet experienced, that I employed myself for some time in repeating its enjoyments: every part of my person I touched, seemed to touch my hand in turn, and gave back sensation for sensation.

I soon found that this faculty was expanded over the whole surface of my body; and I now first began to perceive the limits of my existence, which I had in the beginning supposed spread over all the objects I saw.

Upon casting my eyes upon my body, and surveying my own form, I thought it greater than all the objects that surrounded me. I gazed upon my person with pleasure; I examined the formation of my hand, and all its motions; it seemed to me large or little in proportion as I approached it to my eyes; I brought it very near, and it then hid almost every other object from my sight. I began soon, however, to find that my sight gave me uncertain information, and resolved to depend upon my feeling for redress.

This precaution was of the utmost service; I renewed my motions, and walked forward with my face turned towards the heavens. I happened to strike lightly against a palm-tree, and this renewed my surprise: I laid my hand on this strange body; it seemed replete with new wonders, for it did not return me sensation for sensation, as my former feelings had done. I perceived that there was something external, and which did not make a part of my own existence.

I now, therefore, resolved to touch whatever I saw, and vainly attempted to touch the sun; I stretched forth my arm, and felt only yielding air: at every effort, I fell from one surprise into another, for every object appeared equally near me; and it was not till after an infinity of trials that I found some objects farther removed than the rest

Amazed with the illusions, and the uncertainty of my state, I sat down beneath a tree; the most beautiful fruits hung upon it, within my reach; I stretched forth my hand, and they instantly separated from the branch. I was proud of being able to grasp a substance without me; I held them up, and their weight appeared to me like an animated power that endeavoured to draw them to the earth. I found a pleasure in conquering their resistance.

I held them near my eye; I considered their form and beauty; their fragrance still more allured me to bring them nearer; I approached them to my lips, and drank in their odours; the perfume invited my sense of tasting, and I soon tried a new sense—How new! how exquisite! Hitherto I had tasted only of pleasure; but now it was luxury. The power of tasting gave me the idea of possession.

Flattered with this new acquisition, I continued its exercise, till an agreeable languor stealing upon my mind, I felt all my limbs become heavy, and all my desires suspended. My sensations were now no longer vivid and distinct; but seemed to lose every object, and presented only feeble images, confusedly marked. At that instant I sunk upon the flowery bank, and slumber seized me. All now seemed once more lost to me. It was then as if I was returning into my former nothing. How long my sleep continued, I cannot tell; as I yet had no perception of time. My awaking appeared like a second birth; and I then perceived that I had ceased for a time to exist. This produced a new sensation of fear; and from this interruption in life, I began to conclude that I was not formed to exist for ever.

In this state of doubt and perplexity, I began to harbour new suspicions; and to fear that sleep had robbed me of some of my late powers; when, turning on one side, to resolve my doubts, what was my amazement, to behold another being, like myself, stretched by my side! New ideas now began to arise: new passions, as yet unperceived, with fears, and pleasures, all took possession of my mind, and prompted my curiosity; love served to complete that happiness which was begun in the individual; and every sense was gratified in all its varieties.

CHAPTER X.

OF OLD AGE AND DEATH.*

Every thing in nature has its improvement and decay. The human form is no sooner arrived at its state of perfection, than it begins to decline. The alteration is, at first, insensible; and, often, several years are elapsed before we find ourselves grown old. The news of this disagreeable change too generally comes from without; and we learn from others that we grow old, before we are willing to believe the report.

When the body has come to its full height, and is extended into its

---

* This chapter is taken from Mr. Buffon, except where it is marked by inverted commas.

just dimensions, it then also begins to receive an additional bulk, which rather loads than assists it. This is formed from fat; which generally, at the age of thirty-five or forty, covers all the muscles, and interrupts their activity. Every action is then performed with greater labour, and the increase of size only serves as a forerunner of decay.

The bones, also, become every day more solid. In the embryo they are as soft almost as the muscles of the flesh; but, by degrees, they harden, and acquire their natural vigour; but still, however, the circulation is carried on through them, and, how hard soever the bones may seem, yet the blood holds its current through them as through all other parts of the body. Of this we may be convinced, by an experiment, which was first accidentally discovered by our ingenious countryman Mr. Belcher. Perceiving at a friend's house, that the bones of hogs, which were fed upon madder, were red, he tried it upon various animals by mixing this root with their usual food; and he found that it tinctured the bones in all; an evident demonstration that the juices of the body had a circulation through the bones. He fed some animals alternately upon madder and their common food, for some time, and he found their bones tinctured with alternate layers, in conformity to their manner of living. From all this he naturally concluded, that the blood circulated through the bones, as it does through every other part of the body; and that, how solid soever they seemed, yet, like the softest parts, they were furnished through all their substance with their proper canals. Nevertheless, these canals are of very different capacities, during the different stages of life. In infancy they are capacious; and the blood flows almost as freely through the bones as through any other part of the body; in manhood their size is greatly diminished; the vessels are almost imperceptible; and the circulation through them is proportionably slow. But, in the decline of life, the blood, which flows through the bones, no longer contributing to their growth, must necessarily serve to increase their hardness. The channels that every where run through the human frame, may be compared to those pipes that we every where see crusted on the inside, by the water, for a long continuance, running through them. Both every day grow less and less, by the small rigid particles which are deposited within them. Thus as the vessels are by degrees diminished, the juices also, which were necessary for the circulation through them, are diminished in proportion; till at length, in old age, those props of the human frame are not only more solid but more brittle.

The cartilages, or gristles, which may be considered as bones beginning to be formed, grow also more rigid. The juices circulating through them, for there is a circulation through all parts of the body, every day contributes to render them harder; so that these substances, which in youth are elastic and pliant, in age become hard and bony As these cartilages are generally placed near the joints, the motion of the joints also must, of consequence, become more difficult. Thus, in old age, every action of the body is performed with labour; and the cartilages, formerly so supple, will now sooner break than bend.

" As the cartilages acquire hardness, and unfit the joints for motion, so also, that mucous liquor, which is always separated between the joints, and which serves, like oil to a hinge, to give them an easy and ready play, is now grown more scanty. It becomes thicker, and more clammy, more unfit for answering the purposes of motion; and from thence, in old age, every joint is not only stiff, but awkward. At every motion, this clammy liquor is heard to crack; and it is not without the greatest efforts of the muscles that its resistance is overcome. I have seen an old person, who never moved a single joint, that did not thus give notice of the violence done to it."

The membranes that cover the bones, the joints, and the rest of the body, become, as we grow old, more dense and more dry. These which surround the bones, soon cease to be ductile. The fibres, of which the muscles or flesh is composed, become every day more rigid; and, while to the touch the body seems, as we advance in years, to grow softer, it is, in reality, increasing in hardness. It is the skin, and not the flesh, that we feel upon such occasions. The fat, and the flabbiness of that, seems to give an appearance of softness, which the flesh itself is very far from having. There are few can doubt this, after trying the difference between the flesh of young and old animals. The first is soft and tender, the last is hard and dry.

The skin is the only part of the body that age does not contribute to harden. That stretches to every degree of tension; and we have horrid instances of its pliancy, in many disorders incident to humanity In youth, therefore, while the body is vigorous and increasing, it still gives way to its growth. But, although it thus adapts itself to our increase, it does not in the same manner conform to our decay. The skin, which in youth was filled and glossy, when the body begins to decline, has not elasticity enough to shrink entirely with its diminution. It hangs, therefore, in wrinkles, which no art can remove. The wrinkles of the body, in general, proceed from this cause. But those of the face seem to proceed from another; namely, from the many varieties of positions into which it is put by the speech, the food, or the passions. Every grimace, and every passion, wrinkles up the visage into different forms. These are visible enough in young persons; but what at first was accidental, or transitory, becomes unalterably fixed in the visage as it grows older. "From hence we may conclude, that a freedom from passions not only adds to the happiness of the mind, but preserves the beauty of the face; and the person that has not felt their influence, is less strongly marked by the decays of nature."

Hence, therefore, as we advance in age, the bones, the cartilages, the membranes, the flesh, the skin, and every fibre of the body, become more solid, more brittle, and more dry. Every part shrinks, every motion becomes more slow; the circulation of the fluids is performed with less freedom; perspiration diminishes; the secretions alter; the digestion becomes slow and laborious; and the juices no longer serving to convey their accustomed nourishment, those parts may be said to live no longer when the circulation ceases. Thus the body dies by little and little; all its functions are diminished by degrees; life is driven from one part of the frame to another; universal rigidity prevails; and death at last seizes upon the little that is left.

As the bones, the cartilages, the muscles, and all other parts of the body, are softer in women than in men, these parts must, of consequence, require a longer time to come to that hardness which hastens death. Women, therefore, ought to be a longer time in growing old than men; and this is actually the case. If we consult the tables which have been drawn up respecting human life, we shall find, that, after a certain age, they are more long-lived than men, all other circumstances the same. A woman of sixty has a better chance than a man of the same age to live till eighty. Upon the whole, we may infer, that such persons as have been slow in coming up to maturity, will also be slow in growing old; and this holds as well with regard to other animals as to man.

The whole duration of the life of either vegetables or animals may be, in some measure, determined from their manner of coming to maturity. The tree, or the animal, which takes but a short time to increase to its utmost pitch, perishes much sooner than such as are less premature. In both, the increase upwards is first accomplished; and not till they have acquired their greatest degree of height do they begin to spread in bulk. Man grows in stature till about the age of seventeen; but his body is not completely developed till about thirty. Dogs, on the other hand, are at their utmost size in a year, and become as bulky as they usually are in another. However, man, who is so long in growing, continues to live fourscore, or a hundred years; but the dog seldom above twelve or thirteen. In general, also, it may be said, that large animals live longer than little ones, as they usually take a longer time to grow. But in all animals, one thing is equally certain, that they carry the causes of their own decay about them; and that their deaths are necessary and inevitable. The prospects which some visionaries have formed of perpetuating life by remedies, have been often enough proved false by their own example. Such unaccountable schemes would, therefore, have died with them, had not the love of life always augmented our credulity.

When the body is naturally well formed, it is possible to lengthen out the period of life for some years by management. Temperance in diet is often found conducive to this end. The famous Cornaro, who lived to above a hundred years, although his constitution was naturally feeble, is a strong instance of the benefit of an abstemious life. Moderation in the passions also may contribute to extend the term of our existence. "Fontenelle, the celebrated writer, was naturally of a very weak and delicate habit of body. He was affected by the smallest irregularities; and had frequently suffered severe fits of illness from the slightest causes. But the remarkable equality of his temper, and his seeming want of passion, lengthened out his life to above a hundred. It was remarkable of him, that nothing could vex or make him uneasy; every occurrence seemed equally pleasing; and no event, however unfortunate, seemed to come unexpected." However, the term of life can be prolonged but for a very little time by any art we can use. We are told of men who have lived beyond the ordinary duration of human existence; such as Parr, who lived to a hundred and forty-four; and Jenkins to a hundred and sixty-five; yet these men used no peculiar arts to prolong life; on the contrary, it appears that these, as well as some others, remarkable for their longevity, were peasants accustomed to the greatest fatigues, who had no settled rules of diet, but who often indulged in accidental excesses. Indeed, if we consider that the European, the Negro, the Chinese, and the American, the civilized man and the savage, the rich and the poor, the inhabitant of the city and of the country, though all so different in other respects, are yet entirely similar in the period allotted them for living; if we consider that neither the difference of race, of climate, of nourishment, of convenience, or of soil, makes any difference in the term of life; if we consider that those men who live upon raw flesh or dried fishes, upon sago or rice, upon cassava or upon roots, nevertheless live as long as those who are fed upon bread and meat, we shall readily be brought to acknowledge, that the duration of life depends neither upon habit, customs, nor the quantity of food; we shall confess, that nothing can change the laws of that mechanism which regulates the number of our years, and which can chiefly be affected only by long fasting or great excess.

If there be any difference in the different periods of man's existence, it ought principally to be ascribed to the quality of the air. It has been observed, that in elevated situations there have been found more old people than in those that were low. The mountains of Scotland, Wales, Auvergne, and Switzerland, have furnished more instances of extreme old age, than the plains of Holland, Flanders, Germany, or Poland. But, in general, the duration of life is nearly the same in most countries. Man, if not cut off by accidental diseases, is often found to live to ninety or a hundred years. Our ancestors did not live beyond that date; and, since the times of David, this term has undergone little alteration.

If we be asked, how in the beginning men lived so much longer than at present, and by what means their lives were extended to nine hundred and thirty, or even nine hundred and sixty years; it may be answered, that the productions of the earth, upon which they fed, might be of a different nature at that time from what they are at present. " It may be answered, that the term was abridged by Divine command, in order to keep the earth from being overstocked with human inhabitants; since, if every person were now to live and generate for nine hundred years, mankind would be increased to such a degree, that there would be no room for subsistence: so that the plan of Providence would be altered; which is seen not to produce life, without providing a proper supply."

But to whatever extent life may be prolonged, or however some may have delayed the effects of age, death is the certain goal to which all are hastening. All the causes of decay which have been mentioned, contribute to bring on this dreaded dissolution. However, nature approaches to this awful period by slow and imperceptible degrees; life is consuming day after day; and some one of our faculties, or vital principles, is every hour dying before the rest; so that death is only the last shade in the picture; and it is probable that man suffers a greater change in going from youth to age, than from age into the grave. When we first begin to live, our lives may scarcely be said to be our own; as the child grows, life increases in the same proportion; and is at its height in the prime of manhood. But as soon as the body begins to decrease, life decreases also; for, as the human frame diminishes, and its juices circulate in smaller quantity, life diminishes and circulates with less vigour; so that as we begin to live by degrees, we begin to die in the same manner.

Why then should we fear death, if our lives have been such as not to make eternity dreadful? Why should we fear that moment, which is prepared by a thousand other moments of the same kind? the first pangs of sickness being probably greater than the last struggles of departure. Death, in most persons, is as calmly endured as the disorder that brings it on. If we inquire from those whose business it is to attend the sick and the dying, we shall find that, except in a very few acute cases, where the patient dies in agonies, the greatest number die quietly, and seemingly without pain: and even the agonies of the former rather terrify the spectators than torment the patient; for how many have we not seen who have been accidentally relieved from this extremity, and yet had no memory of what they then endured? In fact, they had ceased to live, during that time when they ceased to have sensation; and their pains were only those of which they had an idea.

The greatest number of mankind die, therefore, without sensation; and of those few that still preserve their faculties entire to the last moment, there is scarce one of them that does not also preserve the hopes of still out-living his disorder. Nature, for the happiness of man, has rendered this sentiment stronger than his reason. A person dying of an incurable disorder, which he must know to be so, by frequent examples of his case, which he perceives to be so, by the inquietude of all around him, by the tears of his friends, and the departure or the face of the physician, is, nevertheless, still in hopes of getting over it. His interest is so great, that he only attends to his own representations; the judgment of others is considered as a hasty conclusion, and while death every moment makes new inroads upon his constitution, and destroys life in some part, hope still seems to escape the universal ruin, and is the last that submits to the blow.

Cast your eyes upon a sick man, who has a hundred times told you that he felt himself dying—that he was convinced he could not recover, and that he was ready to expire; examine what passes on his visage, when, through zeal or indiscretion, any one comes to tell him that his end is at hand. You will see him change, like one who is told an unexpected piece of news. He now appears not to have thoroughly believed what he had been telling you himself; he doubted much, and his fears were greater than his hopes; but he still had some feeble expectations of living, and would not have seen the approaches of death, unless he had been alarmed by the mistaken assiduity of his attendants.

Death, therefore, is not that terrible thing which we suppose it to be. It is a spectre which frights us at a distance, but which disappears when we come to approach it more closely. Our ideas of its terrors are conceived in prejudice, and dressed up by fancy: we regard it not only as the greatest misfortune, but as also an evil accompanied with the most excruciating tortures: we have even increased our apprehensions, by reasoning on the extent of our sufferings. "It must be dreadful," say some, "since it is sufficient to separate the soul from the body; it must be long, since our sufferings are proportioned to the succession of our ideas; and these being painful, must succeed each other with extreme rapidity." In this manner has false philosophy laboured to augment the miseries of our nature; and to aggravate that period, which Nature has kindly covered with insensibility. Neither the mind nor the body can suffer these calamities; the mind is at that time mostly without ideas, and the body too much enfeebled, to be capable of perceiving its pain. A very acute pain produces either death, or fainting, which is a state similar to death: the body can suffer but to a certain degree; if the torture becomes excessive, it destroys itself; and the mind ceases to perceive, when the body can no longer endure.

In this manner excessive pain admits of no reflection; and wherever there are any signs of it, we may be sure that the sufferings of the patient are no greater than what we ourselves may have remembered to endure.

But, in the article of death, we have many instances in which the dying person has shown that very reflection which presupposes an absence of the greatest pain, and consequently that pang which ends life, cannot even be so great as those which have preceded. Thus, when Charles XII. was shot at the siege of Frederickshaldt, he was seen to clap his hand on the hilt of his sword; and although the blow was great enough to terminate one of the boldest and bravest lives in the world, yet it was not painful enough to destroy reflection. He perceived himself attacked; he reflected that he ought to defend himself, and his body obeyed the impulse of his mind, even in the last extremity. Thus it is the prejudice of persons in health, and not the body in pain, that makes us suffer from the approach of death: we have all our lives contracted a habit of making out excessive pleasures and pains; and nothing but repeated experience shows us how seldom the one can be suffered, or the other enjoyed, to the utmost.

If there be any thing necessary to confirm what we have said concerning the gradual cessation of life, or the insensible approaches of our end, nothing can more effectually prove it, than the uncertainty of the signs of death. If we consult what Winslow or Bruhier have said upon this subject, we shall be convinced, that between life and death, the shade is so very undistinguishable, that even all the powers of art can scarcely determine where the one ends, and the other begins. The colour of the visage, the warmth of the body, the suppleness of the joints, are but uncertain signs of life still subsisting; while, on the contrary, the paleness of the complexion, the coldness of the body, the stiffness of the extremities, the cessation of all motion, and the total insensibility of the parts, are but uncertain marks of death begun. In the same manner, also, with regard to the pulse and the breathing, these motions are often so kept under, that it is impossible to perceive them. By approaching a looking-glass to the mouth of the person supposed to be dead, people often expect to find whether he breathes or not. But this is a very uncertain experiment; the glass is frequently sullied by the vapour of the dead man's body, and often the person is still alive, although the glass is no way tarnished. In the same manner, neither burning nor scarifying, neither noises in the ears nor pungent spirits applied to the nostrils, give certain signs of the discontinuance of life; and there are many instances of persons who have endured them all, and afterwards recovered without any external assistance, to the astonishment of the spectators. How careful, therefore, should we be, before we commit those who are dearest to us to the grave, to be well assured of their departure: experience, justice, humanity, all persuade us not to hasten the funerals of our friends, but to keep their bodies unburied until we have certain signs of their real decease.

CHAPTER XI.

OF THE VARIETIES IN THE HUMAN RACE.

Hitherto we have compared man with other animals; we now come to compare men with each other. We have hitherto considered him as an individual, endowed with excellencies above the rest of the creation; we now come to consider the advantages which men have over men, and the various kinds with which our earth is inhabited.

If we compare the minute differences of mankind, there is scarce one nation upon the earth that entirely resembles another; and there may be said to be as many different kinds of men as there are countries inhabited. One polished nation does not differ more from another, than the merest savages do from those savages that lie even contiguous to them; and it frequently happens that a river, or a mountain, divides two barbarous tribes that are unlike each other in manners, customs, features, and complexion. But these differences, however perceivable, do not form such distinctions as come within a general picture of the varieties of mankind. Custom, accident, or fashion, may produce considerable alterations in neighbouring nations; their being derived from ancestors of a different climate, or complexion, may contribute to make accidental distinctions, which every day grow less; and it may be said, that two neighbouring nations, how unlike soever at first, will assimilate by degrees; and, by long continuance, the difference between them will at last become almost imperceptible. It is not, therefore, between contiguous nations we are to look for any strong marked varieties in the human species: it is by comparing the inhabitants of opposite climates and distant countries; those who live within the polar circle with those beneath the equator; those that live on one side of the globe with those that occupy the other.

Of all animals, the differences between mankind are the smallest. Of the lower races of creatures, the changes are so great as often entirely to disguise the natural animal, and to distort, or to disfigure its shape. But the chief differences in man are rather taken from the tincture of his skin than the variety of his figure; and in all climates he preserves his erect deportment, and the marked superiority of his form. If we look round the world, there seems to be not above six* distinct varieties in the human species, each of which is strongly marked, and speaks the kind seldom to have mixed with any other. But there is nothing in the shape, nothing in the faculties, that shows their coming from different originals; and the varieties of climate, of nourishment, and custom, are sufficient to produce every change.

The first distinct race of men is found round the polar regions. The Laplanders, the Esquimaux Indians, the Samoeid Tartars, the inhabitants of Nova Zembla, the Borandians, the Greenlanders, and the natives of Kamtschatka, may be considered as one peculiar race of people, all greatly resembling each other in their stature, their complexion, their customs, and their ignorance. These nations being under a rigorous climate, where the productions of nature are but few, and the provisions coarse and unwholesome, their bodies have shrunk to the nature of their food; and their complexions have suffered from cold almost a similar change to what heat is known to produce; their colour being a deep brown, in some places inclining to actual blackness. These, therefore, in general, are found to be a race of short stature and odd shape, with countenances as savage as their manners are barbarous. The visage, in these countries, is large and broad, the nose flat and short, the eyes of a yellowish brown, inclining to blackness, the eye-lids drawn towards the temples, the cheek-bones extremely high, the mouth very large, the lips thick, and turned outwards, the voice thin and squeaking, the head large, the hair black and straight, the colour of the skin of a dark grayish.† They are short in stature, the generality not being above four feet high, and the tallest not above five. Among all these nations the women are as deformed as the men, and resemble them so nearly, that one cannot at first distinguish the sexes among them.

These nations not only resemble each other in their deformity, their dwarfishness, the colour of their hair and eyes, but they have, in a great measure, the same inclinations, and the same manners, being all equally rude, superstitious, and stupid. The Danish Laplanders have a large black cat, to which they communicate their secrets, and consult in all their affairs. Among the Swedish Laplanders there is in every family a drum for consulting the devil; and although these nations are robust and nimble, yet they are so cowardly that they never can be brought into the field. Gustavus Adolphus attempted to form a regiment of Laplanders, but he found it impossible to accomplish his design; for it should seem that they can live only in their own country, and in their own manner. They make use of skates, which are made of fir, of near three feet long, and half a foot broad; these are pointed, and raised before, and tied to the foot by straps of leather. With these they skate upon the icy snow with such velocity that they very easily overtake the swiftest animals. They make use also of a pole, pointed with iron at one end, and rounded at the other. This pole serves to push them along, to direct their course, to support them from falling, to stop the impetuosity of their motion, and to kill that

* I have taken four of these varieties from Linnæus; those of the Laplanders and Tartars, from Mr. Buffon.

† Krantz.

game which they have overtaken. Upon these skates they descend the steepest mountains, and scale the most craggy precipices; and in these exercises the women are not less skillful than the men. They have all the use of the bow and arrow, which seems to be a contrivance common to all barbarous nations; and which, however, at first required no small skill to invent. They launch a javelin also, with great force, and some say that they can hit a mark no larger than a crown, at thirty yards distance, and with such force as would pierce a man through. They are all hunters, and particularly pursue the ermine, the fox, the ounce, and the martin, for the sake of their skins. These they barter with their southern neighbours for brandy and tobacco; both which they are fond of to excess. Their food is principally dried fish, the flesh of rein-deer and bears. Their bread is composed of the bones of fishes, pounded and mixed with the inside tender bark of the pine-tree. Their drink is train-oil, or brandy; and when deprived of these, water in which juniper berries have been infused. With regard to their morals, they have all the virtues of simplicity, and all the vices of ignorance. They offer their wives and daughters to strangers, and seem to think it a particular honour if their offer be accepted. They have no idea of religion, or a Supreme Being; the greatest number of them are idolaters; and their superstition is as profound as their worship is contemptible. Wretched and ignorant as they are, yet they do not want pride; they set themselves far above the rest of mankind; and Krantz assures us, that when the Greenlanders are got together, nothing is so customary among them as to turn the Europeans into ridicule. They are obliged, indeed, to yield them the pre-eminence in understanding and mechanic arts; but they do not know how to set any value upon these. They therefore count themselves the only civilized and well-bred people in the world; and it is common with them, when they see a quiet, or a modest stranger, to say that he is almost as well-bred as a Greenlander.

From this description, therefore, this whole race of people may be considered as distinct from any other. Their long continuance in a climate the most inhospitable, their being obliged to subsist on food the most coarse and ill prepared, the savageness of their manners, and their laborious lives, all have contributed to shorten their stature, and to deform their bodies.* In proportion as we approach towards the north pole, the size of the natives appears to diminish, growing less and less as we advance higher, till we come to those latitudes that are destitute of all inhabitants whatsoever.

The wretched natives of these climates seem fitted by Nature to endure the rigours of their situation. As their food is but scanty and precarious, their patience in hunger is amazing.† A man who has eaten nothing for four days, can manage his little canoe in the most furious waves, and calmly subsist in the midst of a tempest that would quickly dash an European boat to pieces. Their strength is not less amazing than their patience; a woman among them will carry a piece of timber, or a stone, near double the weight of what an European can lift. Their bodies are of a dark gray all over; and their faces

* Ellis's Voyage, p. 256.   † Krantz, p. 134. vol. 1

brown, or olive. The tincture of their skins partly seems to arise from their dirty manner of living, being generally daubed with train-oil; and partly from the rigours of climate, as the sudden alterations of cold and raw air in winter, and of burning heats in summer, shade their complexions by degrees, till, in a succession of generations, they at last become almost black. As the countries in which these reside are the most barren, so the natives seem the most barbarous of any part of the earth. Their more southern neighbours of America treat them with the same scorn that a polished nation would treat a savage one; and we may readily judge of the rudeness of those manners, which even a native of Canada can think more barbarous than his own.

But the gradations of nature are imperceptible; and while the north is peopled with such miserable inhabitants, there are here and there to be found upon the edges of these regions, people of larger stature, and completer figure. A whole race of dwarfish breed is often found to come down from the north, and settle more to the southward; and on the contrary it sometimes happens that southern nations are seen higher up, in the midst of these diminutive tribes, where they have continued for time immemorial. Thus the Ostiac Tartars seem to be a race that have travelled down from the north, and to be originally sprung from the minute savages we have been describing. There are also Norwegians and Finlanders, of proper stature, who are seen to inhabit in latitudes higher even than Lapland. These, however, are but accidental migrations, and serve as shades to unite the distinct varieties of mankind.

The second great variety in the human species seems to be that of the Tartar race; from whence, probably, the little men we have been describing originally proceeded. The Tartar country, taken in general, comprehends the greatest part of Asia; and is consequently a general name given to a number of nations of various forms and complexions. But however they seem to differ from each other, they agree in being very unlike the people of any other country. All these nations have the upper part of the visage very broad, and wrinkled even while yet in their youth. Their noses are short and flat; their eyes little and sunk in their heads; and in some of them they are seen five or six inches asunder. Their cheek-bones are high, the lower part of their visage narrow, the chin long and advanced forward, their teeth of an enormous size, and growing separate from each other; their eye-brows thick, large, and covering their eyes; their eye-lids thick, the face broad and flat, the complexion olive-coloured, and the hair black. They are of a middle size, extremely strong, and very robust. They have but little beard, which grows stragglingly on the chin. They have large thighs, and short legs. The ugliest of all are the Calmucks, in whose appearance there seems to be something frightful. They all lead an erratic life, remaining under tents of hair, or skins. They live upon horse flesh and that of camels, either raw or a little sodden between the horse and the saddle. They eat also flesh dried in the sun. Their most usual drink is mare's milk, fermented with millet ground into meal. They all have the head shaven, except a lock of hair on the top, which they let grow sufficiently long to form into tresses on each side of the face. The women, who are as ugly as the men, wear their hair, which they bind up with bits of copper and other ornaments of a like nature. The majority of these nations have no religion, no settled notions of morality, no decency of behaviour. They are chiefly robbers: and the natives of Dagestan, who live near their more polished neighbours, make a traffic of Tartar slaves who have been stolen, and sell them to the Turks and the Persians. Their chief riches consist in horses, of which perhaps there are more in Tartary than in any other part of the world. The natives are taught by custom to live in the same place with their horses: they are continually employed in managing them, and at last bring them to such great obedience that the horse seems actually to understand the rider's intention.

To this race of men also, we must refer the Chinese and the Japanese, however different they seem in their manners and ceremonies. It is the form of the body that we are now principally considering; and there is between these countries a surprising resemblance. It is in general allowed that the Chinese have broad faces, small eyes, flat noses, and scarce any beard; that they are broad and square shouldered, and rather less in stature than Europeans. These are marks common to them and the Tartars, and they may therefore be considered as being derived from the same original. "I have observed," says Chardin, "that in all the people from the east and the north of the Caspian sea, to the peninsula of Malacca, that the lines of the face, and the formation of the visage is the same. This has induced me to believe that all these nations are derived from the same original, however different either their complexions or their manners may appear: for as to the complexion, that proceeds entirely from the climate and the food; and as to the manners, these are generally the result of their different degrees of wealth or power." That they come from one stock, is evident also from this, that the Tartars who settle in China, quickly resemble the Chinese; and, on the contrary, the Chinese who settle in Tartary, soon assume the figure and the manners of the Tartars.

The Japanese so much resemble the Chinese, that one cannot hesitate to rank them in the same class. They only differ in being rather browner, as they inhabit a more southern climate. They are, in general, described as of a brown complexion, a short stature, a broad flat face, a very little beard, and black hair. Their customs and ceremonies are nearly the same; their ideas of beauty similar; and their artificial deformities of blackening the teeth, and bandaging the feet, entirely alike in both countries. They both, therefore, proceed from the same stock; and although they differ very much from their brutal progenitors, yet they owe their civilization wholly to the mildness of the climate in which they reside, and to the peculiar fertility of the soil. To this tribe also, we may refer the Cochin Chinese, the Siamese, the Tonquinese, and the inhabitants of Arracan, Laos, and Pegu, who, though all differing from the Chinese, and each other, nevertheless, have too strong a resemblance, not to betray their common original.

Another, which makes the third variety in the human species, is that of the southern Asiatics; the form of whose features and per sons may be easily distinguished from those of the Tartar races. The nations that inhabit the peninsula of India, seem to be the principal stock from whence the inhabitants of the islands that lie scattered in the Indian ocean have been peopled. They are, in general, of a slender shape, with long straight black hair, and often with Roman noses. Thus they resemble the Europeans in stature and features; but greatly differ in colour and habit of body. The Indians are of an olive colour, and in the more southern parts, quite black; although the word Mogul, in their language, signifies a white man. The women are extremely delicate, and bathe very often: they are of an olive colour, as well as the men: their legs and thighs are long, and their bodies short, which is the opposite to what is seen among the women of Europe. They are, as I am assured, by no means so fruitful as the European women; but they feel the pains of child-birth with much less sensibility, and are generally up and well the day following. In fact, these pains seem greatest in all countries where the women are most delicate, or the constitution enfeebled by luxury or indolence. The women of savage nations seem, in a great measure, exempt from painful labours; and even the hard-working wives of the peasants among ourselves, have this advantage from a life of industry, that their child-bearing is less painful. Over all India, the children arrive sooner at maturity, than with us of Europe. They often marry, and consummate, the husband at ten years old, and the wife at eight; and they frequently have children at that age. However, the women who are mothers so soon, cease bearing before they are arrived at thirty; and at that time, they appear wrinkled, and seem marked with all the deformities of age. The Indians have long been remarkable for their cowardice and effeminacy; every conqueror, that has attempted the invasion of their country, having succeeded. The warmth of the climate entirely influences their manners; they are slothful, submissive, and luxurious; satisfied with sensual happiness alone, they find no pleasure in thinking; and contented with slavery, they are ready to obey any master. Many tribes among them eat nothing that has life; they are fearful of killing the meanest insect; and have even erected hospitals for the maintenance of all kinds of vermin. The Asiatic dress is a loose flowing garment, rather fitted for the purposes of peace and indolence, than of industry or war. The vigour of the Asiatics is in general conformable to their dress and nourishment; fed upon rice, and clothed in effeminate silk vestments, their soldiers are unable to oppose the onset of an European army, and from the times of Alexander to the present day, we have scarce any instances of their success in arms. Upon the whole, therefore, they may be considered as a feeble race of sensualists, too dull to find rapture in any pleasures, and too indolent to turn their gravity into wisdom. To this class we may refer the Persians and the Arabians, and in general the inhabitants of the islands that lie scattered in the Indian ocean.

The fourth striking variety in the human species, is to be found among the negroes of Africa. This gloomy race of mankind is found to blacken all the southern parts of Africa, from eighteen degrees north of the line, to its extreme termination at the Cape of Good Hope I know it is said, that the Caffres, who inhabit the southern extremity of that large continent, are not to be ranked among the negro race : however, the difference between them, in point of colour and features, is so small, that they may very easily be grouped in this general picture; and in the one or two that I have seen, I could not perceive the smallest difference. Each of the negro nations, it must be owned, differ from each other ; they have their peculiar countries, for beauty, like us; and different nations, as in Europe, pride themselves upon the regularity of their features.— Those of Guinea, for instance, are extremely ugly, and have an insupportable scent; those of Mosambique are reckoned beautiful, and have no ill smell whatsoever. The negroes, in general, are of a black colour, with a smooth soft skin. This smoothness proceeds from the downy softness of the hair which grows upon it ; the strength of which gives a roughness to the feel, in those of a white complexion. Their skins, therefore, have a velvet smoothness, and seem less braced upon the muscles than ours. The hair of their heads differs entirely from what we are accustomed to, being soft, woolly, and short. The beard also partakes of the same qualities; but in this it differs, that it soon turns gray, which the hair is seldom found to do ; so that several are seen with white beards, and black hair, at the same time. Their eyes are generally of a deep hazel ; their noses flat and short; their lips thick and tumid ; and their teeth of an ivory whiteness. This their only beauty, however, is set off by the colour of their skin ; the contrast between the black and white being the more observable. It is false to say that their features are deformed by art ; since, in the negro children born in European countries, the same deformities are seen to prevail ; the same flatness in the nose; and the same prominence in the lips. They are, in general, said to be well shaped ; but of such as I have seen, I never found one that might be justly called so; their legs being mostly ill formed, and commonly bending outward on the shin-bone. But it is not only in those parts of their bodies that are obvious, that they are disproportioned ; those parts which among us are usually concealed by dress, with them are large and languid.* The women's breasts, after bearing one child, hang down below the navel; and it is customary with them to suckle the child at their backs, by throwing the breast over the shoulder. As their persons are thus naturally deformed, at least to our imaginations, their minds are equally incapable of strong exertions. The climate seems to relax their mental powers still more than those of the body ; they are, therefore, in general, found to be stupid, indolent, and mischievous. The Arabians themselves, many colonies of whom have migrated southward into the most inland parts of Africa, seem to have degenerated from their ancestors; forgetting their ancient learning, and losing their beauty, they have become a race scarcely any way distinguishable from the original natives. Nor does it seem to have fared otherwise with the Portuguese, who, about two centuries ago, settled along this

* Linnæus, in prima linea sua, fæminas Africanas depingit sicut aliquid deforme in parte genitali gestantes, quod sinum pudoris nuncupat. Attamen nihil differunt a nostratibus in hac parte nisi quod labia pudendæ sint aliquantulum tumidiora. In hominibus etiam penis est longior et multo laxior.

coast. They also are become almost as black as the negroes, and are said by some to be even more barbarous.

The inhabitants of America make a fifth race, as different from all the rest in colour, as they are distinct in habitation. The natives of America (except in the northern extremity, where they resemble the Laplanders) are of a red or copper colour; and although, in the old world, different climates produce a variety of complexions and customs, the natives of the new continent seem to resemble each other in almost every respect. They are all nearly of one colour; all have black thick straight hair, and thin black beards; which, however, they take care to pluck out by the roots. They have, in general, flat noses, with high cheek-bones, and small eyes, and these deformities of nature they endeavour to increase by art: they flatten the nose, and often the whole head of their children, while the bones are yet susceptible of every impression. They paint the body and face of various colours, and consider the hair upon any part of it, except the head, as a deformity, which they are careful to eradicate. Their limbs are generally slighter made than those of the Europeans; and I am assured, they are far from being so strong. All these savages seem to be cowardly; they seldom are known to face their enemies in the field, but fall upon them at an advantage; and the greatness of their fears serves to increase the rigours of their cruelty. The wants which they often sustain, make them surprisingly patient in adversity: distress, by being grown familiar, becomes less terrible; so that their patience is less the result of fortitude than of custom. They have all a serious air, although they seldom think; and, however cruel to their enemies, are kind and just to each other. In short, the customs of savage nations in every country are almost the same; a wild, independent, and precarious life, produces a peculiar train of virtues and vices, and patience and hospitality, indolence and rapacity, content and sincerity, are found not less among the natives of America, than all the barbarous nations of the globe.

The sixth and last variety of the human species, is that of the Europeans, and the nations bordering on them. In this class we may reckon the Georgians, Circassians, and Mingrelians, the inhabitants of Asia Minor, and the northern parts of Africa, together with a part of those countries which lie north-west of the Caspian sea. The inhabitants of these countries differ a good deal from each other; but they generally agree in the colour of their bodies, the beauty of their complexions, the largeness of their limbs, and the vigour of their understandings. Those arts which might have had their invention among the other races of mankind, have come to perfection there. In barbarous countries, the inhabitants go either naked, or are awkwardly clothed in furs or feathers; in countries semi-barbarous, the robes are loose and flowing; but here the clothing is less made for show than expedition, and unites, as much as possible, the extremes of ornament and despatch.

To one or other of these classes we may refer the people of every country: and as each nation has been less visited by strangers, or has had less commerce with the rest of mankind, we find their persons and their manners more strongly impressed with one or other of the characters mentioned above. On the contrary, in those places where trade has long flourished, or where enemies have made many incursions, the races are usually found blended, and properly fall beneath no one character. Thus, in the islands of the Indian ocean, where a trade has been carried on for time immemorial, the inhabitants appear to be a mixture of all the nations upon the earth; white, olive, brown, and black men, are all seen living together in the same city, and propagate a mixed breed, that can be referred to none of the classes into which naturalists have thought proper to divide mankind.

Of all the colours by which mankind is diversified, it is easy to perceive that ours is not only the most beautiful to the eye, but the most advantageous. The fair complexion seems, if I may so express it, as a transparent covering to the soul; all the variations of the passions, every expression of joy or sorrow, flows to the cheek, and, without language, marks the mind. In the slightest change of health also, the colour of the European face is the most exact index, and often teaches us to prevent those disorders that we do not as yet perceive: not but the African black, and the Asiatic olive complexions, admit of their alterations also; but these are neither so distinct, nor so visible as with us: and, in some countries, the colour of the visage is never found to change; but the face continues in the same settled shade in shame and in sickness, in anger and despair.

The colour, therefore, most natural to man, ought to be that which is most becoming; and it is found that in all regions, the children are born fair, or at least red, and that they grow more black or tawny, as they advance in age. It should seem, consequently, that man is naturally white, since the same causes that darken the complexion in infants, may have originally operated, in slower degrees, in blackening whole nations. We could, therefore readily account for the blackness of different nations, did we not see the Americans, who live under the line, as well as the natives of Negroland, of a red colour, and but a very small shade darker than the natives of the northern latitudes, in the same continent. For this reason some have sought for other causes of blackness than the climate; and have endeavoured to prove that the blacks are a race of people, bred from one man, who was marked with accidental blackness. This, however, is but mere ungrounded conjecture: and, although the Americans are not so dark as the negroes, yet we must still continue in the ancient opinion, that the deepness of the colour proceeds from the excessive heat of the climate. For, if we compare the heats of Africa with those of America, we shall find they bear no proportion to each other. In America, all that part of the continent which lies under the line, is cool and pleasant, either shaded by mountains, or refreshed by breezes from the sea. But in Africa, the wide tract of country that lies under the line is very extensive, and the soil sandy; the reflection of the sun, therefore, from so large a surface of earth, is almost intolerable; and it is not to be wondered at, that the inhabitants should bear, in their looks, the marks of the inhospitable climate. In America, the country is but thinly inhabited; and the more torrid tracts are generally left desert by the inhabitants; for which reason they are not so deeply tinged by the beams of the sun. But in Africa the whole face of the country is fully peopled, and the natives are obliged to endure their situation without a power of migration. It is there, consequently, that they are in a manner tied down to feel all the severity of the heat; and their complexions take the darkest hue they are capable of receiving. We need not, therefore, have recourse to any imaginary propagation, from persons accidentally black, since the climate is a cause obvious and sufficient to produce the effect.

In fact, if we examine the complexion of different countries, we shall find them darken in proportion to the heat of their climate, and the shades gradually to deepen as they approach the line. Some nations, indeed, may be found not so much tinged by the sun as others, although they lie nearer the line. But this ever proceeds from some accidental causes; either from the country lying higher, and consequently being colder; or from the natives bathing oftener, and leading a more civilized life. In general, it may be asserted, that as we approach the line, we find the inhabitants of each country grow browner, until the colour deepens into perfect blackness. Thus, taking our standard from the whitest race of people, and beginning with our own country, which I believe bids fairest for the pre-eminence, we shall find the French, who are more southern, a slight shade deeper than we; going farther down, the Spaniards are browner than the French; the inhabitants of Fez darker than they; and the natives of Negroland the darkest of all. In what manner the sun produces this effect, and how the same luminary which whitens wax and linen, should darken the human complexion, is not easy to conceive. Sir Thomas Brown first supposed that a mucous substance, which had something of a vitriolic quality, settled under the reticular membrane, and grew darker with heat. Others have supposed that the blackness lay in the epidermis, or scarf-skin, which was burnt up like leather. But nothing has been satisfactorily discovered upon the subject; it is sufficient that we are assured of the fact; and that we have no doubt of the sun's tinging the complexion in proportion to its vicinity.

But we are not to suppose that the sun is the only cause of darkening the skin; the wind, extreme cold, hard labour, or coarse and sparing nourishment, are all found to contribute to this effect. We find the peasants of every country, who are most exposed to the weather, a shade darker than the higher ranks of people. The savage inhabitants of all places are exposed still more, and therefore contract a still deeper hue, and this will account for the tawny colour of the North American Indians. Although they live in a climate the same, or even more northerly than ours, yet they are found to be of complexions very different from those of Europe. But it must be considered that they live continually exposed to the sun; that they use many methods to darken their skins by art, painting them with red ochre, and anointing them with the fat of bears. Had they taken, for a succession of several generations, the same precautions to brighten their colour that an European does, it is very probable that they would in time come to have similar complexions, and perhaps dispute the prize of beauty.

The extremity of cold is not less productive of a tawny complexion than that of heat. The natives of the arctic circle, as was observed, are all brown; and those that lie most to the north are almost entirely black. In this manner both extremes are unfavourable to the human form and colour, and the same effects are produced under the poles that are found at the line.

With regard to the stature of different countries, that seems chiefly to result from the nature of the food, and the quantity of the supply. Not but that the severity of heat or cold may in some measure diminish the growth, and produce a dwarfishness of make. But in general the food is the great agent in producing this effect; where that is supplied in large quantities, and where its quality is wholesome and nutrimental, the inhabitants are generally seen above the ordinary stature. On the contrary, where it is afforded in a sparing quantity, or very coarse, and void of nourishment in its kind, the inhabitants degenerate, and sink below the ordinary size of mankind. In this respect they resemble other animals, whose bodies, by proper feeding, may be greatly augmented. An ox, on the fertile plains of India, grows to a size four times as large as the diminutive animal of the same kind bred in the Alps. The horses bred in the plains are larger than those of the mountain. So it is with man: the inhabitants of the valley are usually found taller than those of the hill: the natives of the Highlands of Scotland, for instance, are short, broad, and hardy; those of the Lowlands are tall and shapely. The inhabitants of Greenland, who live upon dried fish and seals, are less than those of Gambia or Senegal, where nature supplies them with vegetable and animal abundance.

The form of the face seems rather to be the result of custom. Nations who have long considered some artificial deformity as beautiful, who have industriously lessened the feet, or flattened the nose, by degrees begin to receive the impression they are taught to assume; and nature, in a course of ages, shapes itself to the constraint, and assumes hereditary deformity. We find nothing more common in births, than for children to inherit sometimes even the accidental deformities of their parents. We have many instances of squinting in the father, which he received from fright, or habit, communicated to the offspring; and I myself have seen a child distinctly marked with a scar, similar to one the father had received in battle. In this manner, accidental deformities may become natural ones; and by assiduity may be continued, and even increased, through successive generations. From this, therefore, may have arisen the small eyes and long ears of the Tartar and Chinese nations. From hence originally may have come the flat noses of the blacks, and the flat heads of the American Indians.

In this slight survey, therefore, I think we may see that all the variations in the human figure, as far as they differ from our own, are produced either by the rigour of the climate, the bad quality, or the scantiness of the provisions, or by the savage customs of the country. They are actual marks of the degeneracy in the human form; and we may consider the European figure and colour as standards to which to refer all other varieties, and with which to compare them. In proportion as the Tartar or American approaches nearer to European beauty, we consider the race as less degenerated; in proportion as he differs more widely, he has made greater deviations from his original form.

That we have all sprung from one common parent, we are taught, both by reason and religion, to believe; and we have good reason also to think that the Europeans resemble him more than any of the rest of his children. However, it must not be concealed that the olive-coloured Asiatic, and even the jet-black negro, claim this honour of hereditary resemblance; and assert, that white men are mere deviations from original perfection. Odd as this opinion may seem, they have Linnæus, the celebrated naturalist, on their side, who supposes man a native of the tropical climates, and only a sojourner more to the north. But, not to enter into a controversy upon a matter of a very remote speculation, I think one argument alone will suffice to prove the contrary, and show that the white man is the original source from whence the other varieties have sprung. We have frequently seen white children produced from black parents, but have never seen a black offspring the production of two whites. From hence we may conclude, that whiteness is the colour to which mankind naturally tends: for, as in the tulip, the parent stock is known by all the artificial varieties breaking into it; so in man, that colour must be original which never alters, and to which all the rest are accidentally seen to change. I have seen in London, at different times, two white negroes, the issue of black parents, that served to convince me of the truth of this theory. I had before been taught to believe that the whiteness of the negro's skin was a disease, a kind of milky whiteness, that might be called rather a leprous crust than a natural complexion. I was taught to suppose, that the numberless white negroes, found in various parts of Africa, the white men that go by the name of Chacrelas, in the East-Indies, and the white Americans, near the Isthmus of Darien, in the West-Indies, were all as so many diseased persons, and even more deformed than the blackest of the natives. But, upon examining that negro which was last shown in London, I found the colour to be exactly like that of an European; the visage white and ruddy, and the lips of the proper redness. However, there were sufficient marks to convince me of its descent. The hair was white and woolly, and very unlike any thing I had seen before. The iris of the eye was yellow, inclining to red; the nose was flat, exactly resembling that of a negro; and the lips thick and prominent. No doubt, therefore, remained of the child's having been born of negro parents: and the person who showed it had attestations to convince the most incredulous. From this, then, we see that the variations of the negro colour is into whiteness, whereas the white are never found to have a race of negro children. Upon the whole, therefore, all those changes which the African, the Asiatic, or the American, undergo, are but accidental deformities, which a kinder climate, better nourishment, or more civilized manners, would, in a course of centuries, very probably remove.

CHAPTER XII.

OF MONSTERS.

HITHERTO I have only spoken of those varieties in the human species, that are common to whole nations; but there are varieties of another kind, which are only found in the individual; and being more rarely seen, are therefore called *monstrous*. If we examine into the varieties of distorted nature, there is scarcely a limb of the body, or a feature in the face, that has not suffered some reprobation, either from art or nature; being enlarged or diminished, lengthened or wrested from its due proportion. Linnæus, after having given a catalogue of monsters, particularly adds, the flat heads of Canada, the long heads of the Chinese, and the slender waists of the women of Europe, who by straight lacing take such pains to destroy their health, through a mistaken desire to improve their beauty.* It belongs more to the physician than the naturalist to attend to these minute deformities; and indeed it is a melancholy contemplation to speculate upon a catalogue of calamities, inflicted by unpitying nature, or brought upon us by our own caprice. Some, however, are fond of such accounts; and there have been books filled with nothing else. To these, therefore, I refer the reader, who may be better pleased with accounts of men with two heads, or without any head; of children joined in the middle; of bones turned into flesh, or flesh converted into bones, than I am.† It is sufficient here to observe, that every day's experience must have shown us miserable instances of this kind produced by nature or affection: calamities that no pity can soften, or assiduity relieve.

Passing over, therefore, every other account, I shall only mention the famous instance, quoted by Father Malbranche, upon which he founds his beautiful theory of monstrous productions. A woman of Paris, the wife of a tradesman, went to see a criminal broke alive upon the wheel, at the place of public execution. She was at that time two months advanced in her pregnancy, and no way subject to any disorders to affect the child in her womb. She was, however, of a tender habit of body; and, though led by curiosity to this horrid spectacle, very easily moved to pity and compassion. She felt, therefore, all those strong emotions which so terrible a sight must naturally

---

* Linnæi Syst. vol. i. p. 29. Monorchides ut minus fertiles.

† Vide Phil. Trans. passim. Miscellan. Curioss. Johan. Baptist. Wenck. Dissertatio Physica an ex virilis humani seminis cum brutali per nefarium coitum commixtione aut vicissim ex bruti maris cum muliebri humano seminis commixtione possit verus homo generari. Vide etiam, Johnstoni Thaumatographia Naturalis. Vide Adalberti Disqui sitio Physica ostenti duorum puerorum unus quorum dente aureo, alter cum capite giganteo Biluæ spectabantur. A man without lungs and stomach, Journal de Scavans, 1682, p. 301; another without any brain. Andreas Caroli Memorabilia, p. 167 an. 1676; another without any head. Giornale di Roma, anno 1675, p. 26; another without any arms. New Memoirs of Literature, vol. iv. p. 446. In short, the variety of these accounts is almost infinite; and, perhaps, their use is as much circumscribed as their variety is extensive.

inspire; shuddered at every blow the criminal received, and almost swooned at his cries. Upon returning from this scene of blood, she continued for some days pensive, and her imagination still wrought upon the spectacle she had lately seen. After some time, however, she seemed perfectly recovered from her fright, and had almost forgotten her former uneasiness. When the time of her delivery approached, she seemed no ways mindful of her former terrors, nor were her pains in labour more than usual in such circumstances. But what was the amazement of her friends and assistants when the child came into the world! It was found that every limb in its body was broken like those of the malefactor, and just in the same place. This poor infant had suffered the pains of life, even before its coming into the world: it did not die, but lived in a hospital in Paris for twenty years after, a wretched instance of the supposed powers of imagination in the mother, of altering and distorting the infant in the womb. The manner in which Malbranche reasons upon this fact, is as follows:— the Creator has established such a sympathy between the several parts of nature, that we are led not only to imitate each other, but also to partake in the same affections and desires. The animal spirits are thus carried to the respective parts of the body, to perform the same actions which we see others perform, to receive in some measure their wounds, and take part in their sufferings. Experience tells us, that if we look attentively on any person, severely beaten, or sorely wounded, the spirits immediately flow into those parts of the body, which correspond to those we see in pain. The more delicate the constitution, the more it is thus affected; the spirits making a stronger impression on the fibres of a weakly habit than of a robust one. Strong vigorous men see an execution without much concern, while women of nicer texture are struck with horror and concern. This sensibility in them must, of consequence, be communicated to all parts of their body; and, as the fibres of the child in the womb are incomparably finer than those of the mother, the course of the animal spirits must consequently produce greater alterations. Hence, every stroke given to the criminal, forcibly struck the imagination of the woman; and, by a kind of counter-stroke, the delicate tender frame of the child.

Such is the reasoning of an ingenious man upon a fact, the veracity of which many since have called in question.* They have allowed, indeed, that such a child might have been produced, but have denied the cause of its deformity. "How could the imagination of the mother," say they, "produce such dreadful effects upon her child? She has no communication with the infant; she scarcely touches it in any part; quite unaffected with her concerns, it sleeps in security, in a manner secluded by a fluid in which it swims, from her that bears it With what a variety of deformities," say they, "would all mankind be marked, if all the vain and capricious desires of the mother were thus readily written upon the body of the child?" Yet, notwithstanding this plausible way of reasoning, I cannot avoid giving some credit to the variety of instances I have either read or seen upon this subject If it be a prejudice, it is as old as the days of Aristotle, and

* Buffon. vol. iv. p. 9.

to this day, as strongly believed by the generality of mankind as
ever. It does not admit of a reason; and, indeed, I can give none,
even why the child should, in any respect, resemble the father or the
mother. The fact we generally find to be so. But why it should
take the particular print of the father's features in the womb, is as
hard to conceive, as why it should be affected by the mother's imagination. We all know what a strong effect the imagination has on
those parts in particular, without being able to assign a cause how this
effect is produced; and why the imagination may not produce the
same effect in marking the child that it does in forming it, I see no
reason. Those persons whose employment it is to rear up pigeons
of different colours, can breed them, as their expression is, to a feather.
In fact, by properly pairing them, they can give what colour they will
to any feather in any part of the body. Were we to reason upon this
fact, what could we say? Might it not be asserted, that the egg, being distinct from the body of the female, cannot be influenced by it?
Might it not be plausibly said, that there is no similitude between any
part of the egg and any particular feather, which we expect to propagate; and yet for all this, the fact is known to be true, and what
no speculation can invalidate. In the same manner, a thousand various instances assure us, that the child in the womb is sometimes
marked by the strong affections of the mother; how this is performed
we know not; we only see the effect, without any connexion between
it and the cause. The best physicians have allowed it, and have been
satisfied to submit to the experience of a number of ages; but many
disbelieve it because they expect a reason for every effect. This,
however, is very hard to be given, while it is very easy to appear wise
by pretending incredulity.

Among the number of monsters, dwarfs and giants are usually
reckoned; though not, perhaps, with the strictest propriety, since they
are no way different from the rest of mankind, except in stature. It
is a dispute, however, about words, and therefore scarcely worth contending about. But there is a dispute, of a more curious nature, on
this subject, namely, whether there are races of people thus very diminutive, or vastly large; or whether they be merely accidental varieties, that now and then are seen in the country, in a few persons,
whose bodies some external cause has contributed to lessen or enlarge.

With regard to men of diminutive stature, all antiquity has been
unanimous in asserting their national existence. Homer was the first
who has given us an account of the pigmy nation contending with the
cranes; and what poetical licence might be supposed to exaggerate,
Athenæus has attempted seriously to confirm by historical assertion.[*]
If we attend to these, we must believe, that in the internal parts of
Africa there are whole nations of pigmy beings, not more than a foot
in stature, who continually wage an unequal war with the birds and
beasts that inhabit the plains in which they reside. Some of the ancients, however, and Strabo in particular, have supposed all these accounts to be fabulous, and have been more inclined to think this supposed nation of pigmies, nothing more than a species of apes, well

---

[*] Athenæus, ix. 390.

known to be numerous in that part of the world. With this opinion the moderns have all concurred; and that diminutive race, which was described as human, has been long degraded into a class of animals that resemble us but very imperfectly.

The existence, therefore, of a pigmy race of mankind, being founded in error, or in fable, we can expect to find men of diminutive stature only by accident, among men of the ordinary size. Of these accidental dwarfs, every country, and almost every village, can produce numerous instances. There was a time when these unfavoured children of nature were the peculiar favourites of the great; and no prince or nobleman thought himself completely attended, unless he had a dwarf among the number of his domestics. These poor little men were kept to be laughed at; or to raise the barbarous pleasure of their masters, by their contrasted inferiority. Even in England, as late as the times of King James I. the court was at one time furnished with a dwarf, a giant, and a jester: these the king often took a pleasure in opposing to each other, and often fomented quarrels among them, in order to be a concealed spectator of their animosity. It was a particular entertainment of the courtiers at that time, to see little Jeffery, for so the dwarf was called, ride round the lists, expecting his antagonist; and discovering in his actions, all the marks of contemptible resolution.

It was in the same spirit, that Peter of Russia, in the year 1710, celebrated a marriage of dwarfs. This monarch, though raised by his native genius far above a barbarian, was, nevertheless, still many degrees removed from actual refinement. His pleasures, therefore, were of the vulgar kind, and this was among the number. Upon a certain day, which he had ordered to be proclaimed several months before, he invited the whole body of his courtiers, and all the foreign ambassadors, to be present at the marriage of a pigmy man and woman. The preparations for this wedding were not only very grand, but executed in a style of barbarous ridicule. He ordered that all the dwarf men and women, within two hundred miles, should repair to the capital; and also insisted that they should be present at the ceremony. For this purpose he supplied them with proper vehicles; but so contrived it, that one horse was seen carrying in a dozen of them into the city at once, while the mob followed, shouting and laughing, from behind. Some of them were at first unwilling to obey an order which they knew was calculated to turn them into ridicule, and did not come; but he soon obliged them to obey; and, as a punishment, enjoined, that they should wait upon the rest at dinner. The whole company of dwarfs amounted to seventy, beside the bride and bridegroom, who were richly adorned, and in the extremity of the fashion. For this little company in miniature, every thing was suitably provided; a low table, small plates, little glasses, and in short, every thing was so fitted, as if all things had been dwindled to their own standard. It was his great pleasure to see their gravity and their pride; the contention of the women for places, and the men for superiority. This point he attempted to adjust, by ordering that the most diminutive should take the lead; but this bred disputes, for none would then consent to sit foremost. All this, however, being at last settled, dancing followed the dinner, and the ball was opened with a minuet by the bridegroom, who measured exactly three feet two inches high. In the end, matters were so contrived, that this little company, who met together in gloomy pride, and unwilling to be pleased, being at last familiarized to laughter, joined in the diversion, and became, as the journalist has it,* extremely sprightly and entertaining.

But whatever may be the entertainment such guests might afford when united, I never found a dwarf capable of affording any when alone. I have sometimes conversed with some of these that were exhibited at our fairs about town, and have ever found their intellects as contracted as their persons. They, in general, seemed to me to have faculties very much resembling those of children, and their desires likewise of the same kind; being diverted with the same sports, and best pleased with such companions. Of all those I have seen, which may amount to five or six, the little man, whose name was Coan, that died lately at Chelsea, was the most intelligent and sprightly. I have heard him and the giant, who sung at the theatres, sustain a very ridiculous duet, to which they were taught to give great spirit. But this mirth, and seeming sagacity, were but assumed. He had, by long habit, been taught to look cheerful upon the approach of company; and his conversation was but the mere etiquette of a person that had been used to receive visitors. When driven out of his walk, nothing could be more stupid or ignorant, nothing more dejected or forlorn. But we have a complete history of a dwarf, very accurately related by Mr. Daubenton, in his part of the Histoire Naturelle; which I will here take leave to translate.

This dwarf, whose name was Baby, was well known, having spent the greatest part of his life at Lunenville, in the palace of Stanislaus, the titular king of Poland. He was born near the village of Plaisne, in France, in the year 1741. His father and mother were peasants, both of good constitutions, and inured to a life of husbandry and labour. Baby, when born, weighed but a pound and a quarter. We are not informed of the dimensions of his body at that time; but we may conjecture they were very small, as he was presented on a plate to be baptized, and for a long time lay in a slipper. His mouth, although proportioned to the rest of his body, was not, at that time, large enough to take in the nipple; and he was, therefore, obliged to be suckled by a she-goat that was in the house; and that served as a nurse, attending to his cries with a kind of maternal fondness. He began to articulate some words when eighteen months old; and at two years he was able to walk alone. He was then fitted with shoes that were about an inch and a half long. He was attacked with several acute disorders; but the small-pox was the only one which left any marks behind it. Until he was six years old, he eat no other food but pulse, potatoes, and bacon. His father and mother were, from their poverty, incapable of affording him any better nourishment; and his education was little better than his food, being bred up among the rustics of the place. At six years old he was about

* Die dench wurdige. Iwerg. Hockwelt, &c. Lipsiæ, 1713, vol. viii. page 102. seq

fifteen inches high; and his whole body weighed but thirteen pounds. Notwithstanding this, he was well proportioned, and handsome; his health was good, but his understanding scarce passed the bounds of instinct. It was at that time that the king of Poland, having heard of such a curiosity, had him conveyed to Lunenville, gave him the name of *Baby*, and kept him in his palace.

Baby, having thus quitted the hard condition of a peasant to enjoy all the comforts and conveniences of life, seemed to receive no alteration from his new way of living, either in mind or person. He preserved the goodness of his constitution till about the age of sixteen, but his body seemed to increase very slowly during the whole time; and his stupidity was such, that all instructions were lost in improving his understanding. He could never be brought to have any sense of religion, nor even to shew the least signs of a reasoning faculty They attempted to teach him dancing and music, but in vain; he never could make any thing of music; and as for dancing, although he beat time tolerably exact, yet he could never remember the figure, but while his dancing-master stood by to direct his motions. Notwithstanding, a mind thus destitute of understanding was not without its passions; anger and jealousy harassed it at times; nor was he without desires of another nature.

At the age of sixteen, Baby was twenty-nine inches tall; at this he rested; but having thus arrived at his acme, the alterations of puberty, or rather, perhaps, of old age, came fast upon him. From being very beautiful, the poor little creature now became quite deformed; his strength quite forsook him; his back-bone began to bend; his head hung forward; his legs grew weak; one of his shoulders turned awry; and his nose grew disproportionably large. With his strength, his natural spirits also forsook him; and, by the time he was twenty, he was grown feeble, decrepit, and marked with the strongest impressions of old age. It had been before remarked by some, that he would die of old age before he arrived at thirty; and, in fact, by the time he was twenty-two, he could scarcely walk a hundred paces, being worn with the multiplicity of his years, and bent under the burden of protracted life. In this year he died; a cold, attended with a slight fever, threw him into a kind of lethargy, which had a few momentary intervals; but he could scarce be brought to speak. However, it is asserted, that in the five last years of his life, he shewed a clearer understanding, than in his times of best health: but at length he died, after enduring great agonies, in the twenty-second year of his age.

Opposite to this accidental diminution of the human race, is that of its extraordinary magnitude. Concerning the reality of a nation of Giants, there have been many disputes among the learned. Some have affirmed the probability of such a race; and others, as warmly have denied the possibility of their existence. But it is not from any speculative reasonings, upon a subject of this kind, that information is to be obtained; it is not from the disputes of the scholar, but the labours of the enterprising, that we are to be instructed in this inquiry. Indeed, nothing can be more absurd, than what some learned men have advanced upon this subject. It is very unlikely, says

Grew, that there should either be dwarfs or giants; or if such, they cannot be fitted for the usual enjoyment of life and reason. Had man been born a dwarf, he could not have been a reasonable creature; for to that end, he must have a jolt head, and then he would not have body and blood enough to supply his brain with spirits; or if he had a small head, proportionable to his body, there would not be brain enough for conducting life. But it is still worse with giants; and there could never have been a nation of such, for there would not be food enough found in any country to sustain them; or, if there were beasts sufficient for this purpose, there would not be grass enough for their maintenance. But what is still more, add others, giants could never be able to support the weight of their own bodies; since a man of ten feet high, must be eight times as heavy as one of the ordinary stature; whereas he has but twice the size of muscles to support such a burden, and consequently would be overloaded with the weight of his own body. Such are the theories upon this subject, and they require no other answer, but that experience proves them both to be false; dwarfs are found capable of life and reason; and giants are seen to carry their own bodies. We have several accounts from mariners, that a nation of giants actually exists, and mere speculation should never induce us to doubt their veracity.

Ferdinand Magellan was the first who discovered this race of people along the coast, towards the extremity of South America. Magellan was a Portuguese, of noble extraction, who having long behaved with great bravery, under Albuquerque, the conqueror of India, he was treated with neglect by the court, upon his return. Applying therefore, to the king of Spain, he was intrusted with the command of five ships, to subdue the Molucca islands, upon one of which he was slain. It was in his voyage thither that he happened to winter in St. Julian's Bay, an American harbour, forty-nine degrees south of the line. In this desolate region, where nothing was seen but objects of terror, where neither trees nor verdure drest the face of the country, they remained for some months without seeing any human creature. They had judged the country to be utterly uninhabitable, when one day they saw approaching, as if he had been dropped from the clouds, a man of enormous stature, dancing and singing, and putting dust upon his head, as they supposed in token of peace. This overture for friendship was, by Magellan's command, quickly answered by the rest of his men; and the giant approaching, testified every mark of astonishment and surprise. He was so tall, that the Spaniards only reached his waist; his face was broad, his colour brown, and painted over with a variety of tints; each cheek had the resemblance of a heart drawn upon it; his hair was approaching to whiteness; he was clothed in skins, and armed with a bow. Being treated with kindness, and dismissed with some trifling presents, he soon returned with many more of the same stature; two of whom the mariners decoyed on shipboard: nothing could be more gentle than they were in the beginning; they considered the fetters that were preparing for them as ornaments, and played with them like children with their toys; but when they found for what purpose they were intended, they instantly exerted their amazing strength, and broke them in pieces with a very easy effort. This account, with a variety of other circumstances, has been confirmed by succeeding travellers: Herrara, Sebald Wert, Oliver Van Noort, and James le Maire, all correspond in affirming the fact, although they differ in many particulars of their respective descriptions. The last voyager we have had, that has seen this enormous race, is Commodore Byron. I have talked with the person who first gave the relation of that voyage, and who was the carpenter of the Commodore's ship; he was a sensible, understanding man, and I believe extremely faithful. By him, therefore, I was assured, in the most solemn manner, of the truth of his relation; and this account has since been confirmed by one or two publications; in all which the particulars are pretty nearly the same. One of the circumstances which most puzzled me to reconcile to probability, was that of the horses, on which they are described as riding down to the shore. We know the American horse to be of European breed; and, in some measure, to be degenerated from the original. I was at a loss, therefore, to account how a horse of not more than fourteen hands high, was capable of carrying a man of nine feet; or, in other words, an animal almost as large as itself. But the wonder will cease, when we consider, that so small a beast as an ass, will carry a man of ordinary size tolerably well; and the proportion between this and the former instance is nearly exact. We can no longer, therefore, refuse our assent to the existence of this gigantic race of mankind; in what manner they are propagated, or under what regulations they live, is a subject that remains for future investigation. It should appear, however, that they are a wandering nation, changing their abode with the course of the sun, and shifting their situation, for the convenience of food, climate, or pasture.*

This race of giants are described as possessed of great strength; and, no doubt they must be very different from those accidental giants that are to be seen in different parts of Europe. Stature, with these, seems rather their infirmity than their pride; and adds to their burden, without increasing their strength. Of those I have seen, the generality were ill-formed and unhealthful; weak in their persons, or incapable of exerting what strength they were possessed of. The same defects of understanding that attended those of suppressed stature, were found in those who were thus overgrown: they were heavy, phlegmatic, stupid, and inclined to sadness. Their numbers, however, are but few; and it is thus kindly ordered by Providence, that as the middle is the state best fitted for happiness, so the middle ranks of mankind are produced in the greatest variety.

However, mankind seems naturally to have a respect for men of extraordinary stature; and it has been a supposition of long standing, that our ancestors were much taller, as well as much more beautiful, than we. This has been, indeed, a theme of poetical declamation from the beginning; and man was scarce formed, when he began to deplore an imaginary decay. Nothing is more natural than this progress of the mind, in looking up to antiquity with reverential wonder Having been accustomed to compare the wisdom of our fathers with

---

* Later voyagers have not confirmed this account, in some particulars.

our own, in early imbecility, the impression of their superiority remains when they no longer exist, and when we cease to be inferior. Thus the men of every age consider the past as wiser than the present; and the reverence seems to accumulate as our imaginations ascend. For this reason, we allow remote antiquity many advantages, without disputing their title : the inhabitants of uncivilized countries represent them as taller and stronger; and the people of a more polished nation, as more healthy and more wise. Nevertheless, these attributes seem to be only the prejudices of ingenuous minds; a kind of gratitude, which we hope in turn to receive from posterity. The ordinary stature of men, Mr. Derham observes, is, in all probability, the same now as at the beginning. The oldest measure we have of the human figure, is in the monument of Cheops, in the first pyramid of Egypt. This must have subsisted many hundred years before the times of Homer, who is the first that deplores the decay. This monument, however, scarce exceeds the measure of our ordinary coffins: the cavity is no more than six feet long, two feet wide, and deep in about the same proportion. Several mummies also, of a very early age, are found to be only of the ordinary stature; and shew that, for these three thousand years at least, men have not suffered the least diminution. We have many corroborating proofs of this, in the ancient pieces of armour which are dug up in different parts of Europe. The brass helmet dug up at Medauro, fits one of our men, and yet is allowed to have been left there at the overthrow of Asdrubal. Some of our finest antique statues, which we learn from Pliny and others to be exactly as big as life, still continue to this day, remaining monuments of the superior excellence of their workmen indeed, but not of the superiority of their stature. We may conclude, therefore, that men have been, in all ages, pretty much of the same size they are at present; and that the only difference must have been accidental, or perhaps national.

As to the superior beauty of our ancestors, it is not easy to make the comparison; beauty seems a very uncertain charm; and frequently is less in the object, than in the eye of the beholder. Were a modern lady's face formed exactly like the Venus of Medicis, or the Sleeping Vestal, she would scarce be considered beautiful, except by the lovers of antiquity, whom, of all her admirers, perhaps, she would be least desirous of pleasing. It is true, that we have some disorders among us that disfigure the features, and from which the ancients were exempt; but it is equally true, that we want some which were common among them, and which were equally deforming. As for their intellectual powers, these also were probably the same as ours: we excel them in the sciences, which may be considered as a history of accumulated experience; and they excel us in the poetic arts, as they had the first rifling of all the striking images of nature.

CHAPTER XIII.

OF MUMMIES, WAX-WORKS, &c

" MAN[*] is not content with the usual term of life, but he is willing to lengthen out his existence by art; and although he cannot prevent death, he tries to obviate his dissolution It is natural to attempt to preserve even the most trifling relics of what has long given us pleasure; nor does the mind separate from the body, without a wish, that even the wretched heap of dust it leaves behind, may yet be remembered. The embalming, practised in various nations, probably had its rise in this fond desire: an urn filled with ashes, among the Romans, served as a pledge of continuing affection; and even the grassy graves in our own church-yards, are raised above the surface, with the desire that the body below should not be wholly forgotten. The soul, ardent after eternity for itself, is willing to procure, even for the body, a prolonged duration."

But of all nations, the Egyptians carried this art to the highest perfection: as it was a principle of their religion, to suppose the soul continued only coeval to the duration of the body, they tried every art to extend the life of the one, by preventing the dissolution of the other. In this practice they were exercised from the earliest ages; and the mummies they have embalmed in this manner continue in great numbers to the present day. We are told, in Genesis, that Joseph seeing his father expire, gave orders to his physicians to embalm the body, which they executed in the compass of forty days, the usual time of embalming. Herodotus also, the most ancient of the profane historians, gives us a copious detail of this art, as it was practised, in his time, among the Egyptians. There are certain men among them, says he, who practise embalming as a trade; which they perform with all expedition possible. In the first place, they draw out the brain through the nostrils, with irons adapted to this purpose; and in proportion as they evacuate it in this manner, they fill up the cavity with aromatics: they next cut open the belly, near the sides, with a sharpened stone, and take out the entrails, which they cleanse, and wash in palm oil; having performed this operation, they roll them in aromatic powder, fill them with myrrh, cassia, and other perfumes, except incense, and replace them, sewing up the body again. After these precautions, they salt the body with nitre, and keep it in the salting-place for seventy days, it not being permitted to preserve it so any longer. When the seventy days are accomplished, and the body washed once more, they swathe it in bands made of linen, which have been dipt in a gum the Egyptians use instead of salt. When the friends have taken back the body, they make a hollow trough, something like the shape of a man, in which they place the body; and

---

[*] This chapter I have, in a great measure, translated from Mr. Daubenton. Whatever is added from others, is marked with inverted commas.

this they inclose in a box, preserving the whole as a most precious relic, placed against the wall. Such are the ceremonies used with regard to the rich; as for those who are contented with a humbler preparation, they treat them as follows: they fill a syringe with an odoriferous liquor extracted from the cedar-tree, and, without making an incision, inject it up the body of the deceased, and then keep it in nitre, as long as in the former case. When the time is expired, they evacuate the body of the cedar liquor which had been injected; and such is the effect of this operation, that the liquor dissolves the intestines, and brings them away: the nitre also serves to eat away the flesh, and leaves only the skin and the bones remaining. This done, the body is returned to the friends, and the embalmer takes no farther trouble about it. The third method of embalming those of the meanest condition, is merely by purging and cleansing the intestines by frequent injections, and preserving the body for a similar term in nitre, at the end of which it is restored to the relations.

Diodorus Siculus also makes mention of the manner in which these embalmings are performed. According to him, there were several officers appointed for this purpose: the first of them, who was called the scribe, marked those parts of the body, on the left side, which were to be opened; the cutter made the incision; and one of those that were to salt it, drew out all the bowels, except the heart and the kidneys; another washed them in palm wine and odoriferous liquors; afterwards they anointed for above thirty days with cedar, gum, myrrh, cinnamon, and other perfumes. These aromatics preserved the body entire for a long time, and gave it a very agreeable odour. It was not in the least disfigured by this preparation; after which it was returned to the relations, who kept it in a coffin, placed upright against a wall.

Most of the modern writers who have treated on this subject, have merely repeated what has been said by Herodotus; and if they add any thing of their own, it is but merely from conjecture. Dumont observes, that it is very probable, that aloes, bitumen, and cinnamon, make a principal part of the composition which is used on this occasion: he adds, that, after embalming, the body is put into a coffin, made of the sycamore-tree, which is almost incorruptible. Mr. Grew remarks, that in an Egyptian mummy, in the possession of the Royal Society, the preparation was so penetrating, as to enter into the very substance of the bones, and rendered them so black that they seemed to have been burnt. From this he is induced to believe that the Egyptians had a custom of embalming their dead, by boiling them in a kind of liquid preparation, until all the aqueous parts of the body were exhaled away; and until the oily or gummy matter had penetrated throughout. He proposes, in consequence of this, a method of macerating, and afterwards of boiling the dead body in oil of walnut.

I am, for my own part, of opinion that there were several ways of preserving dead bodies from putrefaction; and that this would be no difficult matter, since different nations have all succeeded in the attempt. We have an example of this kind among the Guanches, the ancient inhabitants of the island of Teneriff. Those who survived the general destruction of this people by the Spaniards, when they conquered this island, informed them that the art of embalming was still preserved there; and that there was a tribe of priests among them possessed of the secret, which they kept concealed as a sacred mystery. As the greatest part of the nation was destroyed, the Spaniards could not arrive at a complete knowledge of this art; they only found out a few of the particulars. Having taken out the bowels, they washed the body several times in a lee, made of the dried bark of the pine-tree, warmed during the summer by the sun, or by a stove in the winter. They afterwards anointed it with butter, or the fat of bears, which they had previously boiled with odoriferous herbs, such as sage and lavender. After this unction, they suffered the body to dry, and then repeated the operation as often as it was necessary, until the whole substance was impregnated with the preparation. When it was become very light, it was then a certain sign that it was fit and properly prepared. They then rolled it up in the dried skins of goats; which, when they had a mind to save expense, they suffered to remain with the hair still growing upon them. Purchas assures us, that he has seen mummies of this kind in London; and mentions the name of a gentleman who had seen several of them in the island of Teneriff, which were supposed to have been two thousand years old; but without any certain proofs of such great antiquity. This people, who probably came first from the coasts of Africa, might have learned this art from the Egyptians, as there was a traffic carried on from thence into the most internal parts of Africa.

Father Acosta and Garcilasso de la Vega make no doubt but that the Peruvians understood the art of preserving their dead for a very long space of time. They assert their having seen the bodies of several incas, that were perfectly preserved. They still preserved their hair and their eye-brows; but they had eyes made of gold, put in the places of those taken out. They were clothed in their usual habits, and seated in the manner of the Indians, their arms placed on their breasts. Garcilasso touched one of their fingers, and found it apparently as hard as wood; and the whole body was not heavy enough to overburden a weak man who should attempt to carry it away. Acosta presumes that these bodies were embalmed with bitumen, of which the Indians knew the properties. Garcilasso, however, is of a different opinion, as he saw nothing bituminous about them; but he confesses that he did not examine them very particularly, and he regrets his not having inquired into the methods used for that purpose. He adds, that being a Peruvian, his countrymen would not have scrupled to inform him of the secret, if they really had it still among them.

Garcilasso, thus being ignorant of the secret, makes use of some inductions to throw light upon the subject; he asserts that the air is so dry and so cold at Cusco, that flesh dries there like wood, without corrupting; and he is of opinion that they dried the body in snow, before they applied the bitumen: he adds, that in the times of the incas, they usually dried the flesh which was designed for the use of the army; and that when they had lost their humidity, they might be kept without salt, or any other preparation.

It is said, that at Spitsbergen, which lies within the arctic circle and, consequently, in the coldest climate, bodies never corrupt, nor suffer any apparent alteration, even though buried for thirty years; nothing corrupts or putrefies in that climate; the wood which has been employed in building those houses where the train-oil is separated, appears as fresh as the day it was first cut.

If excessive cold, therefore, be thus capable of preserving bodies from corruption, it is not less certain that a great degree of dryness, produced by heat, produces the same effect. It is well known that the men and animals that are buried in the sands of Arabia, quickly dry up, and continue in preservation for several ages, as if they had been actually embalmed. It has often happened, that whole caravans have perished in crossing those deserts, either by the burning winds that infest them, or by the sands which are raised by the tempest, and overwhelm every creature in certain ruin. The bodies of those persons are preserved entire; and they are often found in this condition by some accidental passenger. Many authors, both ancient and modern, make mention of such mummies as these; and Shaw says that he has been assured that numbers of men, as well as other animals, have been thus preserved, for times immemorial, in the burning sands of Saibah, which is a place, he supposes, situate between Rasem and Egypt.

The corruption of dead bodies being entirely caused by the fermentation of the humours, whatever is capable of hindering or retarding this fermentation, will contribute to their preservation. Both heat and cold, though so contrary in themselves, produce similar effects in this particular, by drying up the humours. The cold in condensing and thickening them, and the heat in evaporating them before they have time to act upon the solids. But it is necessary that these extremes should be constant; for if they succeed each other so as that cold shall follow heat, or dryness humidity, it must then necessarily happen, that corruption must ensue. However, in temperate climates, there are natural causes capable of preserving dead bodies, among which we may reckon the quality of the earth in which they are buried. If the earth be drying and astringent, it will imbibe the humidity of the body; and it may probably be for this reason that the bodies buried in the monastery of the Cordeliers, at Thoulouse, do not putrefy, but dry in such a manner that they may be lifted up by one arm.

The gums, resins, and bitumens, with which dead bodies are embalmed, keep off the impressions which they would else receive from the alteration of the temperature of the air; and still more, if a body thus prepared be placed in a dry or burning sand, the most powerful means will be united for its preservation. We are not to be surprised, therefore, at what we are told by Chardin, of the country of Chorosan, in Persia. The bodies which have been previously embalmed, and buried in the sands of that country, as he assures us, are found to petrify; or, in other words, to become extremely hard, and are preserved for several ages. It is asserted that some of them have continued for a thousand years.

The Egyptians, as has been mentioned above, swathed the body with linen bands, and inclosed it in a coffin; however, it is probable that with all these precautions, they would not have continued at now, if the tombs, or pits, in which they were placed, had not been dug in a dry, chalky soil, which was not susceptible of humidity; and which was, besides, covered over with a dry sand of several feet thickness.

The sepulchres of the ancient Egyptians subsist to this day. Most travellers who have been in Egypt have described those of ancient mummies, and have seen the mummies interred there. These catacombs are within two leagues of the ruins of this city, nine leagues from Grand Cairo, and about two miles from the village of Zaccara. They extend from thence to the Pyramids of Pharaoh, which are about eight miles distant. These sepulchres lie in a field, covered with a fine running sand, of a yellowish colour. The country is dry and hilly; the entrance of the tomb is choaked up with sand; there are many open, but several more that are still concealed. The inhabitants of the neighbouring village have no other commerce, or method of subsisting, but by seeking out mummies, and selling them to such strangers as happen to be at Grand Cairo. " This commerce, some years ago, was not only a very common, but a very gainful one. A complete mummy was often sold for twenty pounds: but it must not be supposed that it was bought at such a high price from a mere passion for antiquity; there were much more powerful motives for this traffic. Mummy, at that time, made a considerable article in medicine; and a thousand imaginary virtues were ascribed to it, for the cure of most disorders, particularly of the paralytic kind. There was no shop, therefore, without mummy in it; and no physician thought he had properly treated his patient, without adding this to his prescription. Induced by the general repute, in which this supposed drug was at that time, several Jews, both of Italy and France, found out the art of imitating mummy so exactly, that they, for a long time, deceived all Europe. This they did by drying dead bodies in ovens, after having prepared them with myrrh, aloes, and bitumen. Still, however, the request for mummies continued, and a variety of cures were daily ascribed to them. At length, Paræus wrote a treatise on their total inefficacy in physic; and showed their abuse in loading the stomach, to the exclusion of more efficacious medicines. From that time, therefore, their reputation began to decline; the Jews discontinued their counterfeits, and the trade returned entire to the Egyptians, when it was no longer of value. The industry of seeking after mummies is now totally relaxed, their price merely arbitrary, and just what the curious are willing to give.

In seeking for mummies, they first clear away the sand, which they may do for weeks together, without finding what is wanted. Upon coming to a little square opening of about eighteen feet in depth, they descended into it, by holes for the feet, placed at proper intervals, and there they are sure of finding what they seek for. These caves, or wells, as they call them, are hollowed out of a white freestone, which is found in all this country, a few feet below the covering of sand. When one gets to the bottom of these, which are sometimes forty feet below the surface, there are several square openings, on each side, into passages of ten or fifteen feet wide, and these lead to chambers of fifteen or twenty feet square. There are all hewn out of the rock; and in each of the catacombs are to be found several of these apartments, communicating with each other. They extend a great way under ground so as to be under the city of Memphis, and in a manner to undermine its environs.

In some of the chambers, the walls are adorned with figures and hieroglyphics; in others, the mummies are found in tombs round the apartment hollowed out in the rock. These tombs are upright, and cut into the shape of a man, with his arms stretched out. There are others found, and these in the greatest number, in wooden coffins, or in cloths covered with bitumen. These coffins, or wrappers, are covered all over with a variety of ornaments. There are some of them painted, and adorned with figures, such as that of Death, and the leaden seals, on which several characters are engraven. Some of these coffins are carved into the human shape; but the head alone is distinguishable; the rest of the body is all of a piece, and terminated by a pedestal, while there are some with their arms hanging down; and it is by these marks that the bodies of persons of rank are distinguished from those of the meaner order. These are generally found lying on the floor, without any profusion of ornaments; and, in some chambers the mummies are found indiscriminately piled upon each other, and buried in the sand.

Many mummies are found lying on their backs; their heads turned to the north, and their hands placed on the belly. The bands of linen, with which these were swathed, are found to be more than a thousand yards long; and, of consequence, the number of circumvolutions they make about the body must have been amazing. These were performed by the beginning at the head, and ending at the feet; but they contrived it so as to avoid covering the face. However, when the face is entirely uncovered, it moulders into dust immediately upon the admission of the air. When, therefore, it is preserved entire, a slight covering of cloth is so disposed over it, as that the shape of the eyes, the nose, and the mouth, are seen under it. Some mummies have been found with a long beard, and hair that reached down to the mid-leg, nails of a surprising length, and some gilt, or at least painted of a gold colour. Some are found with bands upon the breast, covered with hieroglyphics, in gold, silver, or in green: and some with tutelary idols, and other figures of jasper, within their body. A piece of gold, also, has often been found under their tongues, of about two pistoles value; and, for this reason, the Arabians spoil all the mummies they meet with, in order to get at the gold.

But though art, or accident, has thus been found to preserve dead bodies entire, it must by no means be supposed that it is capable of preserving the exact form and lineaments of the deceased person. Those bodies which are found dried away in the deserts, or in some particular church-yards, are totally deformed, and scarce any lineaments remain of their external structure. Nor are the mummies preserved by embalming, in a better condition. The flesh is dried away, hardened, and hidden under a variety of bandages; the bowels, as we have seen, are totally removed: and from hence, in the most perfect of them, we see only a shapeless mass of skin discoloured and even the features scarce distinguishable. The art is, therefore, an effort rather of preserving the substance than the likeness of the deceased; and has, consequently, not been brought to its highest pitch of perfection. It appears from a mummy, not long since dug up in France, that the art of embalming was more completely understood in the western world than even in Egypt. This mummy, which was dug up at Auvergne, was an amazing instance of their skill, and is one of the most curious relics in the art of preservation. As some peasants, in that part of the world, were digging in a field near Rion, within about twenty-six paces of the highway, between that and the river Artier, they discovered a tomb, about a foot and a half beneath the surface. It was composed only of two stones; one of which formed the body of the sepulchre, and the other the cover. This tomb was of free-stone; seven feet and a half long, three feet and a half broad, and about three feet high. It was of rude workmanship; the cover had been polished, but was without figure or inscription: within this tomb was placed a leaden coffin, four feet seven inches long, fourteen inches broad, and fifteen high. It was not made coffin-fashion, but oblong, like a box, equally broad at both ends, and covered with a lid that fitted on like a snuff-box, without a hinge. This cover had two holes in it, each of about two inches long, and very narrow, filled with a substance resembling butter; but for what purpose intended remains unknown. Within this coffin was a mummy, in the highest and most perfect preservation. The internal sides of the coffin were filled with an aromatic substance, mingled with clay. Round the mummy was wrapped a coarse cloth, in form of a napkin; under this were two shirts, or shrouds, of the most exquisite texture; beneath these a bandage, which covered all parts of the body, like an infant in swaddling clothes; still under this general bandage there was another, which went particularly round the extremities, the hands, and the legs. The head was covered with two caps; the feet and hands were without any particular bandages; and the whole body was covered with an aromatic substance, an inch thick. When these were removed, and the body exposed naked to view, nothing could be more astonishing than the preservation of the whole, and the exact resemblance it bore to a body that had been dead a day or two before. It appeared well proportioned, except that the head was rather large, and the feet small. The skin had all the pliancy and colour of a body lately dead: the visage, however, was of a brownish hue. The belly yielded to the touch; all the joints were flexible, except those of the legs and feet; the fingers stretched forth of themselves when bent inwards. The nails still continued entire; and all the marks of the joints, both in the fingers, the palms of the hands, and the soles of the feet, remained perfectly visible. The bones of the arms and legs were soft and pliant; but, on the contrary, those of the skull preserved their rigidity; the hair, which only covered the back of the head, was of a chesnut colour, and about two inches long. The pericranium at top was separated from the skull by an incision, in order to open it for the introducing proper aromatics in the place of the brain, where they were found mixed with clay. The teeth, the tongue, and the ears, were all preserved in perfect form. The intestines were not taken out of the body, but remained pliant and entire, as in a fresh subject; and the breast was made to rise and fall like a pair of bellows. The embalming preparation had a very strong and pungent smell, which the body preserved for more than a month after it was exposed to the air. This odour was perceived wherever the mummy was laid, although it remained there but a very short time, it was even pretended that the peasants of the neighbouring villages were incommoded by it. If one touched either the mummy, or any part of the preparation, the hands smelled of it for several hours after, although washed with water, spirit of wine, or vinegar. This mummy, having remained exposed for some months to the curiosity of the public, began to suffer some mutilations. A part of the skin of the forehead was cut off, the teeth were drawn out, and some attempts were made to pull away the tongue. It was therefore, put into a glass case, and shortly after transmitted to the king of France's cabinet at Paris.

There are many reasons to believe this to be the body of a person of the highest distinction; however, no marks remain to assure us either of the quality of the person, or the time of his decease. There only are to be seen some irregular figures on the coffin, one of which represents a kind of star. There were also some singular characters upon the bandages, which were totally defaced by those who had torn them away. However, it should seem that it had remained for several ages in this state, since the first years immediately succeeding the interment, are usually those in which the body is most liable to decay. It appears also to be a much more perfect method of embalming than that of the Egyptians; as in this the flesh continues with its natural elasticity and colour, the bowels remain entire, and the joints have almost the pliancy which they had when the person was alive. Upon the whole, it is probable that a much less tedious preparation than that used by the Egyptians would have sufficed to keep the body from putrefaction; and that an injection of petreoleum inwardly, and a layer of asphaltum without, would have sufficed to have made a mummy; and it is remarkable that Auvergne, where this was found, affords these two substances in sufficient plenty. This art, therefore, might be brought to greater perfection than it has arrived at hitherto, were the art worth preserving. But mankind have long since grown wiser in this respect, and think it unnecessary to keep by them a deformed carcass, which, instead of aiding their magnificence, must only serve to mortify their pride.

CHAPTER XIV.

OF ANIMALS.

LEAVING man, we now descend to the lower ranks of animated nature, and prepare to examine the life, manners, and characters of these our humble partners in the creation. But, in such a wonderful variety as is diffused around us, where shall we begin! The number of beings endued with life, as well as we, seems, at first view, infinite. Not only the forest, the waters, the air teems with animals of various kinds; but almost every vegetable, every leaf, has millions of minute inhabitants, each of which fill up the circle of its allotted life, and some of which are found objects of the greatest curiosity. In this seeming exuberance of animals, it is natural for ignorance to lie down in hopeless uncertainty, and to declare what requires labour to particularize to be utterly inscrutable. It is otherwise, however, with the active and searching mind; no way intimidated with the immense variety, it begins the task of numbering, grouping, and classing all the various kinds that fall within its notice; finds every day new relations between the several parts of the creation; acquires the art of considering several at a time under one point of view, and at last begins to find that the variety is neither so great nor so inscrutable as was at first imagined. As in a clear night, the number of the stars seems infinite; yet, if we sedulously attend to each in his place, and regularly class them, they will soon be found to diminish, and come within a scanty computation.

Method is one of the principal helps in natural history, and without it very little progress can be made in this science. It is by that alone we can hope to dissipate the glare, if I may so express it, which arises from a multiplicity of objects at once presenting themselves to the view. It is method that fixes the attention to one point, and leads it, by slow and certain degrees, to leave no part of nature unobserved.

All naturalists, therefore, have been very careful in adopting some method of classing or grouping the several parts of nature; and some have written books of natural history with no other view. These methodical divisions some have treated with contempt,* not considering that books, in general, are written with opposite views; some to be read, and some only to be occasionally consulted. The methodists in natural history, seem to be content with the latter advantage; and have sacrificed to order alone, all the delights of the subject, all the arts of heightening, awakening, or continuing curiosity. But they certainly have the same use in science that a dictionary has in language: but with this difference, that in a dictionary we proceed from the name to the definition; in a system of natural history, we proceed from the definition to find out the thing. Without the aid of system, nature must still have lain undistinguished, like furniture in a lumber-room: every thing we wish for is there, indeed; but we know not where to find it. If, for instance, in a morning excursion, I find a plant, or an insect, the name of which I desire to learn; or, perhaps, am curious to know whether already known; in this inquiry I can expect information only from one of these systems, which, being couched in a methodical form, quickly directs me to what I seek for. Thus we will suppose that our inquirer has met with a spider, and that he has never seen such an insect before. He is taught by the writer of a system† to examine whether it has wings, and he finds that it has none. He

---

* Mr. Buffon, in his Introduction, &c. † Linnæus

therefore is to look for it among the wingless insects, or the Aptera, as Linnæus calls them: he then is to see whether the head and breast make one part of the body, or are disunited; he finds they make one ne is then to reckon the number of feet and eyes, and he finds that it has eight of each. The insect, therefore, must be either a scorpion or a spider; but he lastly examines its feelers, which he finds clavated, or clubbed; and, by all these marks, he at last discovers it to be a spider. Of spiders, there are forty-seven sorts; and, by reading the description of each, the inquirer will learn the name of that which he desires to know. With the name of the insect, he is also directed to those authors that have given any account of it, and the page where that account is to be found; by this means he may know at once what has been said of that animal by others, and what there is of novelty in the result of his own researches.

From hence, it will appear how useful these systems in natural history are to the inquirer; but, having given them all their merit, it would be wrong not to observe, that they have, in general, been very much abused. Their authors, in general, seem to think that they are improvers of natural history, when in reality they are but guides; they seem to boast that they are adding to our knowledge, while they are only arranging it. These authors, also, seem to think that the reading of their works and systems, is the best method to attain a knowledge of nature; but, setting aside the impossibility of getting through whole volumes of a dry long catalogue, the multiplicity of whose contents is too great for even the strongest memory, such works rather tell us the names than the history of the creature we desire to inquire after. In these dreary pages, every insect, or plant, that has a name, makes as distinguished a figure as the most wonderful, or the most useful. The true end of studying nature is to make a just selection, to find those parts of it that most conduce to our pleasure or convenience, and to leave the rest in neglect. But these systems, employing the same degree of attention upon all, give us no opportunities of knowing which most deserves attention; and he who has made his knowledge from such systems only, has his memory crowded with a number of trifling, or minute particulars, which it should be his business and his labour to forget. These books, as was said before, are useful to be consulted, but they are very unnecessary to be read; no inquirer into nature should be without one of them; and, without any doubt, Linnæus deserves the preference.

One fault more, in almost all these systematic writers, and that which leads me to the subject of the present chapter, is, that seeing the necessity of methodical distribution in some parts of nature, they have introduced it into all. Finding the utility of arranging plants, birds, or insects, they have arranged quadrupeds also with the same assiduity; and although the number of these is so few as not to exceed two hundred,* they have darkened the subject with distinctions and divisions, which only serve to puzzle and perplex. All method is only useful in giving perspicuity, where the subject is either dark

---

* In Dr. Shaw's General Zoology, the number of quadrupeds, not including the cetaceous and seal tribes, amount to five hundred and twelve, besides their varieties.

or copious: but with regard to quadrupeds, the number is but few; many of them we are well acquainted with by habit; and the rest may very readily be known, without any method. In treating of such, therefore, it would be useless to confound the reader with a multiplicity of divisions; as quadrupeds are conspicuous enough to obtain the second rank in nature, it becomes us to be acquainted with, at least, the names of them all. However, as there are naturalists who have gained a name from the excellence of their methods in classing these animals, some readers may desire to have a knowledge of what has been laboriously invented for their instruction. I will just take leave, therefore, to mention the most applauded methods of classing animals, as adopted by Ray, Klein, and Linnæus; for it often happens, that the terms which have been long used in a science, though frivolous, become by prescription a part of the science itself.

Ray, after Aristotle, divides all animals into two kinds; those which have blood, and those which are bloodless. In the last class, he places all the insect tribes. The former he divides into such as breathe through the lungs, and such as breathe through gills: these last comprehend the fishes. In those which breathe through the lungs, some have the heart composed of two ventricles, and some have it of one. Of the last are all animals of the cetaceous kind, all oviparous quadrupeds, and serpents. Of those that have two ventricles, some are oviparous, which are the birds; and some viviparous, which are quadrupeds. The quadrupeds he divides into such as have a hoof, such as are claw-footed. Those with the hoof, he divides all such such as have it undivided, such as have it cloven, and such as have the hoof divided into more parts, as the rhinoceros, and hippopotamos. Animals with the cloven hoof, he divides into such as chew the cud, such as the cow, and the sheep; and such as are not ruminant, as the hog. He divides those animals that chew the cud, into four kinds: the first have hollow horns, which they never shed, as the cow; the second is of a less species, and is of the sheep kind; the third is of the goat kind; and the last, which have solid horns, and shed them annually, are of the deer kind. Coming to the claw-footed animals, he finds some with large claws, resembling the fingers of the human hand; and these he makes the ape kind. Of the others, some have the foot divided into two, and have a claw to each division; these are the camel kind. The elephant makes a kind by itself, as its claws are covered over by a skin. The rest of the numerous tribe of claw-footed animals he divides into two kinds; the analogous, or such as resemble each other; and the anomalous, which differ from the rest. The analogous claw-footed animals, are of two kinds; they have more than two cutting teeth in each jaw, such as the lion and the dog, which are carnivorous; or they have but two cutting teeth in each jaw; and these are chiefly fed upon vegetables. The carnivorous kinds are divided into the great and little. The great carnivorous animals are divided into such as have a short snout, as the cat and the lion; and such as have it long and pointed, as the dog and the wolf. The little claw-footed carnivorous animals, differ from the great, in having a proportionably smaller head, and a slender body, that fits them for creeping into holes, in pursuit of their prey, like worms; and they are therefore called the vermin kind.

We see, from this sketch of division and sub-division, how a subject, extremely delightful and amusing in itself, may be darkened and rendered disgusting. But notwithstanding, Ray seems to be one of the most simple distributors; and his method is still, and not without reason, adopted by many. Such as have been at the trouble to learn this method, will certainly find it useful; nor would we be thought, in the least, to take from its merits; all we contend for is, that the same information may be obtained by a pleasanter and an easier method.

It was the great success of Ray's method, that soon after produced such a variety of attempts in the same manner; but almost all less simple, and more obscure. Mr. Klein's method is briefly as follows: he makes the power of changing place, the characteristic mark of animals in general; and he takes their distinctions from their aptitude and fitness for such a change. Some change place by means of feet, or some similar contrivance; others have wings and feet: some can change place only in water, and have only fins: some go upon earth, without any feet at all: some change place, by moving their shell; and some move only at a certain time of the year. Of such, however, as do not move at all, he takes no notice. The quadrupeds that move chiefly by means of four feet upon land, he divides into two orders. The first are the hoofed kind; and the second, the claw kind. Each of these orders is divided into four families. The first family of the hoofed kind, are the singled hoofed, such as the horse, ass, &c. The second family are such as have the hoof cloven into two parts, such as the cow, &c. The third family have the hoof divided into three parts; and in this family is found only the rhinoceros. The fourth family have the hoof divided into five parts; and in this is only to be found the elephant. With respect to the clawed kind, the first family comprehends those that have but two claws on each foot, as the camel; the second family have three claws; the third, four; and the fourth, five. This method of taking the distinctions of animals from the organs of motion, is ingenious; but it is, at the same time, incomplete; and, besides, the divisions into which it must necessarily fall, is inadequate; since, for instance, in his family with two claws, there is but one animal; whereas, in his family with five claws, there are above a hundred.

Brison, who has laboured this subject with great accuracy, divides animated nature into nine classes: namely, quadrupeds; cetaceous animals, or those of the whale kind; birds; reptiles, or those of the serpent kind; cartilaginous fishes; spinous fishes; shelled animals; insects; and worms. He divides the quadrupeds into eighteen orders; and takes their distinctions from the number and form of their teeth.

But of all those whose systems have been adopted and admired Linnæus is the foremost; as, with a studied brevity, his system comprehends the greatest variety in the smallest space.

According to him, the first distinction of animals is to be taken from their internal structure. Some have the heart with two ventricles, and hot red blood; namely, quadrupeds and birds. The quadrupeds are viviparous, and the birds oviparous.

Some have the heart with one ventricle, and cold red blood; namely, amphibia and fishes. The amphibia are furnished with lungs; the fishes, with gills.

Some have the heart with one ventricle, and cold white serum; namely, insects and worms: the insects have feelers; and the worms, holders.

The distinctions of quadrupeds, or animals with paps, as he calls them, are taken from their teeth. He divides them into seven orders; to which he gives names that are not easy of translation: Primates or principles, with four cutting teeth in each jaw; Bruta, or brutes, with no cutting teeth; Feræ, or wild beasts, with generally six cutting teeth in each jaw; Glires, or dormice, with two cutting teeth, both above and below; Pecora, or cattle, with many cutting teeth above, and none below; Belluæ, or beasts, with the fore-teeth blunt; Cetæ, or those of the whale kind, with cartilaginous teeth. I have but just sketched out this system, as being, in its own nature, the closest abridgment; it would take volumes to dilate it to its proper length. The names of the different animals, and their classes, alone make two thick octavo volumes; and yet nothing is given but the slightest description of each. I have omitted all criticism also upon the accuracy of the preceding systems: this has been done both by Buffon and Daubenton, not with less truth than humour; for they had too much good sense not to see the absurdity of multiplying the terms of science to no end, and disappointing our curiosity rather with a catalogue of nature's varieties, than a history of nature.

Instead, therefore, of taxing the memory and teazing the patience with such a variety of divisions and subdivisions, I will take leave to class the productions of nature in the most obvious, though not in the most accurate manner. In natural history, of all other sciences, there is the least danger of obscurity. In morals, or in metaphysics, every definition must be precise, because those sciences are built upon definitions; but it is otherwise in those subjects where the exhibition of the object itself is always capable of correcting the error. Thus, it may often happen that in a lax system of natural history, a creature may be ranked among quadrupeds that belongs more properly to the fish or the insect classes. But that can produce very little confusion, and every reader can thus make a system the most agreeable to his own imagination. It will be of no manner of consequence whether we call a bird or an insect a quadruped, if we are careful in marking all its distinctions: the uncertainty in reasoning, or thinking, that those approximations of the different kinds of animals produce, is but very small, and happens but very rarely; whereas the labour that naturalists have been at to keep the kinds asunder, has been excessive. This, in general, has given birth to that variety of systems which we have just mentioned, each of which seems to be almost as good as the preceding.

Taking, therefore, this latitude, and using method only where it contributes to conciseness or perspicuity, we shall divide animated nature into four classes; namely Quadrupeds, Birds, Fishes, and Insects. All these seem in general pretty well distinguished from each other by nature; yet there are several instances in which we can scarce tell whether it is a bird or a quadruped that we are about to examine; whether it is a fish or an insect that offers to our curiosity. Nature is varied by imperceptible gradations, so that no line can be drawn between any two classes of its productions, and no definition made to comprehend them all. However, the distinctions between these classes are sufficiently marked, and their encroachments upon each other are so rare, that it will be sufficient particularly to apprize the reader when they happen to be blended.

There are many quadrupeds that we are well acquainted with; and of those we do not know, we shall form the most clear and distinct conceptions, by being told wherein they differ, and wherein they resemble those with which we are familiar. Each class of quadrupeds may be ranged under some one of the domestic kinds, that may serve for the model by which we are to form some kind of idea of the rest. Thus we may say that a tiger is of the cat kind, a wolf of the dog kind, because there are some rude resemblances between each; and a person who has never seen the wild animals, will have some incomplete knowledge of their figure from the tame ones. On the contrary, I will not, as some systematic writers have done,* say that the bat is of the human kind, or a hog of the horse kind, merely because there is some resemblance in their teeth, or their paps. For although this resemblance may be striking enough, yet a person who has never seen a bat or a hog, will never form any just conception of either, by being told of this minute similitude. In short, the method in classing quadrupeds should be taken from their most striking resemblances and where these do not offer, we should not force the similitude, but leave the animal to be described as a solitary species. The number of quadrupeds is so few, that indeed, without any method whatever, there is no great danger of confusion.

All quadrupeds, the number of which, according to Buffon, amounts to but two hundred, may be classed in the following manner.

First, those of the Horse kind. This class contains the Horse, the Ass, and the Zebra. Of these none have horns, and their hoof is of one solid piece.

The second class are those of the Cow kind; comprehending the Urus, the Buffalo, the Bison, and the Bonassus. These have cloven hoofs, and chew the cud.

The third class is that of the Sheep kind: with cloven hoofs, and chewing the cud, like the former. In this is comprehended the Sheep, the Goat, the Lama, the Vigogne, the Gazelle, the Guinea Deer, and all of a similar form.

The fourth class is that of the Deer kind, with cloven hoofs, and with solid horns, that are shed every year. This class contains the Elk, the Rein-deer, the Stag, the Buck, the Roebuck, and the Axis.

The fifth class comprehends all those of the Hog kind, the Peccari and the Babyrouessa.

* Linnæi Syst.

The sixth class is that numerous one of the Cat kind. This comprehends the Cat, the Lion, the Panther, the Leopard, the Jaguar, the Cougar, the Jaguarette, the Lynx, the Ounce, and the Catamountain. These are all carnivorous, and furnished with crooked claws, which they can sheath and unsheath at pleasure.

The seventh class is that of the Dog kind, carnivorous, and furnished with claws like the former, but which they cannot sheath. This class comprehends the Dog, the Wolf, the Fox, the Jackall, the Isatis, the Hyæna, the Civet, the Gibet, and the Genet.

The eighth class is that of the Weasel kind, with a long small body, with small toes, or claws, on each foot; the first of them separated from the rest like a thumb. This comprehends the Weasel, the Martin, the Pole-cat, the Ferret, the Mangoust, the Vansire, the Ermine, with all the varieties of the American Moufettes.

The ninth class is that of the Rabbit kind, with two large cutting teeth in each jaw. This comprehends the Rabbit, the Hare, the Guinea-pig, all the various species of the Squirrel, the Dormouse, the Marmotte, the Rat, the Mouse, Agouti, the Paca, the Aperea, and the Tapeti.

The tenth class is that of the Hedge-hog kind, with claw feet, and covered with prickles, comprehending the Hedge-hog, and the Porcupine, the Couando, and the Urson.

The eleventh class is that of the Tortoise kind, covered with a shell, or scales. This comprehends the Tortoise, the Pangolin, and the Phataguin.

The twelfth is that of the Otter, or amphibious kind, comprehending the Otter, the Beaver, the Desman, the Morse, and the Seal.

The thirteenth class is that of the Ape and Monkey kinds, with hands, and feet resembling hands.

The fourteenth class is that of winged quadrupeds, or the Bat kind, containing the Bat, the Flying-Squirrel, and some other varieties.

The animals which seem to approach no other kind, either in nature, or in form, but to make each a distinct species in itself, are the following: the Elephant, the Rhinoceros, the Hippopotamus, the Cameleopard, the Camel, the Bear, the Badger, the Tapir, the Cabria, the Conti, the Antbear, the Tatou, and lastly, the Sloth.

All other quadrupeds, whose names are not set down, will be found among some of the above mentioned classes, and referred to that which they most resemble. When, therefore, we are at a loss to know the name of any particular animal, by examining which of the known kinds it most resembles, either in shape, or in hoofs, or claws; and then examining the particular description, we shall be able to discover not only its name, but its history. I have already said that all methods of this kind are merely arbitrary, and that nature makes no exact distinction between her productions. It is hard, for instance, to tell whether we ought to refer the civet to the dog, or the cat kind; but, if we know the exact history of the civet, it is no great matter to which kind we shall judge it to bear the greatest resemblance. It is enough, that a distribution of this kind excites in us some rude outlines of the make, or some marked similitudes in the nature of these animals; but, to know them with any precision, no system, or even description, will serve, since the animal itself, or a good print of it must be seen, and its history be read at length, before it can be said to be known. To pretend to say that we have an idea of a quadruped, because we can tell the number or the make of its teeth, or its paps, is as absurd as if we should pretend to distinguish men by the buttons of their clothes. Indeed, it often happens that the quadruped itself can be but seldom seen, that many of the more rare kinds do not come into Europe above once in an age, and some of them have never been able to bear the removal; in such a case, therefore, there is no other substitute but a good print of the animal, to give an idea of its figure; for no description whatsoever can answer this purpose so well. Mr. Locke, with his usual good sense, has observed, that a drawing of the animal, taken from the life, is one of the best methods of advancing natural history; and yet, most of our modern systematic riters are content rather with describing. Descriptions, no doubt, will go some way towards giving an idea of the figure of an animal; but they are certainly much the longest way about, and, as they are usually managed, much the most obscure. In a drawing we can, at a single glance, gather more instruction than by a day's painful investigation of methodical systems, where we are told the proportions with great exactness, and yet remain ignorant of the totality. In fact, this method of describing all things is a fault that has infected many of our books that treat on the meaner arts, for this last age. They attempt to teach by words, what is only to be learnt by practice and inspection. Most of our dictionaries, and bodies of arts and sciences, are guilty of this error. Suppose, for instance, it be requisite to mention the manner of making shoes; it is plain that all the verbal instructions in the world will never give an adequate idea of this humble art, or teach a man to become a shoemaker. A day or two in a shoemaker's shop will answer the end better than a whole folio of instruction, which only serves to oppress the learner with the weight of its pretended importance. We have lately seen a laborious work carried on at Paris, with this only intent of teaching all the trades by description; however, the design at first blush seems to be ill considered; and it is probable that very few advantages will be derived from so laborious an undertaking. With regard to the descriptions in natural history, these, without all question, under the direction of good sense, are necessary; but still they should be kept within proper bounds; and, where a thing may be much more easily shown than described, the exhibition should ever precede the account.

CHAPTER XV.

OF QUADRUPEDS IN GENERAL, COMPARED TO MAN.

Upon comparing the various animals of the globe with each other, we shall find that quadrupeds demand the rank immediately next ourselves, and consequently come first in consideration. The similitude between the structure of their bodies and ours, those instincts which they enjoy in a superior degree to the rest, their constant services, or their unceasing hostilities, all render them the foremost objects of our curiosity, and the most interesting parts of animated nature. These, however, although now so completely subdued, very probably, in the beginning, were nearer upon an equality with us, and disputed the possession of the earth. Man, while yet savage himself, was but ill qualified to civilize the forest. While yet naked, unarmed, and without shelter, every wild beast was a formidable rival; and the destruction of such was the first employment of heroes. But when he began to multiply, and arts to accumulate, he soon cleared the plains of the most noxious of these his rivals; a part was taken under his protection and care, while the rest found a precarious refuge in the burning desert, or the howling wilderness.

From being rivals, quadrupeds have now become the assistants of man; upon them he devolves the most laborious employments, and finds in them patient and humble coadjutors, ready to obey, and content with the smallest retribution. It was not, however, without long and repeated efforts that the independent spirit of these animals was broken; for the savage freedom, in wild animals, is generally found to pass down through several generations before it is totally subdued. Those cats and dogs that are taken from a state of natural wildness in the forest, still transmit their fierceness to their young; and, however concealed in general, it breaks out upon several occasions. Thus the assiduity and application of man in bringing them up, not only alters their disposition, but their very forms; and the difference between animals in a state of nature and domestic tameness, is so considerable, that Mr. Buffon has taken this as a principal distinction in classing them.

In taking a cursory view of the form of quadrupeds, we may easily perceive, that of all the ranks of animated nature, they bear the nearest resemblance to man. This similitude will be found more striking when erecting themselves on their hinder feet, they are taught to walk forward in an upright posture. We then see that all their extremities in a manner correspond with ours, and present us with a rude imitation of our own. In some of the ape kind the resemblance is so striking, that anatomists are puzzled to find in what part of the human body man's superiority consists; and scarce any but the metaphysician can draw the line that ultimately divides them.

But if we compare their internal structure with our own, the likeness will be found still to increase, and we shall perceive many advantages they enjoy in common with us, above the lower tribes of nature. Like us, they are placed above the class of birds, by bringing forth their young alive; like us, they are placed above the class of fishes, by breathing through the lungs; like us, they are placed above the class of insects, by having red blood circulating through their veins; and lastly, like us, they are different from almost all the other classes of animated nature, being either wholly or partly covered with hair. Thus nearly are we represented in point of conformation to the class of animals immediately below us; and this shows what little reason we have to be proud of our persons alone, to the perfection of which quadrupeds make such very near approaches.

The similitude of quadrupeds to man obtains also in the fixedness of their nature, and their being less apt to be changed by the influence of climate or food, than the lower ranks of nature.* Birds are found very apt to alter both in colour and size; fishes, likewise, still more; insects may be quickly brought to change and adapt themselves to the climate; and if we descend to plants, which may be allowed to have a kind of living existence, their kinds may be surprisingly and readily altered, and taught to assume new forms. The figure of every animal may be considered as a kind of drapery, which it may be made to put on or off by human assiduity; in man the drapery is almost invariable; in quadrupeds it admits of some variation; and the variety may be made greater still, as we descend to the inferior classes of animal existence.

Quadrupeds, although they are thus strongly marked, and in general divided from the various kinds around them, yet some of them are often of so equivocal a nature, that it is hard to tell whether they ought to be ranked in the quadruped class, or degraded to those below them. If, for instance, we were to marshal the whole group of animals round man, placing the most perfect next him, and those most equivocal near the classes they most approach, we should find it difficult, after the principal had taken their stations near him, where to place many that lie at the out-skirts of this phalanx. The bat makes a near approach to the aerial tribe, and might, by some, be reckoned among the birds. The porcupine has not less pretensions to that class, being covered with quills, and showing that birds are not the only part of nature that are furnished with such a defence. The armadillo might be referred to the tribe of insects, or snails, being, like them, covered with a shell; the seal and the morse might be ranked among the fishes, like them being furnished with fins, and almost constantly residing in the same element. All these, the farther they recede from the human figure, become less perfect, and may be considered as the lowest kinds of that class to which we have referred them.

But although the variety in quadrupeds is thus great, they all seem well adapted to the stations in which they are placed. There is scarce one of them, how rudely shaped soever, that is not formed to enjoy a state of happiness fitted to its nature. All its deformities are only relative to us, but all its enjoyments are peculiarly its own. We may superficially suppose the sloth, that takes up months in climbing a single tree, or the mole, whose eyes are too small for distinct vision, are wretched and helpless creatures; but it is probable that their life, with respect to themselves, is a life of luxury; the most pleasing food is easily obtained; and as they are abridged in one pleasure, it may be doubled in those which remain. Quadrupeds, and all the lower kinds of animals, have, at worst, but the torments of immediate evil to encounter, and this is but transient and accidental; man has two sources of calamity, that which he foresees, as well as that which he feels; so that if

---

* Buffon, vol. xviii. p. 179.

his reward were to be in this life alone, then indeed would he be of all beings the most wretched.

The heads of quadrupeds, though differing from each other, are, in general, adapted to their way of living. In some it is sharp, the better to fit the animal for turning up the earth in which its food lies. In some it is long, in order to give a greater room for the olfactory nerves, as in dogs, who are to hunt and find out their prey by the scent. In others it is short and thick, as in the lion, to increase the strength of the jaw, and to fit it the better for combat. In quadrupeds that feed upon grass, they are enabled to hold down their heads to the ground, by a strong tendinous ligament, that runs from the head to the middle of the back. This serves to raise the head, although it has been held to the ground for several hours, without any labour, or any assistance from the muscles of the neck.

The teeth of all animals are entirely fitted to the nature of their food. Those of such as live upon flesh differ in every respect from such as live upon vegetables. In the latter, they seem entirely made for gathering and bruising their simple food, being edged before, and fitted for cutting; but broad towards the back of the jaw, and fitted for pounding. In the carnivorous kinds, they are sharp before, and fitted rather for holding than dividing. In the one the teeth serve as grindstones; in the other as weapons of defence; in both, however, the surface of those teeth which serve for grinding, are unequal; the cavities and risings fitting those of the opposite, so as to tally exactly when the jaws are brought together. These inequalities better serve for comminuting the food; but they become smooth with age; and, for this reason, old animals take a longer time to chew their food than such as are in the vigour of life.

Their legs are not better fitted than their teeth to their respective wants or enjoyments. In some they are made for strength only, and to support a vast unwieldy frame, without much flexibility or beautiful proportion. Thus the legs of the elephant, the rhinoceros, and the sea-horse, resemble pillars; were they made smaller, they would be unfit to support the body; were they endowed with greater flexibility, or swiftness, that would be needless, as they do not pursue other animals for food; and, conscious of their own superior strength, there are none that they deign to avoid. Deers, hares, and other creatures, that are to find safety only in flight, have their legs made entirely for speed; they are slender and nervous. Were it not for this advantage, every carnivorous animal would soon make them a prey, and their races would be entirely extinguished. But, in the present state of nature, the means of safety are rather superior to those of offence; and the pursuing animal must owe success only to patience, perseverance, and industry. The feet of some, that live upon fish alone, are made for swimming. The toes of these animals are joined together with membranes, being web-footed, like a goose or a duck, by which they swim with great rapidity. Those animals that lead a life of hostility, and live upon others, have their feet armed with sharp claws, which some can sheath and unsheath at will. Those, on the contrary who lead peaceful lives, have generally hoofs, which serve some as weapons of defence; and which, in all, are better fitted for traversing extensive tracts of rugged country, than the claw-foot of their pursuers.

The stomach is generally proportioned to the quality of the animal's food, or the ease with which it is obtained. In those that live upon flesh, and such nourishing substances, it is small and glandular, affording such juices as are best adapted to digest its contents; their intestines, also, are short, and without fatness. On the contrary, such animals as feed entirely upon vegetables, have the stomach very large; and those who chew the cud have no less than four stomachs, all which serve as so many laboratories, to prepare and turn their coarse food into proper nourishment. In Africa, where the plants afford greater nourishment than in our temperate climates, several animals, that with us have four stomachs, have there but two.* However, in all animals the size of the intestines are proportioned to the nature of the food; where that is furnished in large quantities, the stomach dilates to answer the increase. In domestic animals, that are plentifully supplied, it is large; in the wild animals, that live precariously, it is much more contracted, and the intestines are much shorter.

In this manner, all animals are fitted by nature to fill up some peculiar station. The greatest animals are made for an inoffensive life, to range the plains and the forest without injuring others; to live upon the productions of the earth, the grass of the field, or the tender branches of trees These, secure in their own strength, neither fly from any other quadrupeds nor yet attack them: nature, to the greatest strength, has added the most gentle and harmless dispositions; without this, those enormous creatures would be more than a match for all the rest of the creation; for what devastation might not ensue were the elephant, or the rhinoceros, or the buffalo, as fierce and as mischievous as the tiger or the rat? In order to oppose these larger animals, and in some measure to prevent their exuberance, there is a species of the carnivorous kind, of inferior strength indeed, but of greater activity and cunning. The lion and the tiger generally watch for the larger kinds of prey, attack them at some disadvantage, and commonly jump upon them by surprise. None of the carnivorous kinds, except the dog alone, will make a voluntary attack, but with the odds on their side. They are all cowards by nature, and usually catch their prey by a bound from some lurking place, seldom attempting to invade them openly; for the larger beasts are too powerful for them, and the smaller too swift.

A lion does not willingly attack a horse; and then only when compelled by the keenest hunger. The combats between a lion and a horse are frequent enough in Italy; where they are both inclosed in a kind of amphitheatre, fitted for that purpose. The lion always approaches wheeling about, while the horse presents his hinder parts to the enemy. The lion in this manner goes round and round, still narrowing his circle, till he comes to the proper distance to make his spring; just at the time the lion springs, the horse lashes with both legs from behind, and, in general, the odds are in his favour: it more often happening that the lion is stunned, and struck motionless by the

---

* Buffon

blow, than that he effects his jump between the horse's shoulders. If the lion is stunned, and left sprawling, the horse escapes, without attempting to improve his victory; but if the lion succeeds, he sticks to his prey, and tears the horse in pieces in a very short time.

But it is not among the larger animals of the forest alone, that these hostilities are carried on; there is a minuter, and a still more treacherous contest between the lower rank of quadrupeds. The panther hunts for the sheep and the goat; the catamountain for the hare or the rabbit; and the wild-cat for the squirrel or the mouse. In proportion as each carnivorous animal wants strength, it uses all the assistance of patience, assiduity, and cunning. However, the arts of these to pursue, are not so great as the tricks of their prey to escape; so that the power of destruction in one class is inferior to the power of safety in the other. Were this otherwise, the forest would soon be dispeopled of the feebler races of animals; and beasts of prey themselves, would want, at one time, that subsistence which they lavishly destroyed at another.

Few wild animals seek their prey in the day-time; they are then generally deterred by their fears of man in the inhabited countries, and by the excessive heat of the sun in those extensive forests that lie towards the south, and in which they reign the undisputed tyrants. As soon, therefore, as the morning appears, the carnivorous animals retire to their dens; and the elephant, the horse, the deer, and all the hare kinds, those inoffensive tenants of the plain, make their appearance. But again, at night-fall, the state of hostility begins; the whole forest then echoes to a variety of different howlings. Nothing can be more terrible than an African landscape at the close of evening; the deep-toned roarings of the lion; the shriller yellings of the tiger; the jackall, pursuing by the scent, and barking like a dog; the hyena, with a note peculiarly solitary and dreadful; but above all, the hissing of the various kinds of serpents that then begin their call, and, as I am assured, make a much louder symphony than the birds in our groves in a morning.

Beasts of prey seldom devour each other; nor can any thing but the greatest degree of hunger induce them to it. What they chiefly seek after, is the deer, or the goat; those harmless creatures, that seem made to embellish nature. These are either pursued or surprised, and afford the most agreeable repast to their destroyers. The most usual method with even the fiercest animals, is to hide and crouch near some path frequented by their prey; or some water where cattle come to drink; and seize them at once with a bound. The lion and the tiger leap twenty feet at a spring; and this, rather than their swiftness or strength, is what they have most to depend upon for a supply. There is scarce one of the deer or hare kind, that is not very easily capable of escaping them by its swiftness; so that whenever any of these fall a prey, it must be owing to their own inattention.

But there is another class of the carnivorous kind, that hunt by the scent, and which it is much more difficult to escape. It is remarkable that all animals of this kind pursue in a pack; and encourage each other by their mutual cries. The jackall, the syagush, the wolf, and the dog, are of this kind; they pursue with patience rather than swiftness; their prey flies at first, and leaves them for miles behind; but they keep on with a constant steady pace, and excite each other by a general spirit of industry and emulation, till at last they share the common plunder. But it too often happens, that the larger beasts of prey, when they hear a cry of this kind begun, pursue the pack, and when they have hunted down the animal, come in and monopolize the spoil. This has given rise to the report of the jackall's being the lion's provider; when the reality is, that the jackall hunts for itself, and the lion is an unwelcome intruder upon the fruit of his toil.

Nevertheless, with all the powers which carnivorous animals are possessed of, they generally lead a life of famine and fatigue. Their prey has such a variety of methods for escaping, that they sometimes continue without food for a fortnight together: but nature has endowed them with a degree of patience equal to the severity of their state; so that as their subsistence is precarious, their appetites are complying. They usually seize their prey with a roar, either of seeming delight, or perhaps to terrify it from resistance. They frequently devour it, bones and all, in the most ravenous manner; and then retire to their dens, continuing inactive till the calls of hunger again excite their courage and industry. But as all their methods of pursuit are counteracted by the arts of evasion, they often continue to range without success, supporting a state of famine for several days, nay, sometimes weeks together. Of their prey, some find protection in holes, in which nature has directed them to bury themselves; some find safety by swiftness; and such as are possessed of neither of these advantages, generally herd together, and endeavour to repel invasion by united force. The very sheep, which to us seem so defenceless, are by no means so in a state of nature; they are furnished with arms of defence, and a very great degree of swiftness; but they are still further assisted by their spirit of mutual defence: the females fall into the centre; and the males, forming a ring round them, oppose their horns to the assailants. Some animals that feed upon fruits which are to be found only at one time of the year, fill their holes with several sorts of plants, which enable them to lie concealed during the hard frosts of the winter, contented with their prison, since it affords them plenty and protection. These holes are dug with so much art, that there seems the design of an architect in the formation. There are usually two apertures, by one of which the little inhabitant can always escape, when the enemy is in possession of the other. Many creatures are equally careful of avoiding their enemies, by placing a sentinel to warn them of the approach of danger. These generally perform this duty by turns; and they know how to punish such as have neglected their post, or have been unmindful of the common safety. Such are a part of the efforts that the weaker races of quadrupeds exert, to avoid their invaders; and, in general, they are attended with success. The arts of instinct are most commonly found an overmatch for the invasions of instinct. Man is the only creature against whom all their little tricks cannot prevail. Wherever he has spread his dominion, scarce any flight can save, or any retreat harbour; wherever he comes, terror seems to follow, and all society ceases among the inferior ten ants of the plain; their union against him can yield them no protection, and their cunning is but weakness. In their fellow brutes, they have an enemy whom they can oppose with an equality of advantage they can oppose fraud or swiftness to force; or numbers to invasion but what can be done against such an enemy as man, who finds them out, though unseen, and though remote, destroys them? Wherever he comes, all the contest among the meaner ranks seems to be at an end, or is carried on only by surprise. Such as he has thought proper to protect, have calmly submitted to his protection; such as he has found it convenient to destroy, carry on an unequal war, and their numbers are every day decreasing.

The wild animal is subject to few alterations; and, in a state of savage nature, continues for ages the same, in size, shape, and colour. But it is otherwise when subdued, and taken under the protection of man; its external form, and even its internal structure, are altered by human assiduity: and this is one of the first and greatest causes of the variety that we see among the several quadrupeds of the same species. Man appears to have changed the very nature of domestic animals, by cultivation and care. A domestic animal is a slave that seems to have few other desires but such as man is willing to allow it. Humble, patient, resigned, and attentive, it fills up the duties of its station; ready for labour, and content with subsistence.

Almost all domestic animals seem to bear the marks of servitude strong upon them. All the varieties in their colour, all the fineness and length of their hair, together with the depending length of their ears, seem to have arisen from a long continuance of domestic slavery. What an immense variety is there to be found in the ordinary race of dogs and horses! the principal differences of which have been effected by the industry of man, so adapting the food, the treatment, the labour, and the climate, that nature seems almost to have forgotten her original design; and the tame animal no longer bears any resemblance to its ancestors in the woods around them.

In this manner nature is under a kind of constraint, in those animals we have taught to live in a state of servitude near us. The savage animals preserve the marks of their first formation; their colours are generally the same; a rough dusky brown, or a tawny, seem almost their only varieties. But it is otherwise in the tame; their colours are various, and their forms different from each other. The nature of the climate, indeed, operates upon all; but more particularly on these. That nourishment which is prepared by the hand of man, not adapted to their appetites, but to suit his own convenience, that climate, the rigours of which he can soften, and that employment to which they are sometimes assigned, produce a number of distinctions that are not to be found among the savage animals. These at first were accidental, but in time became hereditary; and a new race of artificial monsters are propagated, rather to answer the purposes of human pleasure, than their own convenience. In short, their very appetites may be changed; and those that feed only upon grass may be rendered carnivorous. I have seen a sheep that would eat flesh, and a horse that was fond of oysters.

But not their appetites, or their figure alone, but their very dispositions, and their natural sagacity, are altered by the vicinity of man In those countries where men have seldom intruded, some animals have been found, established in a kind of civil state of society. Remote from the tyranny of man, they seem to have a spirit of mutual benevolence, and mutual friendship. The beavers, in these distant solitudes, are known to build like architects, and rule like citizens. The habitations that these have been seen to erect, exceed the houses of the human inhabitants of the same country, both in neatness and convenience. But as soon as man intrudes upon their society, they seem impressed with the terrors of their inferior situation, their spirit of society ceases, the bond is dissolved, and every animal looks for safety in solitude, and there tries all its little industry to shift only for itself.

Next to human influence, the climate seems to have the strongest effects both upon the nature and form of quadrupeds. As in man, we have seen some alterations, produced by the variety of his situation; so in the lower ranks, that are more subject to variation, the influence of climate is more readily perceived. As these are more nearly attached to the earth, and in a manner connected to the soil; as they have none of the arts of shielding off the inclemency of the weather, or softening the rigours of the sun, they are consequently more changed by variations. In general, it may be remarked, that the colder the country, the larger and the warmer is the fur of each animal; it being wisely provided by nature, that the inhabitant should be adapted to the rigours of its situation. Thus, the fox and wolf, which in temperate climates have but short hair, have a fine long fur in the frozen regions, near the pole. On the contrary, those dogs which with us have long hair, when carried to Guinea, or Angola, in a short time cast their thick covering, and assume a lighter dress, and one more adapted to the warmth of the country. The beaver, and the ermine, which are found in the greatest plenty in the cold regions, are remarkable for the warmth and delicacy of their furs; while the elephant, and the rhinoceros, that are natives of the line, have scarce any hair. Not but that human nature can, in some measure, co-operate with, or repress the effects of climate in this particular. It is well known what alterations are produced, by proper care, in the sheep's fleece, in different parts of our own country; and the same industry is pursued with a like success in Syria, where many of their animals are clothed with a long and beautiful hair, which they take care to improve, as they work it into that stuff called camblet, so well known in different parts of Europe.

The disposition of the animal seems also not less marked by the climate than the figure. The same causes that seem to have rendered the human inhabitants of the rigorous climates savage and ignorant, have also operated upon their animals. Both at the line and the pole, the wild quadrupeds are fierce and untameable In these latitudes, their savage dispositions having not been quelled by any afforts from man, and being still farther stimulated by the severity of the weather, they continue fierce and untractable. Most of the attempts which have hitherto been made to tame the wild beasts brought home from the pole or the equator, have proved ineffectual. They are gentle and harmless enough while young; but as they grow up, they acquire their natural ferocity, and snap at the hand that feeds them. It may indeed, in general, be asserted, that in all countries where the men are most barbarous, the beasts are most fierce and cruel: and this is but a natural consequence of the struggle between man and the more savage animals of the forest; for in proportion as he is weak and timid, they must be bold and intrusive; in proportion as his dominion is but feebly supported, their rapacity must be more obnoxious. In the extensive countries, therefore, lying round the pole, or beneath the line, the quadrupeds are fierce and formidable. Africa has ever been remarked for the brutality of its men, and the fierceness of its animals: its lions and its leopards are not less terrible than its crocodiles and its serpents; their dispositions seem entirely marked with the rigours of the climate, and being bred in an extreme of heat, they show a peculiar ferocity, that neither the force of man can conquer, nor his arts allay. However, it is happy for the wretched inhabitants of those climates, that its most formidable animals are all solitary ones; that they have not learnt the art of uniting, to oppress mankind; but each depending on its own strength, invades without any assistant.

The food also is another cause in the variety, which we find among quadrupeds of the same kind. Thus the beasts which feed in the valley are generally larger than those which glean a scanty subsistence on the mountain. Such as live in the warm climates, where the plants are much larger and more succulent than with us, are equally remarkable for their bulk. The ox fed in the plains of Indostan, is much larger than that which is more hardly maintained on the side of the Alps. The deserts of Africa, where the plants are extremly nourishing, produce the largest and fiercest animals; and, perhaps for a contrary reason, America is found not to produce such large animals as are seen in the ancient continent. But, whatever be the reason, the fact is certain, that while America exceeds us in the size of its reptiles of all kinds, it is far inferior in its quadruped productions. Thus, for instance, the largest animal of that country is the tapir, which can by no means be compared to the elephant of Africa. Its beasts of prey also, are devested of that strength and courage which is so dangerous in this part of the world. The American lion, tiger, and leopard, if such diminutive creatures deserve these names, are neither so fierce nor so valiant as those of Africa and Asia. The tiger of Bengal has been seen to measure twelve feet in length, without including the tail; whereas the American tiger seldom exceeds three. This difference obtains still more in the other animals of that country, so that some have been of opinion* that all the quadrupeds in Southern America are of a different species from those most resembling them in the old world; and that there are none which are common to both but such as have entered America by the north; and which, being able to bear the rigours of the frozen pole, have travelled from the ancient continent, by that passage, into the new. Thus the bear, the wolf, the elk,

* Buffon.

the stag, the fox, and the beaver, are known to the inhabitants as well of North America as of Russia; while most of the various kinds to the southward, in both continents, bear no resemblance to each other. Upon the whole, such as peculiarly belong to the new continent, are without any marks of the quadruped perfection. They are almost wholly destitute of the power of defence; they have neither formidable teeth, horns, or tail; their figure is awkward, and their limbs ill proportioned. Some among them, such as the ant-bear, and the sloth, appear so miserably formed, as scarce to have the power of moving and eating. They seemingly drag out a miserable and languid existence in the most desert solitude, and would quickly have been destroyed in a country where there were inhabitants, or powerful beasts to oppose them.

But, if the quadrupeds of the new continent be less, they are found in much greater abundance; for it is a rule that obtains through nature, that the smallest animals multiply the fastest. The goat, imported from Europe to South America, soon begins to degenerate; but as it grows less it becomes more prolific; and, instead of one kid at a time, or two at the most, it generally produces five, and sometimes more. What there is in the food, or the climate, that produces this change, we have not been able to learn; we might be apt to ascribe it to the heat, but that on the African coast, where it is still hotter, this rule does not obtain; for the goat, instead of degenerating there, seems rather to improve.

However, the rule is general among all quadrupeds, that those which are large and formidable produce but few at a time; while such as are mean and contemptible are extremely prolific. The lion, or tiger, have seldom above two cubs at a litter; while the cat, that is of a similar nature, is usually seen to have five or six. In this manner the lower tribes become extremely numerous; and, but for this surprising fecundity, from their natural weakness they would quickly be extirpated. The breed of mice, for instance, would have long since been blotted from the earth, were the mouse as slow in the production as the elephant. But it has been wisely provided that such animals as can make but little resistance, should at least have a means of repairing the destruction which they must often suffer by their quick reproduction; that they should increase even among enemies, and multiply under the hand of the destroyer. On the other hand, it has as wisely been ordered by Providence, that the larger kinds should produce but slowly; otherwise, as they require proportional supplies from nature, they would quickly consume their own store; and, of consequence, many of them would soon perish through want; so that life would thus be given without the necessary means of subsistence. In a word, Providence has most wisely balanced the strength of the great against the weakness of the little. Since it was necessary that some should be great and others mean, since it was expedient that some should live upon others, it has assisted the weakness of the one by granting it fruitfulness; and diminished the number of the other by infecundity

In consequence of this provision, the larger creatures, which bring forth few at a time, seldom begin to generate till they have nearly acquired their full growth. On the contrary, those which bring many, reproduce before they have arrived at half their natural size. Thus the horse and the bull are nearly at their best before they begin to breed; the hog and the rabbit scarce leave the teat before they become parents in turn. Almost all animals likewise continue the time of their pregnancy in proportion to their size. The mare continues eleven months with foal, the cow nine, the wolf five, and the bitch nine weeks. In all, the intermediate litters are the most fruitful; the first and the last generally producing the fewest in number, and the worst of the kind.

Whatever be the natural disposition of animals at other times, they all acquire new courage when they consider themselves as defending their young. No terrors can then drive them from the post of duty; the mildest begin to exert their little force, and resist the most formidable enemy. Where resistance is hopeless, they then incur every danger, in order to rescue their young by flight, and retard their own expedition by providing for their little ones. When the female opossum, an animal of America, is pursued, she instantly takes her young into a false belly, with which nature has supplied her, and carries them off, or dies in the endeavour. I have been lately assured of a she-fox, which, when hunted, took her cub in her mouth, and run for several miles without quitting it, until at last she was forced to leave it behind, upon the approach of a mastiff, as she ran through a farmer's yard. But, if at this period the mildest animals acquire new fierceness, how formidable must those be that subsist by rapine! At such times no obstacles can stop their ravage, nor no threats can terrify; the lioness then seems more hardy than even the lion himself. She attacks men and beasts indiscriminately, and carries all she can overcome reeking to her cubs, whom she thus early accustoms to slaughter. Milk, in the carnivorous animals, is much more sparing than in others; and it may be for this reason that all such carry home their prey alive, that, in feeding their young, its blood may supply the deficiencies of nature, and serve instead of that milk with which they are so sparingly supplied.

Nature, that has thus given them courage to defend their young, has given them instinct to choose the proper times of copulation, so as to bring forth when the provision suited to each kind is to be found in the greatest plenty. The wolf, for instance, couples in December, so that the time of pregnancy continuing five months, it may have its young in April. The mare, who goes eleven months, admits the horse in summer, in order to foal about the beginning of May. On the contrary, those animals which lay up provisions for the winter, such as the beaver and the marmotte, couple in the latter end of autumn, so as to have their young about January, against which season they have provided a very comfortable store. These seasons for coupling, however, among some of the domestic kinds, are generally in consequence of the quantity of provisions with which they are at any time supplied. Thus we may, by feeding any of these animals, and keeping off the rigour of the climate, make them breed whenever we please. In this manner, those contrive who produce lambs all the year round

The choice of situation in bringing forth is also very remarkable. In most of the rapacious kinds, the female takes the utmost precautions to hide the place of her retreat from the male, who otherwise, when pressed by hunger, would be apt to devour her cubs. She seldom, therefore, strays far from the den, and never approaches it while he is in view, nor visits him again till her young are capable of providing for themselves. Such animals as are of tender constitutions, take the utmost care to provide a place of warmth, as well as safety, for their young; the rapacious kinds bring forth in the thickest woods; those that chew the cud, with the various tribes of the vermin kind, choose some hiding-place in the neighbourhood of man. Some dig holes in the ground; some choose the hollow of a tree; and all the amphibious kinds bring up their young near the water, and accustom them betimes to their proper element.

Thus nature seems kindly careful for the protection of the meanest of her creatures: but there is one class of quadrupeds that seems entirely left to chance, that no parent stands forth to protect, nor no instructor leads, to teach the arts of subsistence. These are the quadrupeds that are brought forth from the egg, such as the lizard, the tortoise, and the crocodile. The fecundity of all other animals, compared with these, is sterility itself. These bring forth above two hundred at a time; but, as the offspring is more numerous, the parental care is less exerted. Thus the numerous brood of eggs are, without farther solicitude, buried in the warm sands of the shore, and the heat of the sun alone is left to bring them to perfection. To this perfection they arrive almost as soon as disengaged from the shell. Most of them, without any other guide than instinct, immediately make to the water. In their passage thither, they have numberless enemies to fear. The birds of prey that haunt the shore, the beasts that accidentally come there, and even the animals that give them birth, are known, with a strange rapacity, to thin their numbers as well as the rest.

But it is kindly ordered by Providence, that these animals, which are mostly noxious, should thus have many destroyers; were it not for this, by their extreme fecundity, they would soon overrun the earth and cumber all our plains with deformity.

END OF THE FIRST VOLUME.

www.ingramcontent.com/pod-product-compliance
Lightning Source LLC
Chambersburg PA
CBHW021958220426
43663CB00007B/864